Studies in Systems, Decision and Control

Volume 155

Series editor

Janusz Kacprzyk, Polish Academy of Sciences, Warsaw, Poland
e-mail: kacprzyk@ibspan.waw.pl

The series "Studies in Systems, Decision and Control" (SSDC) covers both new developments and advances, as well as the state of the art, in the various areas of broadly perceived systems, decision making and control- quickly, up to date and with a high quality. The intent is to cover the theory, applications, and perspectives on the state of the art and future developments relevant to systems, decision making, control, complex processes and related areas, as embedded in the fields of engineering, computer science, physics, economics, social and life sciences, as well as the paradigms and methodologies behind them. The series contains monographs, textbooks, lecture notes and edited volumes in systems, decision making and control spanning the areas of Cyber-Physical Systems, Autonomous Systems, Sensor Networks, Control Systems, Energy Systems, Automotive Systems, Biological Systems, Vehicular Networking and Connected Vehicles, Aerospace Systems, Automation, Manufacturing, Smart Grids, Nonlinear Systems, Power Systems, Robotics, Social Systems, Economic Systems and other. Of particular value to both the contributors and the readership are the short publication timeframe and the world-wide distribution and exposure which enable both a wide and rapid dissemination of research output.

More information about this series at http://www.springer.com/series/13304

Aleksander Sładkowski
Editor

Transport Systems and Delivery of Cargo on East–West Routes

 Springer

Editor
Aleksander Sładkowski
Faculty of Transport
Silesian University of Technology
Katowice
Poland

ISSN 2198-4182   ISSN 2198-4190   (electronic)
Studies in Systems, Decision and Control
ISBN 978-3-030-08667-1   ISBN 978-3-319-78295-9   (eBook)
https://doi.org/10.1007/978-3-319-78295-9

This Springer imprint is published by the registered company Springer International Publishing AG part of Springer Nature
The registered company address is: Gewerbestrasse 11, 6330 Cham, Switzerland

Preface

The monograph presented to the attention of the readers is devoted to the consideration of various problems connected with the delivery of goods from the regions of East and South-East Asia to Europe. It is no secret to anyone that a significant part of world production of goods is currently concentrated in these areas, and it is not surprising therefore that the export of goods from these regions is an essential part of the world economy. In particular, according to World Integrated Trade Solution (WITS)[1] exports of goods from the regions of East Asia and the Pacific in 2016 accounted for 26.6% of total world exports. More than half of these exported goods are delivered to the countries of Europe, the Mediterranean and the countries of Central Asia. Thus, the key issue for these countries is the transport of these goods, as well as the return of empty containers, wagons, etc.

The monograph deals with the problems of delivering goods from the point of view of transit countries or recipient countries. It is obvious that different countries have different transport problems, and they approach there in different ways. From the editor's point of view, it was important to acquaint readers with different approaches to solving transport problems in different countries. This could be solved with the involvement of transport scientists from different countries. Therefore, it is very valuable to exchange experience in solving global and regional transport problems, which is described by representatives of Italy, Slovakia, Russia, Georgia, Kazakhstan, Uzbekistan and Poland.

The monograph is structurally divided into two parts. The first part is devoted to the analysis of various problems of global logistics and regional transport, which operates in conditions of transport corridors. In particular, the impact of transport on the economies of Russia, Poland, Slovakia and Kazakhstan is considered. Nevertheless, the authors do not confine themselves to examining regional transport problems, but they consider the impact of cargo transportation on the main transcontinental corridors along the routes connecting Asia and Europe. The initiative of the Chinese leadership to create a new silk road (One Belt One Road) is

---

[1] https://wits.worldbank.org/CountryProfile/en/Country/WLD/Year/2016/TradeFlow/Export.

very positively evaluated. The problems associated with the operation of accelerated container trains are considered.

The problems connected with the organization of transportations of cargoes on various routes are considered also. These problems have not only technical or technological limitations. Most often, the main difficulties are associated with the organization of such deliveries. And if a single one-time transportation can be arranged, then already regular delivery service becomes an irresistible problem. This may be due to financial, information and even political problems. For example, such problems include inconsistencies in the transport and customs legislation of the countries involved in transport processes.

The second part of the monograph is devoted to the solution of some technical and informatics problems related to the organization of transportation on the East–West routes. For example, issues related to the organization of the operation of cargo terminals in Italy and Poland were considered. At the same time, technological, technical, informatics and even financial problems that may arise in this case are considered. Of great importance for the operation of railway transport on these routes are the technical aspects associated with the difference in the standards of construction and the exploration of the railway track. At the same time, it would be advisable to use single wagons without the need to reload goods. In connection with this, the issues of interaction between the track and the rolling stock, including wheels and rails, are very significant. These issues were considered in the section prepared by Georgian scientists.

Very important are also aspects of the aerodynamics of train traffic. This is due to the fact that so far, unfortunately, not all countries located on these routes of transportation of goods can allow individual rail lines for high-speed trains. Most often, these sites are designed simultaneously for the transport of goods and passengers. On the other hand, in such areas it would be necessary to organize the carriage of goods also at higher speeds. It is quite obvious that in this case the aerodynamics of trains is of key importance. In the corresponding section, the aerodynamics of trains is considered using the example of Uzbek railways, where high-speed traffic has already appeared.

The editor of the monograph understands that the subject of the book in question is very broad and it is impossible to cover all aspects. Nevertheless, thanks to the interaction with the publishing house, it is possible to expand the field of knowledge covered by publishing a third monograph in this series. The previous two monographs[2,3] were also devoted to transport problems and could contribute to a more multilateral review of the achievements of modern science in the field of various modes of transport. Despite the fact that most of the authors are working in the universities, the monograph is directly aimed at solving of essential problems

[2] Sładkowski A, Pamuła W (eds.) (2015) Intelligent Transportation Systems – Problems and Perspectives. Studies in Systems, Decision and Control 32. Cham, Heidelberg, New York, Dordrecht, London: Springer. 316 p. ISBN 978-3-319-19149-2.

[3] Sładkowski A (ed.) (2017) Rail transport—systems approach. Studies in systems, Decision and control 87. Cham: Springer. 456 p. ISBN 978-3-319-51502-1.

facing logistics and transport in different countries. Some part of the problems was solved, realizing ideas into concrete technical, economic or organizational solutions. For some problems identified ways for solutions.

The book is written primarily for professionals involved in various problems of cargo delivery on the East–West routes. Nevertheless, the authors hope that this book may be useful for manufacturers, for the technical staff of logistics companies, for managers, for students of transport specialties, as well as for a wide range of readers, who are interested in the current state of transport in different countries.

Katowice, Poland   Aleksander Sładkowski
December, 2017

Contents

Part I Global Logistics and Regional Aspects of Cargo Delivery

Sustainable Development of Transport Systems for Cargo Flows on the East-West Direction 3
Aleksandr Rakhmangulov, Aleksander Sładkowski, Nikita Osintsev, Olesya Kopylova and Natalja Dyorina

Analysis and Development Perspective Scenarios of Transport Corridors Supporting Eurasian Trade 71
Aleksander Sładkowski and Maria Cieśla

The Role of Railway Transport in East-West Traffic Flow Conditions of the Slovak Republic 121
Anna Dolinayova, Lenka Cerna and Vladislav Zitricky

Transnational Value of the Republic of Kazakhstan in International Container Transportation 171
Aleksander Sładkowski, Zhomart Abdirassilov and Amangeldy Molgazhdarov

Economic Aspects of Freight Transportation Along the East-West Routes Through the Transport and Logistics System of Kazakhstan 205
Zhanarys Raimbekov, Bakyt Syzdykbayeva and Kunduz Sharipbekova

Part II Technical and Informatics Issues of Transport Systems

Present and Future Operation of Rail Freight Terminals 233
Marco Antognoli, Luigi Capodilupo, Cristiano Marinacci, Stefano Ricci, Luca Rizzetto and Eros Tombesi

Development of the Silesian Logistic Centres in Terms of Handling Improvement in Intermodal Transport on the East-West Routes 275
Damian Gąska and Jerzy Margielewicz

**Perfection of Technical Characteristics of the Railway Transport
System Europe-Caucasus-Asia (TRACECA)** 303
George Tumanishvili, Tamaz Natriashvili and Tengiz Nadiradze

**Potential and Problems of the Development of Speed Traffic
on the Railways of Uzbekistan** 369
Saidburkhan Djabbarov, Makhamadjan Mirakhmedov
and Aleksander Sładkowski

Part I
Global Logistics and Regional Aspects of Cargo Delivery

Sustainable Development of Transport Systems for Cargo Flows on the East-West Direction

Aleksandr Rakhmangulov, Aleksander Sładkowski, Nikita Osintsev, Olesya Kopylova and Natalja Dyorina

Abstract The implementation of the project "The Economic Belt of the Silk Road" and the creation of new transport links of the international communication make it important to place logistics centers on the road to the flow of freight flows in the East-West direction. The purpose of this study is to develop a methodology for assessing the variants of logistics centers location. The authors proposed the methodology and model allowing us to determine the necessity of placing logistics objects on the basis of a system analysis of transport performance indicators, infrastructural, socio-economic factors. The paper analyzed trends in international trade, reviewed existing and prospective international Euro-Asian transport corridors. The researchers systematized the factors that have the greatest impact on the transport and logistics infrastructure development and on the logistics centers location. Based on the proposed system of factors, the authors developed an assessing methodology for locating variants of logistics centers. The chapter performs the results of the attractiveness assessment of the regions for the logistics infrastructure facilities location according to the developed methodology for the conditions in the Russian Federation as in one of the countries pretending to promote cargo flows through its territory within the framework of the project "The Economic Belt of the Silk Road".

A. Rakhmangulov (✉) · N. Osintsev · O. Kopylova · N. Dyorina
Nosov Magnitogorsk State Technical University,
Lenin Av., 38, 455000 Magnitogorsk, Russia
e-mail: ran@magtu.ru

N. Osintsev
e-mail: osintsev@magtu.ru

O. Kopylova
e-mail: olesya.k863@yandex.ru

N. Dyorina
e-mail: nataljapidckaluck@yandex.ru

A. Sładkowski
Department of Logistics and Aviation Technologies, Faculty of Transport,
Silesian University of Technology, Krasinskiego 8, 40-019 Katowice, Poland
e-mail: aleksander.sladkowski@polsl.pl

© Springer International Publishing AG, part of Springer Nature 2018
A. Sładkowski (ed.), *Transport Systems and Delivery of Cargo on East–West Routes*, Studies in Systems, Decision and Control 155,
https://doi.org/10.1007/978-3-319-78295-9_1

Keywords Sustainable development · Transport systems · Logistics centers
Quartering · Socio-economic factors · Simulation model

1 Trend of International Trade and Their Impact on the Transport Systems Development

The effectively functioning and steadily developing world and national transport systems are the most important factors of countries' economic growth and foreign economic activity expansion. The development of international transport corridors and the creation of an appropriate infrastructure for servicing the East-West freight traffic is becoming a priority in the context of the increasing foreign trade turnover between the countries of Europe and Asia. According to the General Administration of Customs of China, the volume of foreign trade in the first half of 2017 amounted to 13 trillion Yuan, which is 19.5% more than in the same period last year. China is one of the largest trading partners of the European Union. The restoration of the world economy, the huge potential of the European market will contribute to the further growth of the foreign trade turnover and development of the Asia-Europe transport communication.

Over the past decades, the main exporters and importers in the world market are Germany, the United States and the People's Republic of China (PRC). According to ITC statistics, in 2016 the total volume of exports amounted to 15862 billion dollars, of which 13.2% in the PRC, 9.16% in the USA, and 8.45% in Germany [1]. In the world export market, the positions of Asian countries demonstrate strength. So, from 2005 to 2016 the PRC increased its exports by 2.75 times and is currently considered the leader. The exports share of the Republic of Korea increased from 2.75% in 2005 to 3.12% in 2016 from the total, India—from 0.97 to 1.64%, Vietnam—from 0.31 to 1.35%. For Europe, it is typical, both the exports share reduction, on average by 0.5–1% (Germany, Italy, France, Finland), and a slight increase of 0.1–0.5%, mainly for Eastern Europe (Poland, Romania, Slovakia), as well as for the Netherlands. A similar trend in the countries distribution and the trade dynamics is observed in the import market. For the period 2005–2016 significant difference of the world export and import market is to reduce the Russian Federation exports share by 0.5% with an increase in the share of imports in the Russian Federation by 0.2%. Exports and imports data by major countries for 2016 are shown in Table 1.

Since 2014, the growth rate of world foreign trade turnover decelerated a little, which is due to various economic crisis phenomena, but in general, the dynamics for the period of 2005–2016 is positive (Fig. 1, 2) [1]. In 2009 the sharp decline is justified by the global economic crisis, after which, since 2010, there has been an increase in the transportation volumes of both export and import products. In this period, China made the largest contribution to the increase in the world external turnover.

Table 1 The exports and imports volume by major world countries in 2016 [1]

No.	Exporting/importing country	The volume for 2016 (billion US dollars)	Total volume share (%)	
			2005	2016
Export				
1	China	2097.637	7.37	13.22
2	United States of America	1453.167	8.74	9.16
3	Germany	1340.752	9.45	8.45
4	Japan	644.932	5.75	4.07
5	Netherlands	569.384	3.38	3.59
6	Hong Kong, China	516.588	2.82	3.26
7	Korea, Republic of	495.466	2.75	3.12
8	France	488.885	4.20	3.08
9	Italy	461.529	3.61	2.91
10	United Kingdom	415.856	3.8	2.62
…				
16	Russian Federation	285.491	2.33	1.8
Import				
1	United States of America	2249.661	16.33	14.01
2	China	1587.921	6.22	9.89
3	Germany	1060.672	7.35	6.60
4	United Kingdom	635.5699	4.98	3.96
5	Japan	606.924	4.86	3.78
6	France	560.555	4.49	3.49
7	Hong Kong, China	547.124	2.83	3.41
8	Netherlands	504.185	2.93	3.14
9	Korea, Republic of	406.06	2.46	2.53
10	Italy	404.578	3.63	2.52
…				
25	Russian Federation	182.257	0.93	1.13

Currently, the fastest growth rates of trade and economic relations are observed between the PRC and European countries. The EU's import in 2016 amounted to 5218.6 billion US dollars. This is almost 1/3 of the total volume of the world import market. However, the volume of China's exports has doubled to the European Union since 2005 and is equal to 452.891 billion US dollars, which is about 8.7% of the total volume of the EU's import market for 2016 [1]. In 2005, the share of China's imports to the EU countries was only 5%. For comparison, the volume of exports from the United States, as one of the main trade partners of the European Union, increased by 35% and amounted to 302.673 billion US dollars during the period under review.

The growth of the number and welfare of the Chinese population, the active urbanization of the country make the Chinese import market more attractive for the

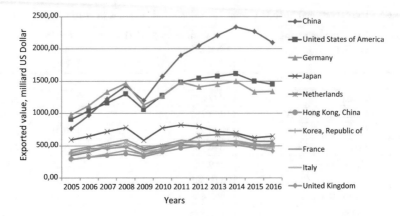

Fig. 1 Exports dynamics of individual countries, billion US dollars

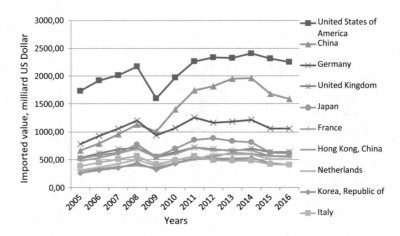

Fig. 2 Imports dynamics of individual countries, billion US dollars

sale of the European Union products. Thus, according to ITC statistics [1], supplies from the EU countries to China (Fig. 3) increased almost by a factor of three.

The largest volume of products exported from China belongs to Germany (16%), to the Netherlands (14%), to Great Britain (14%), to Italy (6%) (Fig. 4). If you consider the goods flow from China to Europe as a whole, its structure will be different—Russia will be located in the fourth place with 10%.

The commodity structure of shipments from China to Europe has practically not changed during the period under review (Fig. 5). The greatest volume of transportations is in various electrical machines and equipment (26%). The share of these goods has increased by almost 4% since 2005, mainly as a result of a share decrease in the mechanical devices and equipment supplies.

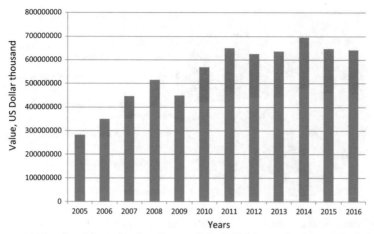

■ Value of supplying markets from European Union (EU 28) for a product imported by China
■ Value of importing markets from European Union (EU 28) for a product exported by China

Fig. 3 External turnover dynamics between the EU and China

Fig. 4 Imports structure
from China to EU countries
(EU 28)

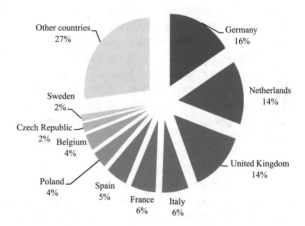

The analysis of the commodity structure of EU exports to China has shown that the main goods are the following commodity groups: machinery, mechanical appliances, nuclear reactors, boilers; parts thereof (18%); vehicles other than railway or tramway rolling stock, and parts and accessories thereof (19%); Electrical machinery and equipment and parts thereof; sound recorders and reproducers, television (11%); optical, photographic, cinematographic, measuring, checking, precision, medical or surgical (8%). At the same time for the period of 2005–2016 the share of machinery and mechanical equipment exports decreased from 31 to 17%, with an increase in the share of vehicles supplies, their parts and accessories by almost 11%, and pharmaceutical products by 5%. There is a rapid increase in the

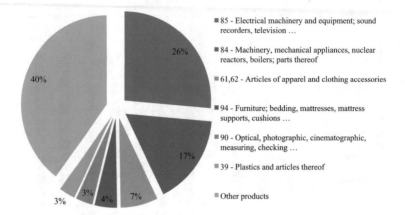

Fig. 5 Imports structure from China to the EU countries by commodity groups

supply of various food products. For example, exports of meat and by-products from the European Union to China increased by a factor of 55 over the period 2005–2016.

It is obvious that such a positive dynamic in the development of international trade, the position strengthening of Asian countries in the import markets of Europe makes it urgent to form a stable transport and logistics system capable of servicing ever increasing freight flows of the East-West direction as efficiently as possible.

The modern transport and logistics infrastructure develops on the basis of the formation of transport corridors and transport hubs with terminals, logistics centers (LC) located in them, where cargoes are being consolidated and divided, temporary storage of goods, transshipment of goods between different vehicles or various modes of transport [2].

The solution to the creation problem of an efficient transport and logistics infrastructure that provides for the handling the cargo flows on the East-West direction is complicated by an increase of the requirements for the environmental aspects of transport and logistics activities [3] by the states where the elements of this infrastructure are located and planned for quartering. This issue is especially urgent for the PRC, which ranks first in the world for carbon dioxide emissions [4] (Fig. 6).

In the context of the growing trend towards specialization of production in countries, of the observed traffic increase in the global transport system, of the shortage of the basic transport infrastructure capacity, the importance of priorities is growing for ensuring the safety and transport environmental friendliness. The world transport system is known to be a sphere of increased risks, one of the main polluters of the environment and consumers of irreplaceable natural resources. Around the world this set of problems is being solved by applying the sustainable development concept principles [2, 5] and green logistics [3], which now constitute the basis for the transport policy of many countries. At the same time, the creation

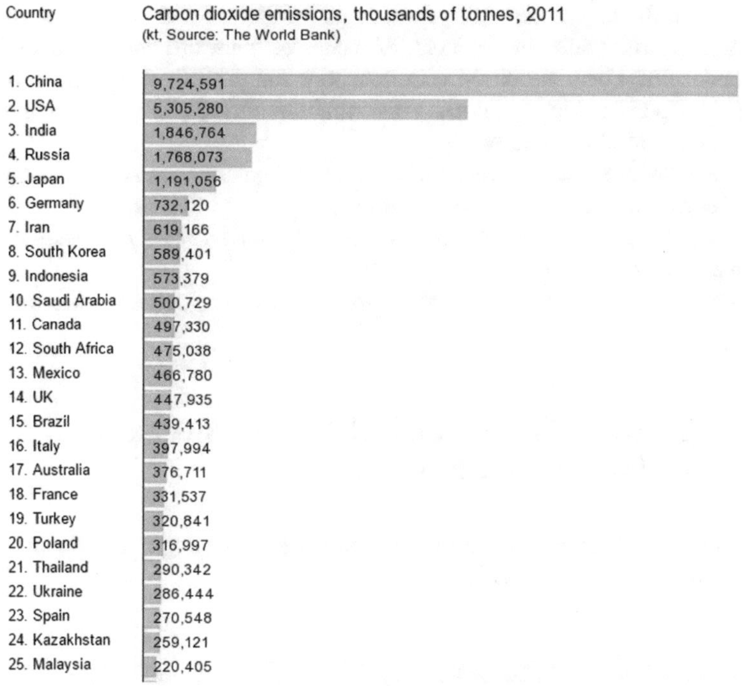

Country	Carbon dioxide emissions, thousands of tonnes, 2011 (kt, Source: The World Bank)
1. China	9,724,591
2. USA	5,305,280
3. India	1,846,764
4. Russia	1,768,073
5. Japan	1,191,056
6. Germany	732,120
7. Iran	619,166
8. South Korea	589,401
9. Indonesia	573,379
10. Saudi Arabia	500,729
11. Canada	497,330
12. South Africa	475,038
13. Mexico	466,780
14. UK	447,935
15. Brazil	439,413
16. Italy	397,994
17. Australia	376,711
18. France	331,537
19. Turkey	320,841
20. Poland	316,997
21. Thailand	290,342
22. Ukraine	286,444
23. Spain	270,548
24. Kazakhstan	259,121
25. Malaysia	220,405

Fig. 6 CO_2 emission level in countries [4]

and development of international transport corridors play a special role in the interaction of the countries-participants in the trade flows promotion.

Balanced accounting of economic, ecological and socio-cultural aspects of the regions with infrastructure elements of international transport corridors will ensure the sustainability of the world transport system.

International transport corridors are the main elements of the world transport system. They connect two or more neighboring states and can pass through several transit states, in particular, to ensure maritime trade for landlocked countries [2].

International transport corridors (ITC) contribute to the acceleration of the goods movement process, they simplify (with effective organization) the passage of border points, contribute to the increase of cargo safety in the case of intermodal transportation. ITCs are the subject of close cooperation and the creation of states unions to ensure spatial accessibility and goods movement freedom. They provide competition between countries, as in the zones with international transport corridors, there is an increase in investment attractiveness and socio-economic development of the regions.

In the transport systems creation and development, the use of the transport corridors concept allows [2]:

- to ensure the coordination of priorities and projects for the transport and economic infrastructure development, of modes of transport, and of territories;
- to reduce the costs associated with transportation directly or indirectly, through the transport and cargo flows concentration, through the reduction of the required land allocation, etc.;
- to develop intermodal transportations, providing the interaction of transport modes at junctions of transport corridors;
- to localize environmental effects by placing different types of transport in the same communication flat;
- to ensure a clear system of priorities for infrastructure projects selection.

2 The Project "The Economic Belt of the Silk Road" and Its Implementation Factors

Currently, the main international transport corridors providing international trade in the East-West direction are (Fig. 7):

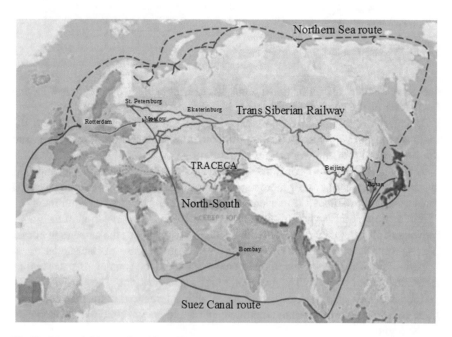

Fig. 7 Scheme of international East-West transport corridors

(1) water routes:

- South Sea Route around Africa;
- South Sea Route through the Suez Canal;
- Arctic Sea route, which includes the Northern Sea Route.

(2) land transport corridors:

- The Trans-Siberian Railway: Europe (Pan-European Transport Corridors II, III and IX)–the Russian Federation–Korean Peninsula–Japan, with two branches from the Russian Federation: Kazakhstan–China and Mongolia–China;
- TPACECA (Transport Corridor Europe-Caucasus-Asia): Eastern Europe (Pan-European Transport Corridors IV, VII, VIII and IX)–through the Black Sea–the Caucasus–through the Caspian Sea–Central Asia;
- South: South-Eastern Europe (Pan-European Transport Corridor IV)–Turkey–the Islamic Republic of Iran, with two branches: to Central Asia–China and South Asia–South-East Asia/Southern China;
- North-South: Northern Europe (Pan-European Transport Corridor IX)–the Russian Federation, with two branches: the Caucasus–Persian Gulf and Central Asia–Persian Gulf [6].

At present, the bulk of cargo flows between Asia and Europe follows the South Sea Route through the Suez Canal. According to [7], the share of sea transport in the total volume of cargo transportation between Asia and Europe is more than 95%. Significant advantages of this route are low cost and relatively high discharge capabilities. In conditions of increasing traffic volume the shortages of this route include: a high load of ports; insufficient capacity of the Suez Canal; High fees for the admission of sea vessels through the Suez Canal; the need for armed guards to accompany ships along the route. These disadvantages lead to significant down-time, an increase in freight handling costs in ports and an increase in the delivery time, and the additional time required to distribute the goods in European consumption centers [6].

The delivery time and its impact on the economic performance of cargo owners and transport companies are important factors in choosing the cargo transportation way in the East-West international communication and the main shortage of water international transport corridors. Travel time reaches 30–45 days depending on navigation conditions and on ports charge capacity.

For example, as for the conditions in the Russian Federation, if the order for goods is located in Asian countries, the delivery time to final consumers (retail outlets) is 60–80 days [8]. In addition, the long term delivery of goods by sea leads to the necrosis of working capital, to the emergence of unnecessary "supplies in transit" and reserves in places of transshipment [8, 9]. Considering that a significant amount of cargo from South-East Asia goes to the countries of South and South-East Europe and Russia, it becomes obvious that the goods transportation by sea to traditional ports—hubs of Northern Europe is an ineffective solution [9]. It is necessary to create more and more effective logistics schemes for the goods delivery.

The need to find the ways to reduce the transportation time and cost, as well as to improve their environmental friendliness appears in the current conditions of increasing trade between the countries of Asia and Europe, as well as rising costs when delivering goods by sea. The solution of the problem is the goods transportation on the Europe-Asia land transport corridors with shorter delivery times.

A promising development of land transport links between Europe and Asia is the implementation of the project "The Economic Belt of the Silk Road" in the framework of the "Belt and Road Initiative" (B&R) initiative, which was first announced in 2013 by the Chairman of the People's Republic of China Xí Jínpíng [10]. The initiative implies the organization of transport, energy, trade corridors between the countries of Asia and Europe, cultural exchange and development of tourism, as well as the creation of a free trade zone [11]. B&R's main objective is to accelerate trade between China and European countries as a result of new transport corridors creation that are alternative to transport connection through the Suez Canal [12].

B&R concept includes two main projects (Fig. 8):

- project "Sea Silk Road of the XXI century", which will pass through the South China Sea and the Indian Ocean from the east coast of China to Europe. The main objective of this project is the creation of an effective network of transport routes between the largest ports of different countries, as well as the

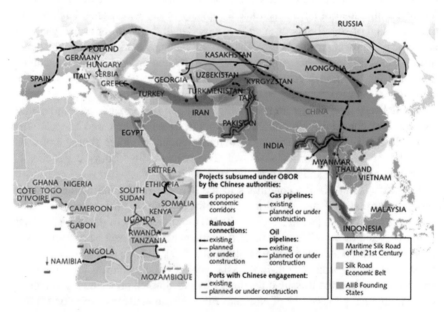

Fig. 8 Prospective transport corridors within the framework of the concept "Belt and Road Initiative" [103]

development of an economic corridor through the Indian Ocean, which will connect China with South Asia, the Middle East and the Mediterranean Sea;

• the project "The Economic Belt of the Silk Road" is a land route that will connect China with Eastern and Western Europe through Central and South-West Asia. The construction of the "Eurasian land bridge" which is a global logistics chain from the eastern part of China to Western Europe (to Rotterdam) is planned [13].

According to experts the forecast of transit and multimodal transportation development from B&R initiative implementation is shown in Fig. 9 [14].

Within the framework of the project "The Economic Belt of the Silk Road", it is planned to build three railway transport corridors (Fig. 8): northern (China–Central Asia–Russia–Europe–the Baltic Sea), central (China–Central and West Asia–Persian Gulf and the Mediterranean Sea) and southern (China–South-East and South Asia–the Indian Ocean).

The central transport corridor runs from the ports of Eastern China (Shanghai, Lianyungang) through the countries of Central Asia (Kyrgyzstan, Uzbekistan, Turkmenistan), and also through Iran. In this region there are currently troubled political circumstances, which is the reason for high risks for transportation in this direction. The southern corridor implies the further use of sea transport for the goods transportation. In this case the need for transshipment of goods in ports leads to an increase in the delivery time and cost.

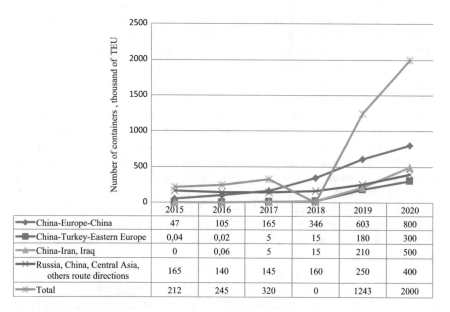

	2015	2016	2017	2018	2019	2020
China-Europe-China	47	105	165	346	603	800
China-Turkey-Eastern Europe	0,04	0,02	5	15	180	300
China-Iran, Iraq	0	0,06	5	15	210	500
Russia, China, Central Asia, others route directions	165	140	145	160	250	400
Total	212	245	320	0	1243	2000

Fig. 9 The development forecast of transit and multimodal transportation between China and the countries participating in the project "Belt and Road Initiative" [14]

From the point of view of delivery terms, low political risks the Northern Railway Corridor, passing through Kazakhstan and the Russian Federation is considered the most attractive. On the territory of the Russian Federation there are three routes of the northern land railway corridor from China to Europe:

1. The city of Manchuria, along the Trans-Siberian Railway to Moscow (9955 km);
2. Through Ulan-Bator–Ulan-Ude and the Trans-Siberian Railway to Moscow (6926 km);
3. Station Alashankou-Astana-Moscow (7144 km).

The possibility of northern transport corridor integration of the project "The Economic Belt of the Silk Road" into the system of operating transport corridors on the territory of the Russian Federation, as well as a significant reduction in the delivery time for this route, make this direction the most attractive and competitive in comparison with other land-based delivery ways. The project implementation of the northern corridor of the "The Economic Belt of the Silk Road" will allow shortening the goods delivery time from China to Eastern Europe to 11 days by rail [13].

The project "The Economic Belt of the Silk Road" meets both China's foreign economic interests and domestic interests in eliminating the economic and transport development unevenness of the central and western regions of the country. To implement this project, special attention should be paid to the development and improvement of the transport and logistics infrastructure, especially in cross-border and difficult-to-reach areas of the route [15], as well as technologies for cargo delivery and organization of interaction between sea and land transport modes.

A number of constraints prevent the implementation of the project "The Economic Belt of the Silk Road" and, in general, the land transport corridors development between Asia and Europe. The main groups of constraints, to our opinion, include technical, economic, organizational and technological, infrastructure and ecological ones.

The technical aspect of restricting the land transport corridors development between Europe and Asia is associated with different track width in China, Europe, Russia and Kazakhstan. The Chinese and European system uses a width of 1435 mm. The railways of the Russian Federation and Kazakhstan, along which the main part of the route for the existing and prospective East-West transport corridors runs (the Trans-Siberian Railway, the Southern Corridor, the North-South, the Northern Railway Transport Corridor of the Economic Belt of the Silk Road project), have a track gauge of 1520 mm. Overloading loads or using various technical solutions (for example, the Variable Gage Axels system—VGA) increases the transportation time and cost.

The economic aspect. The main economic constraints are related to the need for significant investments in the development of land infrastructure and with long payback periods for these investments. The high cost of construction is explained by difficult climatic conditions. For example, the cost of building one kilometer of the railway in the Far East is about 16.5 million US dollars [16]. If we consider the construction of high-speed railways, the construction costs will be even higher, about 32% more than in China or Europe [17].

Restrictive economic factors also include higher operating costs for the goods delivery by rail, compared with sea transport. For instance, the cost of shipping a 40-foot container from Chongching (PRC) to Duisburg (Germany) along the Yu'Xin'Ou Railway is about 8–9 thousand US dollars/FEU, which is twice the cost of organization of delivery by sea, by transit through the Yangtze River and the seaport of Shanghai [18]. This problem is currently being partially solved by the allocation of government subsidies from the government of several provinces of the PRC. This local solution allows compensating up to 50% of containers shipping cost from China to Europe by rail.

Large operating costs are also associated with the empty freight car miles. According to the Chinese railway carrier's report, from January to November 2016, out of 48 block trains sent from China to Europe, only one returned back loaded. Moreover, according to the information agency of the China Youth Daily, between November 2014 and June 2016, more than 58 block trains or 3800 loaded containers were sent to Yiwu-Madrid, only 296 of which returned back loaded [19].

The reduction in operating costs associated with a large empty freight car mile on Europe–China route may result from an increase in imports from Europe to China. At present, such a trend has emerged in the sector of food products, pharmaceuticals and other types of goods, the delivery of which should be carried out within a short time. It is these delivery conditions that can be provided by the technology of express transportation of cargoes on the East–West direction, based on the use of specialized block trains.

The technological and organizational aspect. The main technological and organizational factor negatively affecting the timeliness of freight traffic on the territory of Russia and on the railways of the "1520 space" is the organization of railroad transportations on the basis of the so-called "train formation plan". The main objective of the plan for the formation of freight trains is to minimize the total transport costs for the accumulation of train trains and their reorganization en route. Low-power jets of car traffic volume are connected into combined trains and their subsequent formation at intermediate sorting stations to reduce the cost of their accumulation. Powerful jets of car traffic volume are allocated to independent destinations and the so-called route trains are formed from them.

The main disadvantage of this organization form of trains based on the use of the train formation plan is its orientation solely on reducing transportation costs–total costs for the accumulation and processing of cars. When forming prefabricated trains and route trains, the requirements of cargo owners for timely delivery of cargo are not taken into account. As a result, there is high unevenness in the supply of railcars and goods to end consumers. This leads to an increase in storage costs, as well as to the total logistics costs.

The technology of container block trains being used recently for expedited transportation of cargoes on the East-West direction is based on the trains' formation, the composition of which does not change on the route [20].

Traditionally, block train technology is used to transport containers by rail for short distances between sea and dry ports [21]. Dry ports at the same time ensure the formation of container trains, the time of departure and arrival of which is rigidly fixed and linked with the ships incoming to the seaport [21]. In industrial transport-technological systems, such way of organizing railroad transportation is called a contact schedule.

To date, there has been a rapid increase in the volume of trans-shipment using block trains that deliver goods from western provinces of China, such as Sichuan, Henan, and Yunan to European countries [22]. According to experts' data [14, 18] over the past five years, the number of block train departures increased from 17 to 1702 trains per year (Fig. 10). The majority of Chinese logistics companies decided to develop rail transport using block trains that ensure the goods delivery from China to Europe.

To increase the speed of transportation, such trains are skipped over solid (fixed) train paths. The obvious advantage of this transportation organization form is the shortening of the delivery time. An additional positive factor is the minimum dependence of the block train technology on the system of car traffic volume organization on the "1520 space", as they pass through the territory of Russia and Kazakhstan. In particular, the influence of the plan for the trains' formation on rail transport in these countries is almost completely eliminated, because block trains pass through the "1520" railways without processing.

An important task that needs to be solved for the implementation of block train technology is the development of a coherent schedule for the movement of such trains. However, the potential high intensity of the traffic flows in the East-West direction and, as a consequence, their expected low unevenness and high regularity

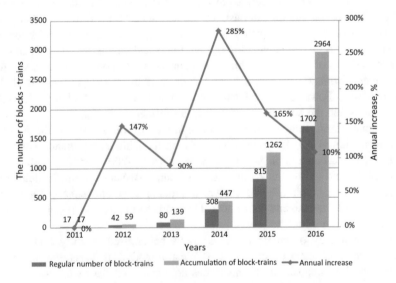

Fig. 10 Dynamics of block trains development between China and Europe

allow predicting low transaction costs for developing and updating existing train schedules [23, 24], operational control of freight trains movement [25] and planning the "1520 space" railways functioning.

The principal shortcomings of the block-trains technology include the need to accumulate powerful streams of car traffic volume to form block trains and also to break them at the final point. On short distances between dry and sea ports, this does not lead to additional costs, since it is possible to flexibly manage the size of the block train, for example, by the criterion of the operating costs minimum for the transportation of such trains [26]. On the contrary, such organization of supplies ensures timely delivery to the seaport of the ship party, rhythmic shipment, coordination of the arrival time of trains and the vessel incoming. As a whole this helps reduce the downtime of vehicles and the costs of goods storage (containers).

However, if it is necessary to transport containers to long distances, in case possibility of regulating the size of the train is limited due to the need to form trains of the maximum permissible size, appears the problem of plan optimizing for the formation of such trains [27]. Reducing transportation costs in the organization of railways on the East-West direction to block trains is associated not only with the passage of these trains throughout the route without processing, but also as a result of a reduction in the cost of such trains accumulation due to the high intensity of the corresponding cargo traffic. To reduce the costs of accumulating block trains, it is necessary to organize large multimodal container terminals or logistic centers both in the accumulation points and in the places of dispersion (distribution) of these carriages. Such terminals must solve two main tasks: to generate and distribute intensive car traffic volume, playing the same role as the sea ports for the Hinterland for the points of production and consumption of cargo flows; the second task is more difficult—it is to ensure the necessary quality (timeliness) of freight traffic. And timeliness should be understood not only as speed and urgency of transport, but also their regularity and the possibility of delivery by the appointed time. Such terminals must have significant processing capabilities and capacities. In order to ensure the timeliness of transportation, the terminals for the formation and breaking block trains should be able not only to carry out large amounts of sorting work for the express formation of the block train composition, but also, if necessary, to delay (accumulate) car traffic volume streams of separate destinations. Such a need arises, for example, if the consumer of the cargo temporarily does not have the possibility of unloading at a minimal cost. As a result of solving this problem, it is possible to achieve a reduction in the total logistics costs for all freight flows from the places of their origin to points of consumption. Such an integrated approach will reduce not only transport costs for the formation and promotion of block trains, but also the storage costs of cargo owners.

The organization of container terminals or logistics centers for cargo operations with block trains actually means "transferring" the reserves of access and processing capacities of railway and cargo infrastructure objects of "1520 space" to the start and end points of the block trains.

Consequently, the issue of forming a network of such container terminals or logistics centers on the territory of China and the European Union is becoming topical. Considering the considerable prospective volumes of freight transportation by rail on the East-West direction, this problem needs to be solved systematically, taking into account many factors that determine not only the economic but also the social and ecological efficiency of the terminal network.

An attempt to systematically assess the efficiency of multimodal transport organization using the so-called LIFT (Long Intermodal Freight Trains) was undertaken in the work [28]. The article presents a comparison of LIFTS with traditional CIFT, and also examines the impact of such a method of organizing freight transport on the environment, the society and the European transport policy in general.

Thus, the organization of freight rail transport on the East-West direction is closely connected with the equally important problem of choosing the location of logistics infrastructure facilities.

The infrastructural factor. In China, the transport and logistics infrastructure is the most developed in the coastal areas. There are sea and dry ports here, eight of which are among the ten largest seaports in the world. In accordance with the strategy for the construction and development of sea freight ports, [29] from 1985 to 1995, China built 217 new deep-sea parking spaces. This allowed increasing the total container turnover in 2016 to 195.9 million of TEU containers. However, the increase in the transport volume led to problems associated with the interaction of rail and sea transport and, as a result, to an increase in the processing time of railcars.

The main reasons for the idle time of wagons at railway port stations are simple railcars awaiting or the process of feeding (cleaning) the railway vehicles to freight fronts; reception of inflated cargo shipments by sea ports; irregularity of shipment within the planned month; violation of the train routing plan in connection with the allocation of certain time windows for the supply of railway vehicles on the way of private use [30].

An effective solution to increase the processing capacity and capacity of seaports is the organization of nearby terminals—dry ports [31]. Dry ports are also an effective tool for reducing the unevenness of cargo flows as a result of managing the formation of ship lots and their coordinated coming to sea ports. In addition, dry ports contribute to reducing the environmental burden on the seaport area as a result of the transfer of the work part for processing cargo and related emissions for redistribution of the building zone [32].

According to experts' value, the creation of a dry port allows to increase the throughput and processing capacity of the sea port by an average of 20–25% [33]. Moreover, such terminals provide not only an increase in the total processing capacity and capacity of the "sea port–dry port" systems, but also play the role of regional distribution centres. For example, in the PRC, where there are over a hundred dry ports, the distance between these terminals can reach 2000 km [34], which considerably exceeds the average effective transport distance between dry and sea ports, typical for European conditions.

Recently, the functions of dry ports have significantly expanded. They provide customs clearance services, a set of logistics services for receiving, storing, handling cargos, their further distribution, etc. The high demand for integrated logistics services in the regions, the expansion of the distribution network for trade networks and the service area for dry ports have contributed to the creation of logistics centres in the regions. The creation of logistics centres in the concentration areas of consignors and consignees at the intersection of transport highways will ensure the formation of large consignments to send goods in international traffic, which will reduce transport costs. And placing them in close proximity to the consumption centres will reduce the costs of processing the goods flow and their distribution.

The location and use of logistic elements (dry ports, logistics centres) along the block trains route will increase the interaction of the participants in the supply chain with more efficient loading of block trains in both directions.

The ecological aspect. The harmful impact of land transport on the environment and the increase in the environmental requirements for the goods transportation are an additional deterrent to the increase in the transport volume of the East-West direction by land transport corridors [35–37].

The fulfilment of key and basic functions [38] in the flow processes of logistics and transport systems is accompanied by a negative impact on nature and the environment. This is expressed [39] in:

- consumption of natural resources;
- contamination of the environment with harmful substances;
- energy and visual pollution of the environment;
- land alienation and degradation;
- injury and death of people, animals, causing harm to health;
- causing material damage as a result of traffic accidents, accidents and road accidents.

The environmental aspect consideration takes particular relevance on the creation and development of international transport corridors, where one of the key tasks is the desire to reduce the environmental load at all stages of the delivery process. This desire is due, on the one hand, to meeting the requirements and restrictions in the field of the environment of the participating countries in promoting commodity flows, on the other hand, to the possibility of gaining economic benefits, competitive advantages, increasing the image and public popularity of the company through the use of green technologies [3].

Throughout the world the solution of these problems is based on the use of the sustainable development concept, and in logistics activity on using the principles of green logistics. In the case of implementing projects to create new transport links of international communication in the East-West direction, the question of the conformity of transport and logistics systems to the principles of the sustainable development concept becomes topical.

3   Factors of Sustainable Development of Transport Systems

Scientific research for 15–20 years emphasize attention at Green Supply Chain Management more and more, as impact of a separate company on the environment is diverse and multidimensional. The opinion of scientists comes down to the fact that ecological aspects need to be considered at all stages of functioning of a supply chain: from production and conversion of raw materials through production and distribution of finished goods before its end use or utilization.

The investigations of the last years [40–42] show the growing interest in Concepts of Sustainability, Livability, Sustainable Development and Sustainable Transport. Fundamental bases of sustainable development are revealed in regulatory legal acts of the international and national legal system in the majority of the countries [3]; they are annually discussed at a set of conferences and the congresses worldwide.

However, the scientific world disputes about the most important factor for stable development. Some authors emphasize importance of preserving functionality of the nature and the environment; others underline the role of social aspects and political institutes or of stable economic growth in society [3]. But, despite differences in approaches and definitions, most of experts speak about balance of economic, ecological and social aspects of stability.

The analysis of the existing determinations of stability in social and economic systems allowed allocating four approaches different from each other [43]:

1. stability as safety, reliability, integrity and durability of a system;
2. stability as a capability of a system to remain invariable during certain time, i.e. the invariance of key parameters of a system;
3. stability as a capability of a system to keep dynamic balance;
4. stability as a capability of a system to develop, i.e. steadily to function, keep movement on the planned trajectory with self-development.

Sustainable development in relation to the transport systems means that meeting transport requirements doesn't contradict priorities of environmental protection and health, doesn't lead to irreversible natural changes and depletion of irreplaceable resources [2].

Reaching reasonable balance of ecological, economic, cultural and social development and of people's needs is the basis for all existing models of sustainable development [5, 44]. However, as the analysis shows, the main shortcomings of the known models of sustainable development are:

- static character and insufficient emphasis on dynamics of development;
- fragmentariness of connections between aspects of sustainable development (ecology, economy and society);
- complexity, taking into accounts various restrictions and the contradictory purposes.

Violation of balance between aspects of sustainable development and achievement priority of economic targets is a consequence of the specified shortcomings in comparison with ecological and sociocultural aspects. It acquires special relevance in economic activity of industrial, trade, transport organizations which function in the difficult, dynamically developing market environment and which by-product of activities is the negative impact on the environment. It motivates the transportation and logistics companies to implement green technologies in their activities. For example, DHL, Schenker AG, Kuehne Nagel, UPS, Group COSCO, etc. in case of activities implementation using green technologies within the concept of sustainable development, determine it as effective approach to management of engineering procedures, resource and energy flows, for the purpose of decrease in an ecological and economic damage to the environment, ensuring social development of workers and effective innovative development of production.

The special role in development and implementation of the sustainable development concept is played by different ecological programs and projects realized with support of public and state institutions, business of structures, research organizations and international associations in the countries of Europe, Asia, North America and BRICS countries. The Global Green Freight Project [45], Lean & Green [46], ECO Stars [47], Green Freight Europe [48], SWIFTLY Green [49] and many others are the most significant ones.

The greatest interest concerning environmental protection and sustainable development of transport corridors is represented by the concept of "Green Transport Corridors" which was firstly mentioned at a congress of EU countries "Prospects and development plans in transport logistics" [49, 50] in 2007. The purpose of "Green channel" is in increasing performance indicators, an energy efficiency and ecological compatibility of the transport systems in all territory of the EU. Reaching this purpose [51–53] is possible by integration of different transport types into a single system on the basis of their merits, use of innovative technologies, methods and tools of green logistics in the process of goods flows and also in a cooperation and coordination of participants of a supply chain at the different levels (municipal, regional, federal and international), including interaction with participants of cargo delivery and state bodies at infrastructure design and constructions stages.

Reaching this purpose is followed by availability of certain risks of that might occur in Green Transport Corridor. According to categories of sustainable development the main risks in development of green transport channels are [54, 55]:

- Economical: missing overall strategy, disharmonized standards, lack in infrastructure, not reached service levels, bad information exchange;
- Environmental: problems of transport modes, violations of environmental requirements;
- Social: safety deficiencies, lack of knowledge, skills shortage.

Risk minimization is reached by use of the sustainable development principles by the companies, applying methods and tools of green logistics now; by forming ecological consciousness and skills of ecological behavior in business and private life, through implementation of actions for training, forming of competences for sustainable development; realization of different ecological programs and projects.

Thus, implementation of the sustainable development concept in supply chain management within functioning of the international transport corridors will demand enhancement of the following aspects [3, 50, 52, 56, 57]:

- social problems and policy are the questions of business ethics and corporate social liability; environmental audits; transport policy harmonization in a certain region and compliance of the legislation requirements in the field of ecology;
- coordination and interaction–cooperation between business sector, academic structures and political bodies both at local and regional levels; creation of coordinating committee for the green channel, development of IT systems for the support of physical infrastructure;
- purchasing and manufacturing are the questions of ecological development, development and processing products to reduce harmful emissions and waste; compliance to the quality standards in the field of the environment; improvement of service quality;
- green logistics is the environmental issues connected with steady transportation, handling and storage of dangerous materials, inventory management, warehousing, packaging and distribution of objects location, directed to decrease CO_2 emissions;
- infrastructure is ecological designing of objects for transport and logistics infrastructure, construction and placement of the logistics centers, of additional infrastructure.

According to the described approach as the generalizing factors influencing sustainable development of the transport systems, transport infrastructure, the authors offer three groups of factors (Fig. 11):

- socio-economic factors characterize the production and use of GRP (Gross Regional Product) rate, level and people's quality of life, investment and labor potential of the subject, a foreign trade turnover;

Fig. 11 Groups of factors for sustainable development of transport systems

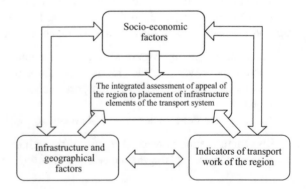

- infrastructure and geographical factors characterize the development level of transport infrastructure, determine topography of the area and a territorial arrangement of the subject, including rather external borders, transport corridors, the markets;
- indicators of the region transport work characterize amount of transportations in the examined subject, structure of a cargo flow.

In detail the provided factors are considered in paragraph 6.

Sustainable development of the transport system for development of cargo flows on the East-West direction should be based on the analysis of the existing infrastructure equipment and social and economic development of regions in which territory the existing transport corridors pass or in the territory of which it is supposed to place new elements of transport and logistics infrastructure. The authors offer to use system accounting of the considered factors for the location choice of the logistics centers as of the most important elements of the transport systems. It will allow reducing capital investments in infrastructure development, providing their usage with a necessary manpower, reducing both resource intensity of cargo delivery in general, and negative impact of this process on the environment.

4 Russian Transit Potential in the Cargo Transport Maintenance on the East-West Direction

One of the important tasks of the Transport Strategy of the Russian Federation until 2030 is the realization of the country transit potential, the creation of competitive transport corridors on the basis of an integrated transport and logistics infrastructure [58]. Russia has a favorable geographic position for the promotion of cargo flows in the direction East-West, demonstrates a stable political situation. According to the data of the Bank of Russia, in January–July 2017, Russia's foreign trade turnover amounted to 318.3 billion US dollars, which is 27% more than in the same period in 2016, including exports—190.9 billion US dollars (126.6% against the previous period), imports—127.4 billion US dollars (127.5%) [59]. External turnover with the EU countries in January–July 2017 increased by 26.3%, with China by 36.4%

However, at present only a small part of Russia's transit potential is used. The volume of transit traffic on the territory of the Russian Federation is less than 1% of the trade turnover between the countries of Asia and Europe. The interest of Asian countries, in particular of China, in alternative routes for the goods transportation on the East-West direction, which pass through the territory of the Russian Federation, has been recently increasing more and more.

The international transport corridors connecting Russia with Asian countries include: Trans-Siberian Railway (TS), North–South (NS), Northern Sea Route (NSR), Primorye-1 (Harbin–Grodekovo–Vladivostok/Nakhodka/East–ports of the Asia-Pacific region)–PR1, Primorye-2 (Hunchun–Kraskino–Posiet/Zarubino–ports of the Asia-Pacific region)–PR2. Also, the system of international transport corridors of Russia includes three Pan-European transport corridors No. 1 (PE1), No. 2 (PE2), No. 9 (PE9).

Recent advances in the icebreaker fleet make the Northern Sea Route (NSR) a rather promising direction. NSR will allow cargo transportation from the northern regions of Russia to the countries of Asia, will provide access to the market of the European part of the Russian Federation through the ports of the Arctic basin. At the same time, for transportation of goods from Rotterdam to the ports of China and Japan via NSR, it will take about 23 days (the distance averages 7610 miles). A similar route through the Suez Canal will take 35 days with a transportation distance of about 10,600 miles. Thus, in the case of using NSR, delivery time, fuel costs are reduced, costs associated with ship protection are reduced, and the amount of carbon dioxide emissions is reduced by optimizing the movement route of ships. For the development of NSR, modernization and reconstruction of infrastructure (of Arctic ports), re-equipping the icebreaker fleet is required, which implies large capital investments [60].

The basis of overland transport communications of the East-West communication is the Trans-Siberian Railway, passing through the territory of the Russian Federation and providing access to the network of railways of North Korea, China, Kazakhstan and Mongolia in the east through the border stations, and in the west via Russian ports and border transfers to European countries [61]. The Trans-Siberian Railway is included as a priority route in the communication between Europe and Asia to the projects of the UNECE UNO (United Nations Economic Commission for Europe UNECE), UNESCAP (United Nations Economic and Social Commission for Asia and the Pacific), OCR (Organization for Cooperation of Railways).

In comparison with the traditional sea route, Trans-Siberian Railway has the following advantages: reduction of delivery time to 10 days (for example, from China to Finland); increasing the safety of goods by reducing the number of cargo transshipments, and as a consequence, reducing costs; transport safety and low political risks. When transporting goods through the Suez Canal, there is a high risk of unauthorized seizure of ships. When choosing Trans-Siberian Railway 90% of the entire route passes through Russia, which has a stable political and economic environment.

The basis of the Trans-Siberian Railway is railways, stretching from Moscow to Vladivostok 9288 km. For the first half of 2017, the volume of transit traffic by rail on the territory of the Russian Federation amounted to 10,688.72 thousand ton, which is 25% more than in the same period in 2016 (8525.83 thousand ton) [62]. However, if we consider the statistics for the last 3 years, then there is a decrease in transit rail freight by 20% (Fig. 12). Basically, this is due to a decrease in the volume of rail transit from Kazakhstan (from 17,337.79 thousand ton in 2014 to 13,798.35 thousand ton in 2016, a decrease of 20%) and from the Ukraine (from 1,495.66 thousand ton, to 384.63 thousand ton), which is connected with anti-Russian sanctions and Russian countermeasures.

Fig. 12 Volumes of transit rail transportation on the territory of the Russian Federation

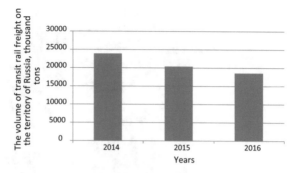

Fig. 13 Structure of transit traffic through the Russian Federation in countries of origin

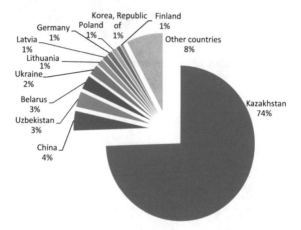

Transit railway transportation from China through the territory of Russia for the period 2014–2016 increased from 1 to 4% of the total volume. Since 2014, transit volumes have increased almost 3 times from 258.83 thousand ton to 756.23 thousand ton (Fig. 13).

The main countries receiving transit railway traffic from China through Russia are Poland (58%), Germany (14.6%), Belarus (8.7%), Uzbekistan (5%). The structure of the recipient countries of transit cargo from China through the territory of the Russian Federation is shown in Fig. 14.

At the same time, Kazakhstan remains one of Russia's key transit partners, accounting for 74% of the total volume of transit rail traffic via Russia (Fig. 13).

Among the factors identified in the second paragraph limiting the development of land transport corridors, the most acute problem in the Russian Federation is the elimination and/or reduction of the negative impact of the technical and technological aspect. In addition to the different track widths, the mismanaged work of rail and maritime transport, the organizational problem of Russian railways is added which is associated with a large number of railcars and owners of railway rolling stock.

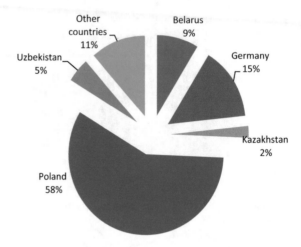

Fig. 14 Structure of recipient countries of transit cargo from China through the territory of the Russian Federation

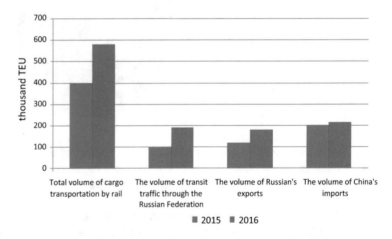

Fig. 15 The volumes of cargo transportation and foreign trade between China and Russia

The total volume of containers transported by rail to China–Russia–Europe also increased in 2016, an increase was 30% (Fig. 15) [63].

At the beginning of the century reforming the railway transport of the Russian Federation, the purpose of which was to form a competitive environment, led to the large quantity of railcar owners in the railway transportation market in Russia. These owners do not co-operate the cars. As a result, the total time of railcars turnover and the time of empty railcars movement increased.

However, the competitiveness exists only in the field of cars operation. The railways infrastructure is owned by JSC Russian Railways. This company defines the basic principles of organizing rail transport in Russia: the first one is the plan for the trains' formation; the second is the train schedule. The complexity of the organization of transit rail transportation through the territory of Russia, connected with the technological and organizational aspect is presented in the second paragraph of the chapter. An effective solution to this problem is connected, in the opinion of the authors, with the implementation of the technology of block trains.

The technology of block trains, which is common in Europe and is increasingly used in China, is poorly developed in the Russian Federation. The practice of putting container trains in the Russian Federation shows that their use is effective if the direction has a stable container flow. The organization of the container trains movement is carried out according to the "firm" threads of the schedule. The average train consists of 57 railcars. When organizing the movement on demanding schedule, the goods and railcars are registered in such a way that the carriages are carried out without processing and reformation in the journey. This ensures the arrival of the train to the destination station on an accelerated schedule.

As a result of the meeting of the authorized representatives of the railway administrations of the Commonwealth of Independent States, Georgia, the Republic of Latvia, the Republic of Lithuania and the Republic of Estonia, held on May 30–June 2, 2017 in Vilnius, it is planned to provide for laying and destination for 94 international container trains [64].

Currently, only the terminals "Voskhod" and "Logistics-Terminal" (railway station of Shushary), "Predportovy" and "Interterminal" (the railway station Predportovaya) can serve for organizing the movement of block trains, for example, in St. Petersburg transport junction [21]. At the same time, 60% of all foreign trade cargoes (exports, imports) passes through the transport hub of St. Petersburg.

In other regions of the Russian Federation, such national terminals (dry ports, logistics centers) capable of providing cargo servicing in international traffic and forming container trains are practically absent. This negatively affects the interaction of participants in the transportation process and leads to inefficient, in terms of common logistics costs, schemes of cargo transportation.

Delivery of goods from China to Russia is most often performed as follows: cargoes from China arrive by traditional sea way to St. Petersburg, and then by road transport are delivered to Moscow for further distribution by regions. Given the proximity of the Far East, Siberia, the Urals, and the Volga Federal District to the border with China, such a scheme leads to an increase in the delivery time, vehicle mileage, the costs of "supplies on the road," etc. The creation of terminals and logistics centers in the regions, near the land transport corridors of the East-West direction, will ensure the servicing of internal and external cargo flows in Russia.

The processes of globalization, changes in foreign trade relations, and the tendency to increase the volume of foreign trade pose a challenge for Russia to rationally use the country's transport sector. The realization of the transit potential of the Russian Federation is possible only in the integrated development of large transport corridors in the directions "West–East" (using the Trans-Siberian Railway), "North–South" (the Baltic Sea coast–Persian Gulf) and the Northern Sea Route [58]. The increase in the volume of transit traffic, the growth of foreign trade turnover requires a qualitatively new development of transport hubs, customs entry points, the construction of modern objects of logistics infrastructure (container terminals, logistics centers, dry ports).

5 Practical Experience of Placing Logistics Centers in Cargo-Flows Maintenance of the East-West Direction

The modern infrastructure for the promotion and maintenance of freight flows is represented by logistics centers (LC). Logistics centers allow to increase not only the quality of provided services and the coordination of participants work in the transportation process, but also to reduce logistics costs in the final price of products. The experience of Western countries has shown that the work of logistics centers makes it possible to reduce transportation costs by 7–20%, the cost of handling and storage of material resources and finished products by 15–30%, allover logistics costs by 12–35% [65].

In Western Europe there are about 120 logistics centers, some of which are integrated within the European Platform, which is coordinated by the cooperation of seven European countries (Denmark, France, Spain, Luxembourg, Germany, Portugal and Italy) [66].

In Europe, the first projects to create logical centers were associated with the development of the seaports infrastructure. As a result of the high growth rates of the world economy, seaports have become unable to meet the demands of regional trade. The construction of new docks required high costs; there was also a problem of free space lack to expand the water area of the ports. The way out of this situation was the change in the seaports concept and the transfer of functions part into the internal territory, where logistics centers or so-called "dry ports" began to be created.

With the growing demand for integrated logistics services, the functions of such logistics elements have become more complex, and the service and distribution areas of domestic freight flows have increased. An important characteristic of logistics centers is the ability to handle the traffic of various types of transport. In such conditions, special attention is paid to the location of the logistics center. World experience demonstrates that the economic feasibility of locating the LC determines the proximity to the junctions of national land routes, navigable routes, international transport corridors, to sales centers (regions with a high population density), and to production centers that export products, provide traffic congestion infrastructure (Fig. 16) [67]. Particular attention is paid to social infrastructure. On the one hand, the logistics center can be located next to a large settlement, which allows solving labor issues; on the other hand, the complex itself can cause the growth of the district population.

At present, the European transport and logistics system is characterized by the following features [68, 69]:

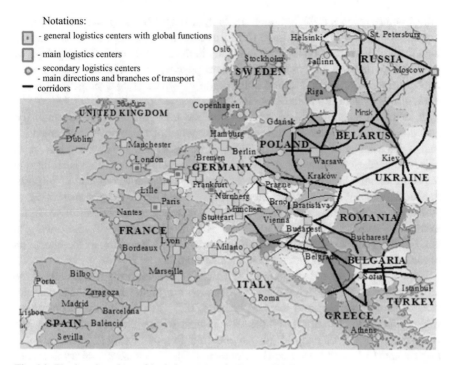

Fig. 16 The location chart of logistics centers in Europe [104]

- integration of regional logistics structures into the national transport system;
- interaction of various modes of transport through the coordinated work of existing hubs, instead of creating new ones;
- high level of cooperation between logistics centers, railways, carriers, authorities;
- strong government support at the design and development stages of logistics centers.

In Russia, until 1995, large, unorganized warehouse complexes operated in the country that did not meet contemporary standards of goods storage and processing of goods traffic. Qualitative changes in the markets of transport and logistics services and logistics real estate in Russia have occurred since 2003–2004 with the arrival of large international trading companies and distributors placing high demands on the quality of services on the basis of European standards on the regional markets.

Currently, the main number of objects of logistics infrastructure is concentrated in Moscow and in Moscow region, in St. Petersburg. In the regions of the Russian Federation, the logistics infrastructure is poorly represented, while in the regional markets there is a high level of demand for rental of warehouse space by end-users —trading and manufacturing companies, as well as by logistics operators providing integrated logistics services. The gap in the equipment of the transport and logistics infrastructure reaches 60%, although the bulk of freight traffic (more than 80%), of manufacturing (70%) and of trade (65%) is accounted for regions. In connection with the high investment risk of the company in the construction LCs are guided by profitability, demand and, as a result, a short payback period of the project. Such logistic centers can be considered only on-object—there is no synergetic effect of the logistics system of commodity circulation.

In a market economy, state regulation of the transport and logistics industry, aimed at harmonizing the interests of various modes of transport, at implementing transit potential and improving the quality of transport and logistics services, is an objective necessity. In Russia, the model of public-private partnership is realized in transport mainly in the construction of highways, the development of airports, transport hubs. When creating the logistics infrastructure (the construction of logistics centers), this form of cooperation just begins to be introduced, for example, "Sviyazhsky Logocenter" in Kazan, "Logopark Tolmachevo" in Novosibirsk (public-private partnership is implemented, basically, as the allocation of land for construction), but these are single objects.

One of the largest objects realized in recent years within the framework of the development of the transport corridor infrastructure of the Europe-Asia communication is the transport and logistics center (TLC) "Yuzhnouralsky". The project was created with the support of the Government of the Chelyabinsk region, the Government of the Xinjiang Uygur Autonomous Region of China, Resource LLC, Logic Land LLC, and Resource Development [70]. TLC Yuzhnouralsky is part of the concept of the "The Economic Belt of the Silk Road" and a key component of the logistics infrastructure of the transport corridor "Xinjiang Uygur Autonomous Region (XUAR)–Kazakhstan–Chelyabinsk Region." Geographically, this logistics center is located at the intersection of the main roads, for the international traffic it is the nearest entry point from the territory of north-west and central China to Russia. The work of the center allows distributing incoming goods to major cities in Russia–Chelyabinsk, Ekaterinburg, Ufa, Tyumen, Perm, Samara are located in a radius of 700 km from TLC.

The total area of the complex after the completion of the construction will be 180 ha, on 60 ha of which there will be distribution warehouses, a customs station with a temporary storage warehouse and a railway container terminal capable of receiving up to 400 containers per day. In April 2015, operations started on the Container site and a modern warehouse complex. Operating capacity of the terminal allows serving 2.5 million ton of cargo per year or 252 thousand TEU per year (20 container equivalents). The daily volume of processing will be 7 trains or 700 TEU. However, at present, the existing capacities of the logistics center are under-utilized [71] because of the long distance from the existing freight flows of the East-West direction, which allows us to speak about insufficient consideration of the factors determining the efficiency of logistics centers when choosing their location. Disclosure of the entire potential of the Yuzhnouralsky TLC is possible when one of the variants of the route is being passed within the framework of the project "The Economic Belt of the Silk Road" in the section Astana–Chelyabinsk.

This practical example demonstrates the importance and necessity of taking into account a multitude of factors when choosing the location of logistics centers as one of the main elements of transport systems.

The study of the practical experience of locating logistics infrastructure facilities for servicing freight flows in the direction of Europe-Asia showed that at present there is no common understanding of the term "logistics center". The terms "logopark", "logistic village", "logistics center", "transport and logistics complex", "dry port", etc. are often found, which are interpreted as synonyms. The most complete terminology and classification of logistics infrastructure objects is revealed in the works [72–75].

In this paper, the logistic center is understood as the "territorial association of independent companies and bodies engaged in cargo transportation (for example, of transport intermediaries, consignors, transport operators, customs authorities) and dealing with related services (for example, storage, maintenance and repair) including at least one terminal" [76].

The authors' representation of the logistics infrastructure objects depending on the management object and logistics integration degree in the business processes of the enterprise is depicted in Fig. 17 [77].

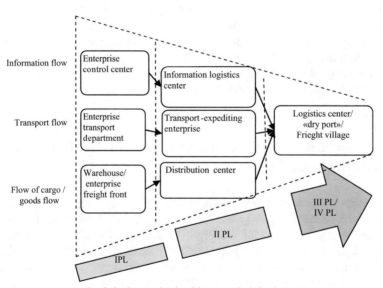

Logistics integration level in enterprise's business-processes

Fig. 17 Level representation of the logistics infrastructure objects

The existing logistics infrastructure objects in world practice can be classified as follows:

- by the nature and scale of the activity: of international (national), regional and local significance;
- by the main brunch of business: wholesale and logistics center, information and logistics center, customs and logistics center and transport and logistics center;
- by type of cargo being processed;
- by the management object (information flow, flow of goods, cargo flow, flow of vehicles).

An extended characteristic of the logistics infrastructure objects is presented in Table 2 [77].

Analysis of the transport and transit potential of Russia, of practical experience in the location of logistics centers showed that the formation of LC becomes a promising area for the region development. Both the business structures and the governments of the regions are interested in developing the logistics infrastructure.

Nevertheless, the construction of new logistics centers in the conditions of absence of a unified methodology and methods of forming a logistics infrastructure leads to an uneven distribution of the logistics infrastructure objects across the country. The formation of a sustainable transport systems development for servicing freight flows to the East-West directions should be based on an analysis of the existing infrastructural equipment and socio-economic development of the regions through which the transport corridors pass. Further this will reduce the capital investment in the infrastructure development, provide the necessary human resources and the required service level.

6 Methodical Estimation Bases for Logistic Centers Placement Options

An important place in logistics planning and creating conditions for the sustainable development of the transport system for servicing freight flows in the Europe-Asia direction belongs to the location analysis of the logistics infrastructure facilities. Typical problems associated with the logistics centers location are characterized by high complexity. The complexity is explained by the numerous variants of placement to be analyzed, the number of which is estimated by the product term of the number of types of logistics centers, the number of possible locations and the number of strategies for using each center. Such analysis needs detailed information of technical and economic indicators variety. The choice of the best alternatives on the basis of such an abundance of data requires an appropriate sophisticated technique of analysis and modeling.

In the past, distribution networks were relatively stable, and the task of locating logistics infrastructure objects was decided from the position of a separate group of

Table 2 Characteristics of the logistics infrastructure objects

Parameter	Distribution center	Dry port	Logistics center		
			Local	Regional	National and international
The main logistics functions	Distribution logistics functions	Transport logistics transport	Limited logistics service	Definite logistics services	Full range of logistics services
Servicing zone	Is determined by enterprise	80–500 km	50–80 km	80–500 km	Over 500 km
Object area	Is determined by the unique features of production	Is determined by process volume	2–10 ha	20–50 ha	50–150 ha
Processed cargo types	Is determined by the unique features of production	Different cargo categories	consumption cargo	General consumption cargo	Different cargo categories
Quantity of transport modes	As a rule, one	One-two	One or more	Two or more	Two or more
Management object	Goods flow	Cargo flow Container flow	Transport flow Cargo flow		

consumers. With the expansion of the markets borders, with the increase in foreign trade relations and the number of intermediaries providing logistical services, it becomes urgent to find alternative channels for distribution and placement of logistics infrastructure objects, taking into account the influence of such factors as foreign trade volume, proximity to transport corridors, etc. [76].

In order to assess the effective operation of the logistics center in the long term, it is necessary to take into consideration the dynamics of the infrastructural equipment development, indicators of transport work and the economy of the region; use a systematic approach to the study of the parameters of potential locations for objects of the logistics infrastructure. In this regard, the issues of defining the main factors influencing the location of the logistics center, determining the statistically significant links between them for a comprehensive assessment of LC sites, and developing a methodology in which the territorial development of the logistics infrastructure is connected with the long-term development prospects of the country (regions) have become the object of the research.

When locating logistics facilities, it is proposed to take into account a number of socio-economic factors, an objective assessment of which, in our opinion, not only creates the conditions for minimizing investor costs, but also increases the competitiveness of the region and the companies owning such logistics capacities [76, 78–83].

In works [84, 85], such criteria are identified for placing logistics infrastructure objects, such as: transport accessibility of the terrain, proximity to markets, availability of land, state support, availability of labor resources, load on the transport network.

Regarding the experience of logistics centers location abroad, world trends in servicing transport corridors and relying on work [86, 87]; it is suggested to take into account the geographical location of the subject relative to transport corridors when choosing the location of the logistics center. The presence of transport corridors on the territory of the Russian Federation subject of will allow expanding the scope of the logistics center, which will serve not only transit freight traffic, but also provide domestic regional and interregional transport links.

Using the proposed level representation of the logistics infrastructure objects, as well as taking into consideration the approaches determining the factors affecting the transport services volumes, freight transportation and the transport infrastructure development, the factors were grouped as follows (Fig. 18):

- socio-economic factors characterize the level of production and use of GRP, the level and quality of the population life, the investment and labor potential of the subject, foreign trade turnover;
- geographical factors determine the topography of the terrain and the territorial location of the subject, including external borders, transport corridors, markets;
- infrastructure factors reflect the development level of transport infrastructure;
- political and regulatory factors demonstrate the level of state support for transport and logistics activities, programs implemented in the transport and logistics complex, changes in transport legislation, customs policy peculiarities

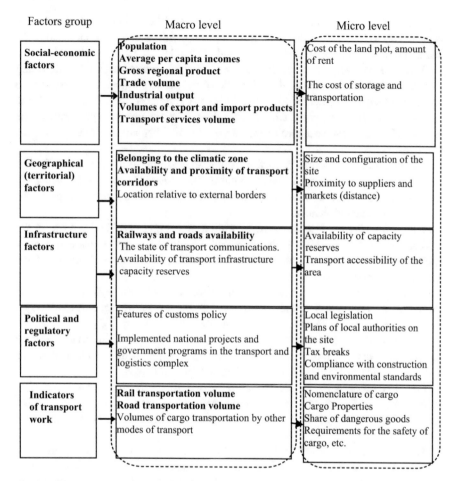

Factors group	Macro level	Micro level
Social-economic factors	**Population** **Average per capita incomes** **Gross regional product** **Trade volume** **Industrial output** **Volumes of export and import products** **Transport services volume**	Cost of the land plot, amount of rent The cost of storage and transportation
Geographical (territorial) factors	**Belonging to the climatic zone** **Availability and proximity of transport corridors** Location relative to external borders	Size and configuration of the site Proximity to suppliers and markets (distance)
Infrastructure factors	**Railways and roads availability** The state of transport communications. Availability of transport infrastructure capacity reserves	Availability of capacity reserves Transport accessibility of the area
Political and regulatory factors	Features of customs policy Implemented national projects and government programs in the transport and logistics complex	Local legislation Plans of local authorities on the site Tax breaks Compliance with construction and environmental standards
Indicators of transport work	**Rail transportation volume** **Road transportation volume** Volumes of cargo transportation by other modes of transport	Nomenclature of cargo Cargo Properties Share of dangerous goods Requirements for the safety of cargo, etc.

Fig. 18 The main factors affecting the placement of logistics infrastructure facilities at macro and micro levels

for the management object (information flow, goods flow, cargo flow, traffic flow);

- indicators of transport work are the data on the traffic volumes in the region, the structure of freight traffic.

In our opinion, it is possible to study the factors influence on the location choice of the logistics infrastructure objects on two basic levels: the macro level and the micro level.

The macro-level factors influence the development of the logistics infrastructure at the level of the country, district, region (II and III level of the object of the logistics infrastructure, depending on the degree of logistics integration); and micro-level factors influence the choice of a specific site for building and have a direct impact on the parameters of the future object of the logistics infrastructure (it is necessary to take them into account for logistic facilities of level I, and also at the last stage of selecting a land plot for II and III levels).

At the micro level, i.e. at the level of the municipal district (city), when choosing a specific site for the location of the logistics infrastructure object, it is advisable to apply methods of "center of gravity", linear programming, etc. In this case, the choice is made from the given alternatives taking into account the topography of the subject (city) area, the structure of the cargo flow, the distances to its main sources and consumers.

The factors that are subject to further research are performed in Table 3. The choice of indicators that are to be included in the system of factors influencing the logistics center location was based on an analysis of existing approaches to the allocation of demand factors.

When choosing the logistics center location, it is suggested to be guided by the assessment of the Russian Federation subject attractiveness, considering as a complex the proposed indicators of transport work, infrastructural, geographical and socio-economic factors.

The most common tool for assessing the attractiveness of the regions today is the formation of investment attractiveness ratings by the rating agencies: Moody's Investors, Standard & Poor's, the Bank of Austria and Expert [88].

The methods used to assess the investment attractiveness of the regions can be conditionally divided into three groups, which use: expert scores; statistical scores and converting particular indicators into an integral indicator.

The first group of methods is based on assigning the points to investment attractiveness of the region. These methods are subjective in nature and smooth the spread of the values of the factors, since experts, in general, tend to exhibit an average estimate.

The next method, which is widely used—statistical scores—is often used by reducing the numerical values of statistical indicators to scores on a scale. In this regard, this method also does not fully reflect the degree of statistical indicators differentiation by subjects because of the limited number of "data breakdown" of used intervals or the range of scores given in advance [88].

The methods based on the summary of the individual indicators are the calculation of the integral indicator, which is defined as the products sum of weight coefficients and the values of factors [89]

$$I = \sum_{i=1}^{n} w_i \cdot x_i, \tag{1}$$

where I—integral criteria; w_i—weight coefficient of i—factor; x_i—value of i—factor.

Disadvantages of such methods is that investment attractiveness assessments of the regions assess only the current state of the subjects, without regarding the projected values of the parameters influencing the investment attractiveness of the

Table 3 System of factors affecting the logistics center location

No	Factor	Observed value	Unit	Designation
Group of social-economic factors				
1	Population	Mid-year population	Thousand person	Population
2	Average per capita incomes of the population	The ratio of the annual volume of money income to the number of months and the average annual population number	rub./ person	Income
3	Gross regional product volume	Gross regional product volume per head	rub./ person	GRP
4	Industrial output	Industrial output per head (processing industry)	rub./ person	Industrial output
5	Trade volume	Retail turnover per head	rub./ person	Trade volume
6	Export products volume	Export products volume per head from a region to near and far abroad	USD/ person	Export
7	Import products volumes	Import products volume per head from a region to near and far abroad	USD/ person	Import
8	Transport services volume	Transport services volume per head	rub./ person	Transport services
Group of infrastructure and geographical factors				
9	Road traffic routes density	Length of hard stand road traffic routes in km on 1000 m^2 of the region area	km/ 1000 m^2	Density of roads
10	Railroad facilities density	Length of public railway tracks in km on 1000 m^2 of the region area	km/ 10,000 m^2	Density of railway
11	Belonging to the climatic zone	Points on the scale from 1 to 5 depending on the region climatic zone. 5 points for location in climatic zone I; 4 points for zone II, etc.	point	Climatic zone
12	Transport corridors on the region area	Point. The region location on the main direction of the transport corridor—1 point for each one; on the transport corridor brunch—0.5 of a point	point	Transport corridors
Indicators of transport work				
13	Rail transportation volume	Cargo weight in tones, established for public railroad transportation	mln ton	Railway transport
14	Road transportation volume	Dispatch outwards sum on the region area	mln ton	Road transport

regions. Indicators of transport work in the regions either are not considered in the rating formation, or are taken with low weighting factors. The calculation of the weights of each chosen factor should be based on functioning efficiency assessment of the facility, which will be determined by the volumes of transport and logistics services and freight traffic in the region.

In this paper, the authors propose realize a comprehensive assessment of the factors impact using the "integrated assessment of the attractiveness of the Russian Federation subject". An integrated assessment of the attractiveness of the Russian Federation subject is understood as a numerical characteristic of the attractiveness (competitiveness) of a subject for placing logistics infrastructure objects on its territory that takes into account the influence of a group of socio-economic, infrastructural and geographical factors, as well as transport performance.

An integrated assessment of the attractiveness of the Russian Federation subject is a relative indicator, therefore, should not depend on the size of the territory or the population of the region. For this reason, the proposed factors influencing LC location are included in the calculation of the integrated indicator with units of measurement expressed by relative values—per capita, share, and in some cases, as an exception, scores.

The calculation of the integrated assessment of the attractiveness of the Russian Federation subject is proposed to be carried out on the basis of the methodology [90, 91]. This methodology was implemented to assess the economic attractiveness of enterprises and was a calculation with the subsequent consolidation of individual indicators (consolidated coefficients of a group of parameters that affect the enterprise attractiveness) into a complex criterion.

In general, the integrated assessment of the attractiveness of the potential location of the logistics center is a values combination of the consolidated coefficients of the group of indicators transport work (K_{TR}), socio-economic (K_{SE}), infrastructure and geographical (K_{INF}) factors:

$$S\{K_{SE}, K_{INF}, K_{TR}\} \rightarrow \max. \tag{2}$$

The higher the integrated assessment of the attractiveness, the more attractive is the Russian Federation subject for placing a logistics center on its territory.

The methodology for evaluating the options for locating regional LCs, based on the system of factors, includes the following main stages:

1. Calculation of the estimation of the partial ith factor for the jth subject (t_{ij}) is defined as the ratio of the actual factor value to the maximum value (t_{max}).

Since each factor influences the choice of LC sites and the efficiency of its operations to varying degrees, the weight coefficients of each of them are determined by the analytical hierarchy method.

Calculation of the estimation of the jth subject, taking into account the weight coefficients for each group of factors: socio-economic (Ω_{SE}), infrastructure and geographical factors (Ω_{INF}) and indicators of transport work of the region (Ω_{TR}).

$$\Omega = \sum_{i=1}^{n} t_{ij} \cdot W_i, \tag{3}$$

where t_{ij}—integral score i of the indicator in each group of factors for the j region; W_i—weighting factor of i index.

2. Determination of consolidated coefficients for each group of factors (K_{SE}, K_{INF}, K_{TR}) as the ratio of the difference between the maximum value of the estimate (Ω_{max}) and the estimation of the jth subject in the group of factors (Ω_{SE}, Ω_{INF}, Ω_{TR}) to the range of values (Ω_{max}, Ω_{min})

$$K_{SE}^j = 1 - \frac{|\Omega_{max} - \Omega_{SE}|}{|\Omega_{max} - \Omega_{min}|}. \tag{4}$$

3. The calculation of the integrated assessment of the attractiveness of LC location options (S_j) is defined as the mean square value of the consolidated coefficients [92]. The value of the integrated assessment of the attractiveness varies in the interval [0,1]. The larger the value of the integrated indicator of evaluation, the more competitive the subject for locating LC.
4. The decision to place a LC is taken on conditions

$$\begin{cases} K_{INF}^j \cdot \frac{K_{SE}^j + K_{TR}^j}{2} \geq K_{INF}^{AVG} \cdot \frac{K_{SE}^{AVG} + K_{TR}^{AVG}}{2} \\ S_j > S_{AVG} \end{cases}, \tag{5}$$

where K_{INF}^{AVG}, K_{SE}^{AVG} and K_{TR}^{AVG}—average values of consolidated coefficients, referring to the group of socio-economic (Ω_{SE}), infrastructure and geographical factors (Ω_{INF}) and indicators of transport work of the region; S_{AVG}—integrated indicator for the studied subjects.

Taking into consideration that the attractiveness of the region is determined on the basis of three consolidated coefficients, which for each subject contribute differently to the overall integrated indicator, it is suggested to compare the values of the consolidated coefficients among themselves. Comparison of consolidated coefficients will allow identifying a group of factors for each entity whose values need to be improved in order to increase the region attractiveness to locate the objects of the logistics infrastructure on its territory and to develop recommendations for the formation of an effective logistics infrastructure [93].

Infrastructural factors reflect the current level of the infrastructural equipment of the subject, while at low values they are a deterrent to the region development, so a comparison of this group of factors should be performed with a combination of consolidated coefficients values for a group of socio-economic and indicators of region transport work.

The matrix of regional grouping by the level of attractiveness for the location of logistics centers when comparing the values of consolidated coefficients among themselves is shown in Fig. 19. In graphical form, it can be represented as applying the value of consolidated coefficients in the XY coordinate system.

Comparison of consolidated coefficients can also be carried out in the XYZ-coordinate system. In this case, the attractiveness of the subject and the

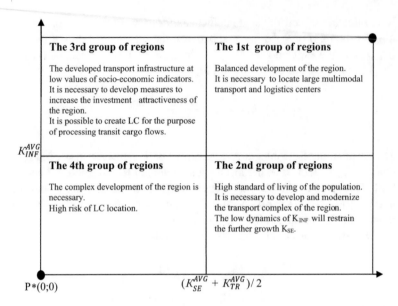

Fig. 19 Matrix of regions grouping in terms of attractiveness for the logistics centers location

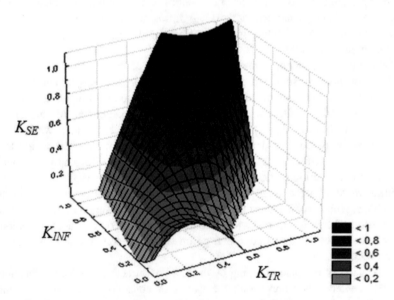

Fig. 20 Grouping of regions by the level of attractiveness for the logistics centers location in the XYZ-coordinate system

direction of development of the logistics infrastructure is determined by the conformity of values of the consolidated coefficients of a certain color area (Fig. 20). The point in the n-dimensional unit cube is set according to the rule $P_j \rightarrow P_j \ (K_{SE}, K_{INF}, K_{TR}) \in P^n$ [94]. Then:

- $P_j > 0.6$—high values of consolidated coefficients, it is advisable to place international and national logistics centers;
- $0.2 < P_j < 0.6$—average values of consolidated coefficients, it is possible to create regional logistics centers;
- $P_j < 0.2$—low values of consolidated coefficients, high risk of LC placement. The "risk of placing LC" means the possibility of the emergence of adverse situations in the creation of LC: the lack of profit, loss, as well as the short-received income or profit.

The developed algorithm for locating the logistics center is shown in Fig. 21.

With the final choice of the LC location, it is necessary to take into account already existing logistics centers, their capacity, as well as the dynamics of the factors affecting the LC location, in order to determine the prospective regions and the operational efficiency of the facility.

7 Formation of Logistic Infrastructure Simulation

In order to obtain predictive values of indicators and study the system of factors in dynamics, a set of models for the logistics infrastructure formation, including a statistical model and a simulation model necessary to form a logistics infrastructure, was developed.

The statistical model is developed using the methods of correlation-regression analysis and is accepted as the basis for the simulation model. It establishes the dependencies between the proposed indicators of the LC allocation factors system and is a system of equations for forecasting LC allocation factors.

In order to obtain more accurate values of the coefficients of the regression equation and to take into account regional differences in socio-economic development, infrastructure equipment and the volume of transport work to be done, clustering of regions should be carried out. In the work, clustering was carried out using the k-averages method of the Russian Federation regions. It consists in constructing k clusters, located at possibly large distances from each other.

As a result of the k-average cluster analysis of the Russian Federation regions, four clusters were obtained, depending on the level of their socio-economic, infrastructural development and geographic location [76, 95]. The indicators that are taken in the final grouping of regions include: population, per capita incomes of the population, GRP, industrial output, export products volume, density of railways, density of roads, belonging to the climatic zone. The distribution of regions into homogeneous groups allows the construction of regression equations for the studied factors interaction, considering the development features of the subjects depending on their belonging to a particular cluster, which contributes to a study accuracy increase of the LC allocation factors system.

The analysis of the LC location factors system was carried out on the basis of a spatial-temporal sample (17 indicators for 20 regions of the Russian Federation, the

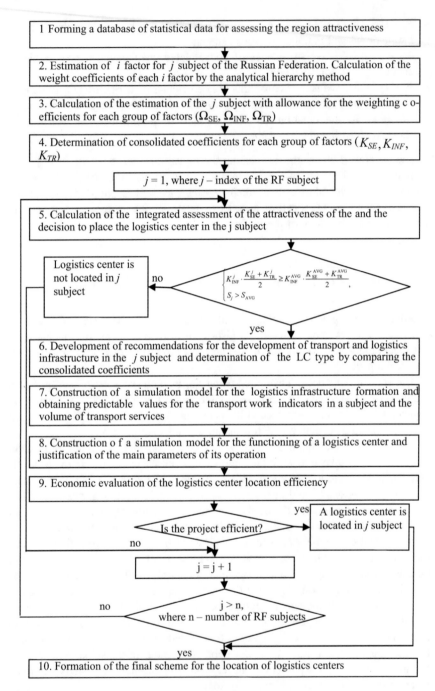

Fig. 21 Algorithm for choosing locations of logistics infrastructure objects

research period 2000–2012), based on Rosstat data [96, 97]. The use of data before 2000 for the Russian Federation conditions is inexpedient because of their incompatibility. Each cluster is represented by 5 regions that have the shortest distance to the cluster center.

For a precise problem statement, a number of simplifications were adopted:

1. Railroad facilities density, belonging to the climatic zone, the presence of transport corridors on the territory of the subject, all these are fixed values for a particular subject.
2. The population is defined as the indicator value sum of the previous period and the natural population increase, calculated as the difference in the number of born people (parameter named "Additions" in simulation model) and deceased (parameter named "Loss" in simulation model) people during a calendar year.
3. From the perspective of a preliminary analysis of the dependencies type between factors, the relationships between variables are defined as linear.
4. The indicators will be included in the statistical model with different regression coefficients. The value of the coefficients of the regression equation depends on the subject's belonging to a particular cluster.

Basing on the previously developed statistical models [98, 99], in order to obtain more accurate predictive values, additional indicators were included in the modeled system of factors: investments in fixed capital per capita, rubles/person ("Investments"), fixed assets per capita, rub/person ("Fixed assets"), the number of people employed in the economy, thousand people (parameter named "Working_population" in simulation model).

To obtain more accurate values of the coefficients of the regression equation and take into account the influence of the Russian Federation subject on a particular cluster, dummy variables are introduced. Based on the results of the cluster analysis of the logistics centers potential location, four dummy variables were adopted:

$$r_1 = \begin{cases} 1\text{---}if\ the\ subject\ belongs\ to\ cluster\ 1; \\ 0\text{---}others. \end{cases}$$

$$r_2 = \begin{cases} 1\text{---}if\ the\ subject\ belongs\ to\ cluster\ 2; \\ 0\text{---}others. \end{cases}$$

$$r_3 = \begin{cases} 1\text{---}if\ the\ subject\ belongs\ to\ cluster\ 3; \\ 0\text{---}others. \end{cases} \tag{6}$$

$$r_4 = \begin{cases} 1\text{---}if\ the\ subject\ belongs\ to\ cluster\ 4; \\ 0\text{---}others. \end{cases}$$

Then, a priori model of the i-factor dependence will represent the form

$$Y_i(X) = f(Y, X_1, X_2, \ldots, X_p, r_1, r_2, r_3, r_4) + \varepsilon_i, \tag{7}$$

where ε_i—random component.

The research database consists of a set of values $X_t^i(j)$, where j is the number of the Russian Federation subject, $j = 1, 2, \ldots, N$; i—the number of the factor (indicator) influencing the location of logistics centers, $i = 1,2, \ldots, P$; $t = 1,2, \ldots, T$.

Multiple regression analysis was used to evaluate the parameters of the regression equation. The parameters of the equations of the statistical (econometric) model were determined by the method of least squares. In this study, the necessary estimates of the coefficients of the regression equation were obtained using the software product Statistica.

Based on the developed statistical model, reflecting the type and tightness of the relationship between the factors, a simulation model imitating the development of the logistics infrastructure in the region has been developed for the formation and location of the LC.

The general view of the statistical model of forecasting the values of the i-parameter (factor) (Y_i) has the following form

$$Y_i(X) = \beta_0 + \sum \alpha_k \cdot r_k, + \sum \beta_j \cdot X_j, + \varepsilon_j \tag{8}$$

where X_j is the predicted value of the j-variable having the strongest influence on the i-indicator; β_0 and β_j are unknown parameters of the regression equation, which are subject to estimation based on the results of sample observations; r_k, is a dummy variable, where k is the cluster number, $k = 1,2,3,4$; α_k is the regression coefficient for the dummy variable; ε_i is a random component.

Traditionally, three basic elements of the simulated system are distinguished in system-dynamic simulation modeling, which reflect the processes taking place in the real world; it is in [100]:

- stock (levels), represent the accumulation of values within the system (for example, population size, infrastructure, level of industrial production);
- flows, characterize the intensity of the drive change. They are divided into incoming and outgoing flows;
- information, determines the change in intensity of the flows.

In this study, 12 main stock were allocated in the construction of the simulation model, these are: socio-economic factors, indicators of transport work, as well as the existing supply of logistics infrastructure facilities in the Russian Federation subject. Representation of these factors and indicators as storage devices will allow changing their values during experiments, i.e. to specify a random event, for example, a decrease in the volume of industrial production, a change in the population, etc.

Infrastructural factors were defined in the model as fixed parameters, which is associated with a low dynamics of the transport network development. Geographical factors (belonging to the climatic zone, the presence of transport corridors in the territory of the region) are set, in our case, for a specific subject of the Russian Federation, which is of interest from the point of view of placing logistics infrastructure objects on its territory.

The relationships between variables, stock and fixed parameters in the simulation model are established based on the developed statistical model of the LC allocation factor system. In this case, there is practically no feedback in the developed simulation model. This circumstance is due to the complexity of such closures. The rationale for this kind of feedback should be based on a greater number of different hypotheses and assumptions, which makes their formalization more difficult [101]. There is a feedback between the stock, characterizing the existing volume of quality warehouse space ("Existing capacities"), and the variable, reflecting the unmet demand for objects of the logistics infrastructure ("Unmet demand").

To build the simulation model, the toolkit of system-dynamic simulation of the AnyLogic program was used. The appearance of the simulation model implemented in the AnyLogic program is shown in Fig. 22. The main stages of building a complex of simulation models for the development of logistics infrastructure in the regions are shown in Fig. 23.

The functioning effectiveness of the logistics infrastructure object will be determined by a reasonable choice of its parameters (area, processing capacity, flow time) [102].

The choice of the parameters of the logistics infrastructure object is determined in the simulation model, depending on the structure of the material flow (the type of

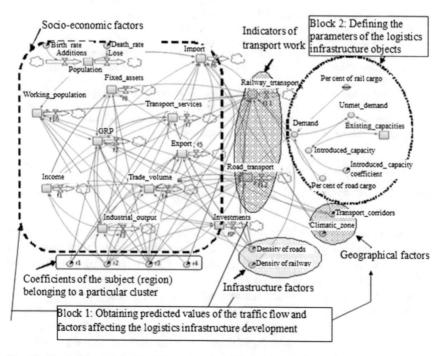

Fig. 22 Simulation model of logistics infrastructure formation

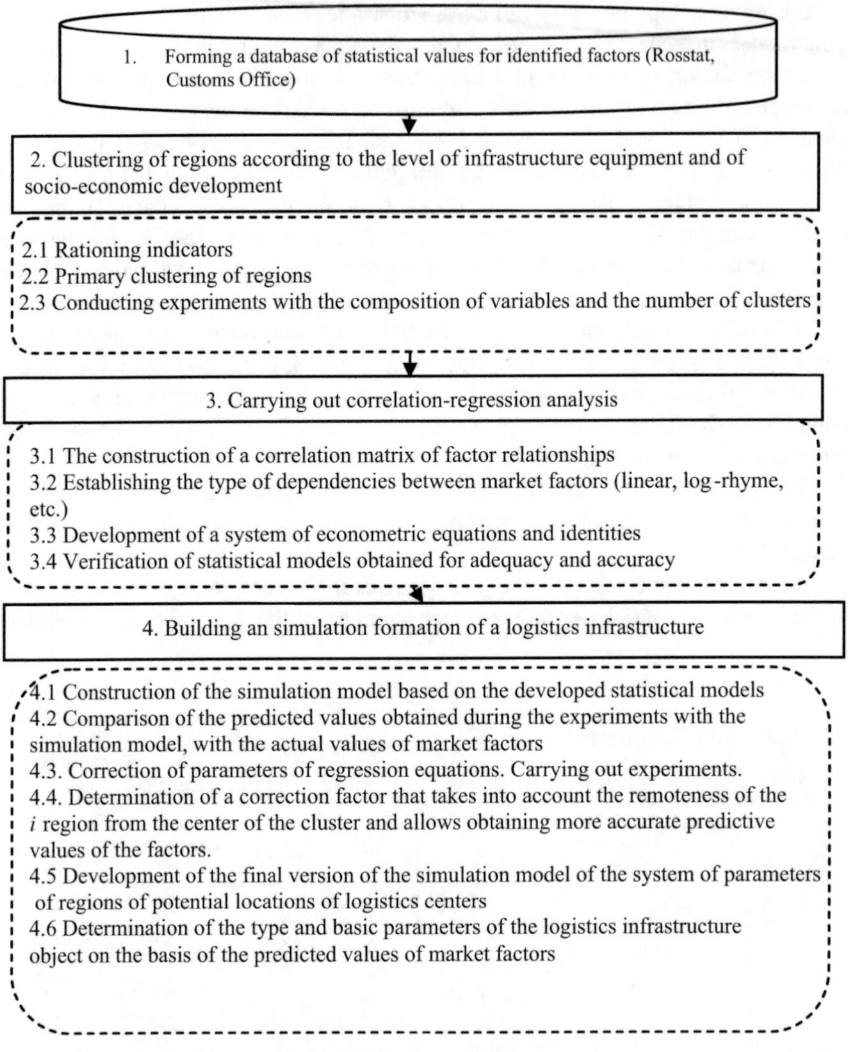

1. Forming a database of statistical values for identified factors (Rosstat, Customs Office)

2. Clustering of regions according to the level of infrastructure equipment and of socio-economic development

2.1 Rationing indicators
2.2 Primary clustering of regions
2.3 Conducting experiments with the composition of variables and the number of clusters

3. Carrying out correlation-regression analysis

3.1 The construction of a correlation matrix of factor relationships
3.2 Establishing the type of dependencies between market factors (linear, log-rhyme, etc.)
3.3 Development of a system of econometric equations and identities
3.4 Verification of statistical models obtained for adequacy and accuracy

4. Building an simulation formation of a logistics infrastructure

4.1 Construction of the simulation model based on the developed statistical models
4.2 Comparison of the predicted values obtained during the experiments with the simulation model, with the actual values of market factors
4.3. Correction of parameters of regression equations. Carrying out experiments.
4.4. Determination of a correction factor that takes into account the remoteness of the i region from the center of the cluster and allows obtaining more accurate predictive values of the factors.
4.5 Development of the final version of the simulation model of the system of parameters of regions of potential locations of logistics centers
4.6 Determination of the type and basic parameters of the logistics infrastructure object on the basis of the predicted values of market factors

Fig. 23 The main stages of building a simulation model for the logistics infrastructure formation

transport in transit, the volume of cargo) and the volumes of freight flow that are subject to processing within the given logistics center. At the same time, it is necessary to take into account existing objects of logistics infrastructure on the territory of the Russian Federation subject.

Demand for logistics infrastructure facilities is determined depending on the volume of cargo processing, the coefficient of turnover of inventory in the warehouse, the load norm in ton per $1 \, \text{m}^2$ of storage space. The parameters "Inventory_turnover_ratio" and "Load norm" are given as fixed values. The volumes of cargo handling depend on the structure of traffic volumes, i.e. the transportation of what goods predominate in the Russian Federation subject (consumer goods, bulk goods, etc.) In the simulation model, this is taken into account by specifying additional parameters that reflect the percentage of goods subject to processing in the warehouse, from the volumes of transport by the road and rail transport (parameters "Per_cent_of_road_cargo" and "Per_cent_of_rail_cargo").

As new objects of the logistics infrastructure are introduced, the demand will decrease. In order to avoid negative values of demand and volumes of input areas, the function limitMin (double min, double X) is specified. This function returns X if its value is greater than or equal to the argument min, that is, the value of X is bounded by the value min to the left.

When carrying out experiments, the quantity of goods subject to processing in a warehouse can be specified either by fixed values (requires additional investigation of the cargo flows of the subject) or by specifying the *random()* function. The *random()* function generates an equidistant random variable in the interval from [0; 1). Using this function allows taking into account the random nature of demand for the cargo processing volume.

The values of the simulation model parameters for the logistics infrastructure formation are presented in Table 4.

To assess the adequacy of the constructed simulation model to real processes, the experiments with the specification of various parameter values were conducted.

The situation of a decline in industrial production was simulated, that imitates the economy state during the economic crisis in 2009 (Fig. 24). For this purpose, an "event" was set in the simulation model—"Economic_crisis_2009". The event "Economic_crisis_2009" is performed in the simulation model for the ninth period of model time and performs the following action:

$$Industrial\_output = Industrial\_output \cdot 0.9.$$

The value of 0.9 in this experiment corresponds to a decrease in industrial production by 10% from the previous period. This value can be set arbitrarily or reflect the real percentage reduction in industrial production in a particular region.

In Fig. 24 it can be seen that with the decline in industrial production, GRP is decreasing and the pace of growth in the volume of freight transportation is slowing down, which confirms the adequacy of the developed simulation model.

The estimation of the accuracy of the developed simulation model was carried out by comparing the actual data with the predicted values by the example of a specific Russian Federation subject. According to the results of the analysis, the relative error was from 6 to 20%, depending on the indicator. To assess the prospects for the establishment of regional logistics centers on the territory of the subject, the experiments with a simulation model were conducted under different scenarios for the development of the region. As the studied region, the Trans-Baikal Territory was chosen.

The subject of the Russian Federation belongs to the first cluster, has low values for infrastructural and medium values for socio-economic factors. The survey of the transport and logistics services market showed that in the Siberian Federal District,

Table 4 Values of the simulation model parameters for the logistics infrastructure formation

No.	Parameter name in the simulation model	Value
Stock		
1	Income	$5764.267 + 0.032 \cdot GRP - 0.013 \cdot Industrial\_output + 2469.706 \cdot r2 - f1$
2	GRP	$23,604.1 + 0.775 \cdot Industrial\_output + 0.301 \cdot Investments + 105,922.9 \cdot r3 + 29,418.4 \cdot r1 + 0.484 \cdot Trade\_volume - 27.4 \cdot Working\_population - 0.010 \cdot Fixed\_assets - f2$
3	Industrial_output	$63,924 + 0.234 \cdot Fixed\_assets + 360,064.5 \cdot r3 + 278,458.9 \cdot r2 - 163.85 \cdot Working\_population - f3$
4	Trade_volume	$-7056.57 + 6.704 \cdot Income - 0.074 \cdot Industrial\_output + 1.904 \cdot Export + 11,294.715 \cdot r2 + 49.484 \cdot Density\_of\_roads - f4$
5	Export	$-7937.07 + 0.01 \cdot Industrial\_output + 8320.25 \cdot r3 + 1002.23 \cdot Transport\_corridors + 35.1 \cdot Density\_of\_railway + 5055.54 \cdot r1 - f5$
6	Import	$-241.979 + 0.007 \cdot Trade\_volume + 337.004 \cdot r1 - 0.001 \cdot GRP - 0.437 \cdot opulation + 0.033 \cdot Export + 1358.516 \cdot r3 + 1334.685 \cdot r2 + 285.1 \cdot Climatic\_zone - f6$
7	Transport_services	$1846.09 + 0.45 \cdot Income - 0.01 \cdot GRP + 0.12 \cdot Export - 16.9 \cdot Density\_of\_roads - 1133.31 \cdot r1 - f7$
8	Fixed_assets	$-372.957 + 66.197 \cdot Income + 1.096 \cdot Industrial\_output + 694,341.201 \cdot r3 - f8$
9	Investments	$-31,538.4 + 1.443 \cdot GRP - 0.946 \cdot Industrial\_output - 0.693 \cdot Trade\_volume + 59,945.711 \cdot r2 - f9$
10	Working_population	$153.217 + 0.529 \cdot Population - 187.355 \cdot Climatic\_zone + 385. 477 \cdot r4 + 95.875 \cdot r3 - f10$
11	Railway_transport	$7018.4 + 12.2 \cdot Working\_population + 23,935.9 \cdot r2 - 181.5 \cdot Density\_of\_roads + 54.8 \cdot Density\_of\_railway - 17,123.6 \cdot r3 + 1.3 \cdot Income + 22,341 \cdot r4 - 0.009 \cdot GRP - 3623.9 \cdot Climatic\_zone + 2.1 \cdot Import - 0.002 \cdot Fixed\_assets - 2057.2 \cdot Transport\_corridors + 0.038 \cdot Road\_transport - 0.099 \cdot Trade\_volume - f11$
12	Road_transport	$5073.3 + 107.7 \cdot Working\_population + 39894 \cdot r3 - 61,430 \times r2 - 18,256 \cdot Transport\_corridors + 0.287 \cdot Trade\_volume - 102 \cdot Density\_of\_roads + 0.034 \cdot Industrial\_output - 3.861 \times Income + 0.005 \cdot Fixed\_assets - f12$
13	Population	$Additions - Loss$
14	Existing_capacities	$Introduced\_capacity$

(continued)

Table 4 (continued)

No.	Parameter name in the simulation model	Value
Variables		
15	Demand	(Per_cent_of_rail_cargo · Railway_transport × Inventory_turnover_ratio · 1,000,000 + Per_cent_of_road_cargo × Road transport × Inventory_turnover_ratio · 1,000,000)/(Number_of_work_days · load_norm)
16	Unmet_demand	limitMin(1,Demand − Existing capacities)
17	Per_cent_of_road_cargo	random()
18	Introduced_capacity	limitMin(0, Unmet demand * Introduced_capacity_coefficient)
Flows		
19	f1	Parameter "Income" value for the previous period
20	f2	Parameter "GRP" value for the previous period
21	f3	Parameter "Industrial_output" value for the previous period
22	f4	Parameter "Trade_volume" value for the previous period
23	f5	Parameter "Export" value for the previous period
24	f6	Parameter "Import" value for the previous period
25	f7	Parameter "Transport_services" value for the previous period
26	f8	Parameter "Fixed_assets" value for the previous period
27	f9	Parameter "Investments" value for the previous period
28	f10	Parameter "Working_population" value for the previous period
29	f11	Parameter "Railway_transport" value for the previous period
30	f12	Parameter "Road_transport" value for the previous period
31	Additions	Birth rate
32	Loss	Death rate

(continued)

Table 4 (continued)

No.	Parameter name in the simulation model	Value
Parameters		
33	Climatic_zone	const
34	Transport_corridors	const
35	Density_of_roads	const
36	Density_of_railway	const
37	Per_cent_of_rail_cargo	const
38	Introduced_capacity_coefficient	const
Dummy variables		
29	r1	=1, if the subject belongs to cluster 1; 0, others
30	r2	=1, if the subject belongs to cluster 2; 0, others
31	r3	=1, if the subject belongs to cluster 3; 0, others
32	r4	=1, if the subject belongs to cluster 4; 0, others

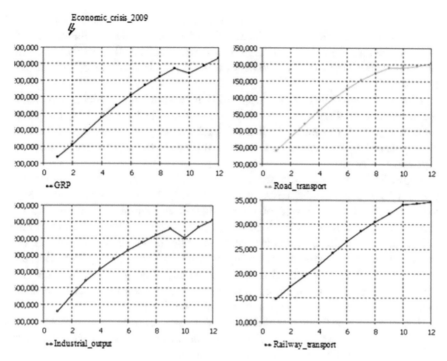

Fig. 24 The results of modeling the decline in industrial production by 10% in 2009

to which the Trans-Baikal Territory belongs, there are currently no high-quality A and B warehousing facilities. Regarding the proximity of this Russian Federation subject to the border with China, the situation of a sharp increase in the volume of transit traffic by rail was modeled.

To simulate this situation, the event "Growth of rail freight volume" was introduced. It is assumed that the volume of cargo transportation by rail increases in geometric progression for 5 years, for this purpose the function *pow(a, b)* is given, it returns the value of the first argument raised to the power of the second argument. The event occurs for the 20 period of model time. The increase in the number of cargoes to be processed in the LC (the parameter "Per_cent_of_rail_cargo") was also set, from 5 to 70% of the total volume of rail transport.

In order to assess the demand for the construction of a logistics center, the experiment simulates the situation of the logistics centers construction, when the level of demand is growing quite high, and the volume of freight transportation is reduced to the same level. Thus, we simulate the situation of a sharp increase and fall in demand for logistics infrastructure objects.

The "Logistics center creation" event simulates the construction of 10% warehouse space from the volume of unsatisfied demand for logistics infrastructure objects. This event is performed in the developed simulation model in the case that the demand for logistics infrastructure objects in the region exceeds 500 thousand m^2.

The decrease in the volume of transportation by rail is modeled by the introduction of the event "Transportation volume decrease", which assumes that for the 25th period of the model time the volumes of cargo transportation will be 40 thousand ton. At the same time, the quantity of cargo to be processed in LC is also reduced to its former low values.

Characteristics of the parameters specified in the simulation model and modeling the scenario of a sharp increase and decrease in demand for the logistics infrastructure objects, for example the Trans-Baikal Territory, are presented in Table 5. The dynamics of the simulation model parameters in the experiment is shown in Fig. 25.

Figure 25 shows that with the increase in traffic volumes by 10 times, the demand for logistics infrastructure increases sharply. However, the construction of logistics centers with an area of about 240,000 m^2 during this period and the simulated decrease in the volumes of freight traffic to the previous values makes the objects of the logistics infrastructure constructed during this period unclaimed. For the conditions of our experiment, the supply will exceed demand by more than 100 thousand m^2. Thus, using the simulation model and playing various scenarios for the region development will make it possible to avoid the construction of unclaimed logistics centers and assess the parameters of such an object.

The developed simulation model makes it possible to trace the dynamics of the development of the system of LC distribution factors and to determine the degree of their influence on the logistics infrastructure formation.

The analysis of the obtained results during the experiments with the model allows to draw a conclusion about the adequacy of the developed model to real processes and the possibility of using it to obtain the predicted values of the factors. The estimation of the exactness of the simulation results was carried out using the example of a specific subject of the Russian Federation. For the indicators of transport work and the volume of transport services, the relative error of the forecast in percent was 11% for the indicator "Transport services volume", for the indicators "Road transportation volume" and "Rail transportation volume"—10.7 and 5.9% respectively. To increase the accuracy of obtaining predictive values, it is possible by adjusting the coefficients of the regression equation in a series of experiments with the simulation model for the studied region.

The simulation model of logistics infrastructure formation allows to take into account the random character of demand for volumes of cargo processing in the region and the influence of various external influences on the formation of demand for the logistics infrastructure objects. The application of the developed complex of models for the logistic infrastructure formation makes it possible to evaluate the options for locating logistics centers, modeling various scenarios for the development of a logistics infrastructure.

Table 5 Characteristics of the parameters that simulate the situation of increasing the freight flow of the East-West direction using the Russian Federation subject (Trans-Baikal Territory) as an example

Parameter name in the model	Varied parameter	Event type	Functioning model time/condition	Done action
Growth of rail freight volume	Railway transport	On time out	20	Railway_transport = Railway_transport × pow (1.1, 3)
Increase in logistics center cargo	Per_cent_of_rail_cargo	On time out	20	Per_cent_of_rail_cargo = 0.7
Logistics center creation	Introduced capacity coefficient	In condition of	Demand ≥ 500	Introduced_capacity_coefficient = 0.1
Transportation volume decrease	Railway transport	On timeout	25	Railway_transport = 40,000
Decrease in logistics center cargo	Per_cent_of_rail_cargo	On timeout	25	Per_cent_of_rail_cargo = 0.1

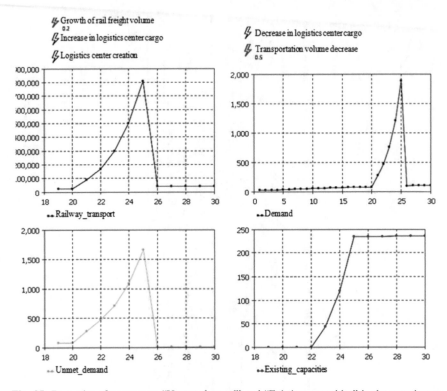

Fig. 25 Dynamics of parameters "Unmet_demand" and "Existing_capacities" in the experiment with the simulation model in the conditions of growth of the cargo transportation volume by rail transport on the example of the Zabaykalsky Krai

8 Calculated Example—Estimation of the Russian Regions Attraction for Placing Logistics Centers

In order to assess the options for locating regional logistics centers on the Russian Federation territory, approbation of the developed methodology was carried out at various administrative and territorial levels, both at the level of the country (subjects of the Russian Federation) and at the city level.

For the adopted system of factors influencing the location of the logistics infrastructure objects (Table 3), a matrix of pairwise comparisons was formed and the weights of each indicator were calculated by the analytical hierarchy method. In Table 6, the indicator number is adopted according to Table 3. In determining the integrated assessment of the attractiveness of the constituent entities of the Russian Federation, data on state and customs statistics of the Russian Federation were used to estimate the magnitude of the factors [96, 97].

According to formulas (1)–(4), consolidated coefficients and an integrated assessment of the region's attractiveness are calculated. The values of the indicators

Table 6 Pairwise comparisons matrix of indicators for calculating weighting coefficients

Factor number	1	2	3	4	5	6	7	8	9	10	12	13	14	15	Estimates of the components of the eigenvector by row	Result normalization
1	1	6	5	1	3	5	5	4	3	5	5	5	1	6	3.07	0.17
2	1/6	1	1/3	1/6	1/5	1/3	1/3	1/4	1/5	1/3	1/3	1/3	1/6	1	0.33	0.02
3	1/5	3	1	1/5	1/4	1	1	1/3	1/4	1	1	1	1/5	4	0.66	0.04
4	1	6	5	1	3	5	5	4	3	5	5	5	1	6	3.07	0.17
5	1/3	5	4	1/3	1	4	4	1/3	1	4	4	4	1/3	5	1.61	0.09
6	1/5	3	1	1/5	1/4	1	1/5	1/3	1/5	1/3	0.5	1/3	1/5	4	0.48	0.03
7	1/5	3	1	1/5	1/4	5	1	1/2	1/4	1	1	1	1/5	4	0.76	0.04
8	1/4	4	3	1/4	3	3	2	1	1/4	1/2	3	3	1/4	4	1.20	0.07
9	1/3	5	4	1/3	1	5	4	4	1	1/4	4	4	1/3	5	1.60	0.09
10	1/5	3	1	1/5	1/4	3	1	2	4	1	1	1	1	4	1.07	0.06
11	1/5	3	1	1/5	1/4	2	1	1/3	1/4	1	1	1/2	1/5	4	0.66	0.04
12	1/5	3	1	1/5	1/4	3	1	1/3	1/4	1	2	1	1/5	4	0.74	0.04
13	1	6	5	1	3	5	5	4	3	1	5	5	1	6	2.76	0.15
14	1/6	1	1/4	1/6	1/5	1/4	1/4	1/4	1/5	1/4	1/4	1/4	1/6	1	0.30	0.02

Table 7 The results of ranking subjects of the Russian Federation in terms of the integrated assessment of the region attractiveness

The territorial entity of the Russian Federation and its position according to integrated assessment		K_{SE}	K_{TR}	K_{INF}	S
The Russian Federation in average		0.145	0.163	0.211	0.194
1	The Moscow Region and Moscow	0.623	0.651	0.973	0.766
2	The Leningrad Region and St. Petersburg	0.303	0.456	0.550	0.448
3	The Tyumen Region	0.420	0.573	0.162	0.421
4	The Sverdlovsk Region	0.236	0.532	0.236	0.362
5	The Khanty-Mansi Autonomous Area—Yugra	0.474	0.414	0.005	0.362
6	Primorski Krai	0.133	0.263	0.509	0.340
7	The Irkutsk Region	0.143	0.514	0.168	0.323
8	Krasnodar Krai	0.195	0.375	0.357	0.319
9	Krasnoyarsk Krai	0.183	0.494	0.159	0.318
10	The Nenets Autonomous Okrug	0.503	0.053	0.155	0.305
11	The Kemerovo Region	0.185	0.454	0.072	0.286
12	The Kaliningrad Region	0.202	0.061	0.448	0.286
13	The Nizhni Novgorod Region	0.182	0.162	0.429	0.285
14	The Vladimir Region	0.098	0.055	0.483	0.285
15	The Smolensk Region	0.111	0.057	0.454	0.272
16	The Tver Region	0.110	0.065	0.453	0.272
17	The Chelyabinsk Region	0.203	0.381	0.176	0.269
18	The Novgorod Region	0.111	0.077	0.448	0.270
19	Perm kray	0.190	0.360	0.215	0.266
20	The Republic of Bashkortostan	0.200	0.374	0.160	0.261
21	The Samara Region	0.213	0.315	0.239	0.259
22	The Yamal-Nenets Autonomous Area	0.387	0.223	0.001	0.258
23	The Republic of Tatarstan	0.211	0.261	0.291	0.256
24	The Rostov Region	0.190	0.244	0.288	0.244
25	The Saratov Region	0.116	0.133	0.353	0.228
26	The Belgorod Region	0.143	0.266	0.249	0.226
27	The Volgograd Region	0.139	0.132	0.332	0.221
28	The Kaluga Region	0.152	0.037	0.354	0.223
29	Khabarovsk Krai	0.123	0.326	0.164	0.222
30	The Sakhalin Region	0.274	0.148	0.170	0.205

for each factor for the Russian Federation subjects under consideration and the results of calculations of the integrated indicator are given in [76]. The ranking according to the integrated indicator magnitude of the constituent entities of the Russian Federation attractiveness assessment is presented in Table 7.

To develop recommendations for the logistics infrastructure formation, the values of the consolidated coefficients of a group of infrastructure and geographic

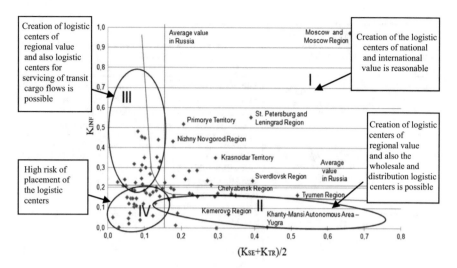

Fig. 26 Distribution of Russian Federation regions by the value of consolidated coefficients in the XY coordinate system

factors were compared with socio-economic factors and indicators of transport work (Fig. 26).

Subjects of the Russian Federation, divided into groups depending on the priority of the construction of the logistics center and the development direction of the logistics infrastructure, are presented in Table 8. The condition of the transportation and logistics center is satisfied by 26 subjects of the Russian Federation. The proposed scheme for locating LCs on the territory of the constituent entities of the Russian Federation is shown in Fig. 27.

In regions where the consolidated coefficients are greater than the average values for each of the factor groups, it is proposed to place national and international logistics centers. On the territory of the regions that have the largest values of consolidated coefficients for two of the three groups of factors, the location of regional LCs is recommended. The conditions for placing international and national logistics centers are met by: Moscow and Moscow region, St. Petersburg and Leningrad region, Sverdlovsk, Samara, Rostov, Lipetsk regions, the Republic of Tatarstan, Perm and Krasnodar Krai. The creation of regional LCs, based on the results of calculations, is advisable in 17 regions of the Russian Federation: Chelyabinsk, Irkutsk, Novosibirsk, Nizhny Novgorod, etc. The remaining subjects of the Russian Federation require an additional assessment of the options for locating logistics centers for the conditions of a particular federal district and determining the location places of local logistics centers.

To determine the specific cities of the regional logistics centers location, it is proposed to test the developed methodology at the level of the municipal formations of a particular district.

Table 8 Subjects of the Russian Federation, distributed by groups of attractiveness for the placement of logistics infrastructure objects

Regions group	Indicator value	Russian Federation entities	
1	$\frac{K_{SE}^j + K_{TR}^j}{2} \geq \frac{K_{SE}^{AVG} + K_{TR}^{AVG}}{2}$ $K_{INF}^j \geq K_{INF}^{AVG}$	The Moscow Region The Leningrad Region The Rostov Region The Sverdlovsk Region TOTAL: 12 entities	The Samara Region Krasnodar Krai Perm Krai The Republic of Tatarstan et al.
2	$\frac{K_{SE}^j + K_{TR}^j}{2} \geq \frac{K_{SE}^{AVG} + K_{TR}^{AVG}}{2}$ $K_{INF}^j < K_{INF}^{AVG}$	Krasnoyarsk Krai The Murmansk Region The Republic of Bashkortostan TOTAL: 18 entities	The Chelyabinsk Region The Tyumen Region The Novosibirsk Region et al.
3	$\frac{K_{SE}^j + K_{TR}^j}{2} < \frac{K_{SE}^{AVG} + K_{TR}^{AVG}}{2}$ $K_{INF}^j \geq K_{INF}^{AVG}$	The Vladimir Region The Voronezh Region The Saratov Region TOTAL: 22 entities	The Chuvash Republic The Novgorod Region The Kurgan Region et al.
4	$\frac{K_{SE}^j + K_{TR}^j}{2} < \frac{K_{SE}^{AVG} + K_{TR}^{AVG}}{2}$ $K_{INF}^j < K_{INF}^{AVG}$	The Orenburg Region Altai Krai The Kursk Region the Buryat Republic TOTAL: 31 entities	The Penza Region Trans-Baikal Territory the Yaroslav Region the Tula Region et al.

RF entities are marked by figures on the map:

1.Moscow and Moscow Region; 2.St.Petersburg and Leningrad Region; 3.Sverdlovsk Region; 4.Krasnodar Krai; 5.Perm Krai; 6.Samara Region; 7.The Tatar Republic; 8.Rostov Region; 9.Lipetsk Region; 10.Tyumen Region; 11.the Khanty-Mansijsk Autonomous District; 12.Primorskii Krai; 13.Irkutsk Region; 14.Krasnoyarsk Krai; 15.Kemerovo Region; 16.Kaliningrad Region;17.the Nizhni Novgorod Region; 18.Chelyabinsk Region; 19.The Republic of Bashkortostan; 20.Belgorod Region; 21.Kaluga Region; 22.Novosibirsk Region; 23.Murmansk Region; 24.Republic of Karelia; 25.Republic of Komi; 26.Khabarovsk Krai

Fig. 27 Calculated scheme for the logistics centers location on the territory of the Russian Federation

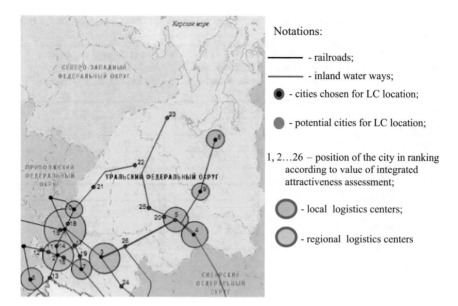

Notations:

────── - railroads;

────── - inland water ways;

● - cities chosen for LC location;

⬤ - potential cities for LC location;

1, 2...26 – position of the city in ranking
according to value of integrated
attractiveness assessment;

◯ - local logistics centers;

◯ - regional logistics centers

Fig. 28 Calculated scheme for the logistics centers location on the territory of the Urals Federal District

It is proposed to develop the scheme for the logistic centers location in the territory of the Urals Federal District, because according to the ranking results (Table 7), five of the six entities that are part of the Urals Federal District organization have a high integrated indicator of the subjects' attractiveness. The Urals Federal District is located on the border with Kazakhstan, the transport corridor "Trans-Siberian Railway" passes through its territory and it is planned to lay the northern transport corridor of the project "The Economic Belt of the Silk Road".

Calculation of the integrated assessment of the attractiveness of LC options for the conditions of the Urals Federal District was carried out among the cities with the largest population (total 26 cities).

Based on the results of the integrated assessment of the attractiveness calculation in the territory of the considered district, it is expedient to create 10 logistics centers (Fig. 28). The work proposed location of one logistics center of international importance (Yekaterinburg), two ones of regional importance (Chelyabinsk, Tyumen) and seven local ones.

The conducted experiments with the developed simulation model of the logistics infrastructure formation allowed to predict the demand for logistics infrastructure objects in the Urals Federal District cities (Table 9). The volume of proposals was determined on the basis of analysis of the warehouse real estate market, as one of the logistics center characteristics is the availability of terminals, class A and B warehouse premises on the territory of the facility.

Thus, the developed methodology for assessing the location options of logistics centers, based on integrated consideration of the transport work indicators,

Table 9 The demand volume for warehouse real estate in cities selected for the logistics centers location

Cities for LCs location	Forecast of demand for warehouse area, m² (forecast)	Supply in the warehouse real estate market, m²	Unmet demand for A and B warehouses, m²	Project capacity, thous.t/ year
Ekaterinburg	1,030,000	766,000	264,000	1584
Chelyabinsk	580,000	170,000	410,000	2940
Tyumen	300,000	13,000	287,000	1722
Surgut	100,000	2000	98,000	588
Magnitogorsk	82,000	0	82,000	492
Nizhnii Tagil	75,000	600	74,400	446
Kurgan	70,000	0	70,000	420
Nizhnevartovsk	50,000	6000	44,000	264
New Urengoi	45,000	0	45,000	270
Nojabrsk	20,000	0	20,000	120

infrastructure, geographical and socio-economic factors, is applicable at various administrative-territorial levels. The use of the simulation model allows forecasting the demand for the logistics infrastructure objects in the region under consideration and determining the parameters of such an object.

9 Conclusion

This research presents scientifically valid topical solutions for the formation of sustainable development of transport systems for servicing freight flows in the East-West direction. It is proposed to take into account the dynamics of transportation, infrastructure and economic development of the region when developing the transport and logistics infrastructure of the East-West direction.

For this purpose the work has developed and justified: a system of factors that determines the attractiveness of the region when choosing the location of the objects of the logistics infrastructure; the integrated assessment of the attractiveness Russian Federation subject when the logistics center is located, based on the integrated accounting of indicators transport work, infrastructure, geographical and socio-economic factors; a statistical model of the system of distribution factors for logistics centers and an simulation model for the logistics infrastructure formation; methodology and algorithm for assessing the options for locating regional logistics centers.

Based on the analysis of foreign economic and transport links between Europe and Asia, China and Russia, objective prerequisites for the development of transport and logistics infrastructure and the creation of logistics facilities (dry ports, logistics centers) on the routes of the Europe-Asia traffic flow are disclosed: growth of volumes of the foreign trade turnover and demand for complex logistical services in regions; expanding the boundaries of trading companies; a long period of delivery of goods by traditional sea way through the Suez Canal and, as a result, an increase in the total logistics costs (costs for supplies in transit and places of consumption, loss of profit); an increase in the number of block trains implying the use of dry ports.

A promising development of transport links is the concept "The Economic Belt of the Silk Road", which affects the development of trade and transport routes of many countries, and also meets the internal interests of the country to develop remote regions of the Chinese People's Republic. The implementation of the northern transport corridor passing through the territory of the Russian Federation, Kazakhstan, will make it possible to shorten the delivery time by rail from 30–45 to 11 days by sea.

It has been established that Europe is characterized by strong state support for the development of transport and logistics infrastructure and the logistics centers integration in the national transport system. In Russia, the formation of a modern logistics infrastructure is mainly due to private investors and large logistics companies that need a branched network of warehouses and terminals. In the Russian Federation regions, where logistics centers construction is carried out by private companies, there are no alternative national terminals. This leads to an increase in overall logistics costs and to a low formation rate of block trains in international traffic, which has a negative impact on the realization of the transit potential in the Russian Federation.

The analysis of practical experience and methodology of logistics infrastructure development revealed the absence of a common understanding of the term "logistics center". The level representation of the logistics infrastructure objects is developed depending on the management object and the integration degree of logistics into the business processes of the enterprise.

Based on the system analysis, the authors proposed a system of factors influencing the logistics centers location, including socio-economic, infrastructural, geographical factors of the region and indicators of transport work.

An integrated assessment of the attractiveness of logistics centers options allocation is introduced, the basis of which is the consolidated coefficients of a group of factors that characterize the socio-economic region development, its infrastructure level, the transport work volume, and also the ones reflecting the territorial advantage of the subject location relative to transport corridors.

The methodology for assessing regional logistics centers location options is proposed, which performs the ranking in terms of their attractiveness for the logistics infrastructure development and allows making an informed decision on the choice of LC location sites, taking into account the indicators of transport operations, infrastructure, geographic and socio-economic factors.

The work suggests to take into consideration the distinctive features of the regions through which the existing and perspective transport corridors of the East-West direction pass by conducting a cluster analysis and introducing a dummy variable into the statistical and simulation model of the logistics infrastructure development. Depending on the level of socio-economic, infrastructure development of the region, the peculiarities of its geographical location and the volume of performed transport work, a cluster analysis was carried out based on the results of which the regions are divided into four groups. The indicators of the final grouping include: population size, per capita incomes of the population, GRP, industrial output volume, export product volume, density of railways, density of highways, belonging to the climatic zone.

Using the methods of correlation-regression analysis allowed developing a statistical model for the logistics centers location, establishing the type and tightness of the connection between the identified factors and taking into account the specifics of the logistics infrastructure development in the region. The determination coefficient for most established dependency functions is greater than 0.8. Thus, the developed statistical model explains more than 80% of the factors majority variations in the studied system.

The established statistical dependencies allowed to develop a simulation model for the logistic infrastructure formation. The adequacy of the model is confirmed as a result of decline modeling in industrial production in 2009, caused by the global economic crisis. An estimation of the accuracy of the modeling results was carried out on the example of the specific Russian Federation subject. Based on the model experiment results, it was found that the discrepancy between the actual values of the factors and the model values varies between 6 and 20%, depending on the indicator.

The simulation model of the logistics infrastructure development developed by the authors allows us to study the dynamics of the factors system, to obtain their forecast values and to assess the prospect of locating logistics centers in the region, and to predict the design capacity of the logistics center in conditions of a random demand and the influence of external effects.

Application of the developed methodology in combination with the simulation model for the logistics infrastructure formation allows analyzing various development scenarios, modeling the change in foreign trade indicators, traffic volumes, the number of investments and other parameters; and also on the basis of calculations to determine the parameters of the logistics facility and to justify the locating option of logistics centers.

References

1. International Trade Centre (2017) http://www.trademap.org
2. Герами ВД, Колик АВ (2015) Управление транспортными системами. Транспортное обеспечение логистики: учебник и практикум. Москва: Издательство Юрайт. 510 p [In Russian: Gerami VD., Kolik AV (2015) Management of transport systems. Transport logistics. Urait Publishing House, Moscow]
3. Osintsev N, Rakhmangulov A, Sladkowski A, Muravev D (2017) Green logistics: element of the sustainable development concept. Part 1. Nase More 64(3):120–126

4. The Global Economy (2017) http://ru.theglobaleconomy.com/rankings/Carbon_dioxide_emissions/

5. Рахмангулов АН, Орехова НН, Осинцев НА (2016) Концепция системы повышения квалификации преподавателей в области экологического образования на основе логистической модели устойчивого развития. Современные проблемы транспортного комплекса России, vol 6. No. 1, pp 4–18 [In Russian: Rakhmangulov AN, Orekhova NN, Osintsev NA (2016) The concept of a system for advanced training teachers in the field of the ecological education on the basis of logistics model of sustainable development. Modern Problems of Russian Transport Complex]

6. Transport Links between Europe & Asia (2006) ECMT. 81 p. ISBN 92-821-1379-5. https://www.itf-oecd.org/sites/default/files/docs/06europe-asia.pdf

7. Пегин НА (2016) Национальная арктическая транспортная линия: проблемы и перспективы. Арктика и Север. 2016. No. 23, pp 32–40 [In Russian: Pegin NA (2016) National Arctic Transport Line: problems and prospects. Arctic and North]

8. Почему торговые сети упорно теряют миллиарды? (2015) Отраслевой портал logistics. ru. http://www.logistics.ru/retail/news/pochemu-torgovye-seti-uporno-teryayut-milliardy [In Russian: Why do retail chains persist in losing billions? (2015) The industry portal logistics. ru]

9. Сладковски А (2011). Контейнерные перевозки Запад – Восток, Восток – Запад. In: Миндур, М. (ed.) Транспорт в товарообмене между Европой и Азией. Варшава – Радом: ИтеЕ – PIB. 2011, pp 254–283 [In Russian: Sladkowski A (2011) Container transportation East–West, West–East]

10. Malle S (2017) Russia and China in the 21st century. Moving towards cooperative behaviour. J Eurasian Stud 8:136–150

11. Aoyama R (2016) "One Belt, One Road": China's new global strategy. J Contemp East Asia Stud 5(2):3–22

12. Ploberger C (2017) One Belt, One Road—China's new grand strategy. J Chin Econ Bus Stud 15:289–305

13. Алиев ТМ и др (2016) Экономический пояс Евразийской интеграции: доклад о путях реализации проекта сопряжения интеграции Евразийского экономического союза и Экономического пояса "Шёлкового пути". Москва: ITI. 200 p [In Russian: Aliev TM (2016) The Economic Belt of the Eurasian Integration: a report on the ways of implementing the integration project of the integration of the Eurasian Economic Union and of the Silk Road Economic Belt. Moscow: ITI]

14. Kuanyshev B (2017) The Role of Kazakhstan in the Development of the Programme 'One Belt and One Road'. https://www.slideshare.net/ciltinternational/the-role-of-kazakhstan-in-the-development-of-the-programme-one-belt-and-one-road

15. Литвинов АИ (2016) Политика современных "Шёлковых путей". Вестник МГИМО. No. 4 (49), pp 176–180 [In Russian: Litvinov AI (2016) Politics of contemporary "Silk Roads". Vestnik MGIMO]

16. Карела П (2010) Мы же не на Луне строим. РЖД-Партнер. No. 5(177), pp 28–29 [In Russian: Karela P (2010) We are not building the moon. RZD-Partner]

17. Железнодорожная инфраструктура в России, проблемы и их решения (2017) Транс-регион. http://trreg.ru/zheleznodorozhnaya-infrastruktura-v-rossii [In Russian: Railway infrastructure in Russia, problems and their solutions]

18. Besharati B et al (2017) The ways to maintain sustainable China-Europe block train operation. Bus Manage Stud 3(3):25–33

19. Loroun BB, Ming X (2017) Drawing the economical balanced line for Railway and Sea way transportation between Iran and China. Int J Bus Manage 12(5):217–231

20. Rotter H (2004) New operating concepts for intermodal transport: the mega hub in Hanover/Lehrte in Germany. Transp Plan Technol 27:347–365

21. Белозеров ВЛ, Грошев ГМ, Ковалёв ВИ, Климова НВ (2013) Использование прогрессивных форм транспортных услуг при организации работы приприпортовых станций. Известия Петербургского университета путей сообщения. No. 2(35), pp 31–43

[In Russian: Belozerov VL, Groshev GM, Kovalyev VI, Klimova NV (2013) The use of advanced forms of transport services in organization of seaport railway station work. Proceedings of Petersburg Transport University]

22. Tian S (2016) "Zheng'ou" freight train services enhance ties with countries along Silk Road. http://news.xinhuanet.com/english/photo/2016-08/19/c_135615759.htm

23. Шапкин ИН (2009) Организация железнодорожных перевозок на основе информационных технологий. Автореферат дисс. … д.т.н., Москва, МИИТ. 48 p [In Russian: Shapkin IN (2009) Organization of rail transportation on the basis of information technology. Doctoral dissertation. MIIT, Moscow]

24. Шапкин ИН (2008) Организация железнодорожных перевозок на основе дискретных методов управления и твердого графика движения поездов. Транспорт: наука, техника, управление. No. 11, pp 14–19 [In Russian: Shapkin IN (2008) Organization of Rail Freightage on the Basis and Stable Time-table of Railway Traffic. Transport: science, technology, management]

25. Суюнбаев ШМ (2010) Оперативное планирование эксплуатационной работы в условиях организации движения грузовых поездов по твердому графику. Известия Петербургского университета путей сообщения. No. 3, pp 15–25 [In Russian: Suyunbaev ShM (2010) Day-to-Day Planning of Freight Trains Operation Under a Rigid Timetable. Proceedings of Petersburg Transport University]

26. Грошев ГМ, Белозеров ВЛ, Кукушкина ЯВ, Климова НВ (2016) Об оптимизации величины состава блок-поездов при доставке контейнеров в морской торговый порт в транспортном узле. Известия Петербургского университета путей сообщения, vol 13. No. 4(49), pp 451–459 [In Russian: Groshev GM, Belozerov VL, Kukushkina YaV, Klimova NV (2016) About optimizing the size of the block train formation when delivering containers to a commercial sea port in a transportation hub. Proceedings of Petersburg Transport University]

27. Xiao J, Lin B (2016) Comprehensive optimization of the one-block and two-block train formation plan. J Rail Transp Plan Manage 6:218–236

28. Janic M (2008) An assessment of the performance of the European long intermodal freight trains (LIFTS). Transp Res Part A: Policy Pract 42:1326–1339

29. Шполянская АА (2014) Основные проблемы морских грузовых портов России и пути их решения. SCI-ARTICLE.RU No. 7. http://sci-article.ru/stat.php?i=1394648802 [In Russian: Shpolyansky AA (2014) The main problems of Russian sea freight ports and ways of their solving]

30. Коровяковский ЕК (2013) Проблемы развития системы логистических центров на железнодорожном транспорте, Логистические системы в глобальной экономике. No. 3-1, pp 121–125 [In Russian: Korovyakovskiy EK (2013) Problems of logistical hubs system development on rail transport. Logistic systems in the global economy]

31. Muravev D, Rakhmangulov A (2016) The development of the regional sea port infrastructure on the basis of dry port. Econ Reg 12(3):924–936

32. Muravev D, Rakhmangulov A (2016) Environmental factors consideration at industrial transportation organization in the "seaport–dry port" system. Open Eng 6(1):476–484

33. Jeevan J, Salleh NHM, Loke KB, Saharuddin AH (2017) Preparation of dry ports for a competitive environment in the container seaport system: a process benchmarking approach. Int J e-Navigation Marit Econ 7:19–33

34. Beresford AKC, Pettit SJ, Xu Q, Williams W (2012) A study of dry port development in China. Marit Econ Logistics 14(1):73–98

35. Zhang Y, Thompson RG, Bao X, Jiang Y (2014) Analyzing the promoting factors for adopting green logistics practices: a case study of road freight industry in Nanjing. Procedia —Soc Behav Sci 125:432–444

36. Ho Y-H, Lin C-Y, Tsai J-S (2014) An empirical study on organizational infusion of green practices in Chinese logistics companies. J Econ Soc Stud 4(2):159–189

37. Guo R, Gui H, Changlei Guo LC (2015) Multiregional economic development in China. Springer, Berlin, p 540

38. Rakhmangulov A, Sładkowski A, Osintsev N (2016) Design of an ITS for Industrial enterprises. Intelligent transportation systems—problems and perspectives: studies in systems. Decis Control 32:161–215

39. Осинцев НА, Казармщикова ЕВ (2017) Факторы устойчивого развития транспортно-логистических систем. Современные проблемы транспортного комплекса России. 1 (7): 13-21. [In Russian: Osintsev NA, Kazarmshchikova EV (2017) Factors of sustainable development of transport and logistics systems. Modern Problems of Russian Transport Complex]

40. Litman T (2015) Well measured: developing indicators for sustainable and livable transport planning. Victoria Transport Policy Institute, 100 p

41. McKinnon A, Browne M, Whiteing A, Piecyk M (eds) (2015) Green logistics: improving the environmental sustainability of logistics, 3rd edn. Kogan Page Limited, 426 p

42. Журавская МА, Лемперт АМ, Маслов АМ, Гашкова ЛВ (2015) Функционирование транспортно-логистических систем с учетом оценки экологических последствий. Инновационный транспорт. Vol. 18. No. 4, pp 31–37 [In Russian: Zhuravskaya MA, Lempert AM, Maslov AM, Gashkova LV (2015) Operation of transport and logistics systems with account to environmental impact assessment. Innovative transport]

43. Ускова ТВ (2009) Управление устойчивым развитием региона. Вологда: ИСЭРТ РАН, 355 p. [In Russian: Uskova TV (2009) Management of sustainable development of the region, Vologda, 355 p]

44. Jutvik G, Liepin I (2010) Education for changes: handbook for teaching and studying the sustainable development. Izd-vo Baltijskogo universiteta, Vides Vestis, Uppsala, p 74

45. Global Green Freight (2017) http://www.globalgreenfreight.org

46. Lean & Green (2017) http://lean-green.nl/en-GB/international/

47. ECO Stars (2017) http://www.ecostars-uk.com/about-eco-stars/introduction/

48. Green Freight Europe (2017) http://www.greenfreighteurope.eu/

49. SWIFTLY Green (2017) http://www.swiftlygreen.eu

50. Psaraftis HN, Panagakos G (2012) Green corridors in European surface freight logistics and the SuperGreen project. Procedia—Soc Behav Sci 48:1723–1732

51. Горяев НК, Циулин СС (2014) Перспективы развития "зелёных транспортных коридоров" в Европе с учётом ключевых показателей эффективности. Вестник Сибирской государственной автомобильно-дорожной академии. No. 6(40), pp 14–20 [In Russian: Goryaev NK, Tsiulin SS (2014) Development perspectives of the green transport corridors in Europe with the key performance indicators. Vestnik SibADI]

52. Miliauskaitė L (2011) Evaluation of the green transport corridor concepts in the EU Area. In: The 7th international conference May 5–6, 2011, Vilnius, Lithuania. Selected papers, pp 112–119

53. Psaraftis HN (2016) Green transportation logistics—the quest for win-win solutions. Springer, International Series in Operations Research & Management Science, Switzerland, vol 226, 558 p

54. Schröder M, Prause G (2015) Risk management for green transport corridors. J Secur Sustain 5(2):229–239

55. Prause G, Schröder M (2015) KPI building blocks for successful green transport corridor implementation. Transp Telecommun 16(4):277–287

56. Green Corridor Manual (Draft)—Purpose, definition and vision for Green Transport Corridors (2011) Report. Danish Transport Authority, 36 p

57. Журавская МА, Мартыненко АВ, Цяо Ц (2016) Оптимизация "зеленых" цепей поставок в условиях неопределенности. Транспорт Урала. 3(50):20–26 [In Russian: Zhuravskaya MG, Martynenko AV, Tsyao T (2016) Optimization of green supply chains under uncertainty. Transport of the Urals]

58. Постановление Правительства РФ No. 848. О федеральной целевой программе "Развитие транспортной системы России (2010-2020 годы)" (5 декабря 2001) http://base.garant.ru/1587083/1/#ixzz4uLAjBWSs. [In Russian: Decree of the Government of the

Russian Federation of December 5, 2001 No. 848. On the Federal Target Program "Development of the Transport System of Russia (2010-2020)" (5 Dec 2001)]

59. О состоянии внешней торговли в январе-июле 2017 года. Федеральная служба государственной статистики (2017) http://www.gks.ru/bgd/free/b04_03/IssWWW.exe/Stg/d02/191.htm [In Russian: On foreign trade in January-July 2017. Federal State Statistics Service]

60. Чижков ЮВ (2015) Международные транспортные коридоры - коммуникационный каркас экономики. Транспорт Российской Федерации. No. 5 (60), pp 9–15 [In Russian: Chizhkov YV (2015). International transport corridors are a communication frame of economy. The Russian Federation transport]

61. Цыденов АС (2014) Восток – Запад: состояние и перспективы развития международного транспортного коридора. Транспорт Российской Федерации. No. 5 (54), pp 4–6 [In Russian: Tsydenov AU (2014) East–West: state and prospects of development of the international transport corridor. The Russian Federation transport]

62. Информационно-справочный портал "Железнодорожные перевозки" (2017) https://cargo-report.info/ [In Russian: Information and reference portal "Railway transport"]

63. Объем контейнерных перевозок между Россией и Китаем за 2016 год вырос на 30% (2017) Информационное агентство РЖД Партнер. http://www.rzd-partner.ru/zhd-transport/news/obem-konteynernykh-perevozok-mezhdu-rossiey-i-kitaem-za-2016-god-vyros-na-30/ [In Russian: The volume of container traffic between Russia and China for 2016 grew by 30% (2017). Information Agency RRR Partner]

64. О назначении международных контейнерных поездов в графике движения грузовых поездов на 2017/2018 год (телеграмма от 30 августа 2017 года No. 15859). http://docs.cntd.ru/document/456096699 [In Russian: On the appointment of international container trains in the schedule of freight trains for 2017/2018 (telegram of August 30, 2017 No. 15859)]

65. Саркисов СВ (2008) Международные логистические системы в условиях глобализации: автореф. дис. ... докт. эк. наук: 08.00.14. Москва, 53 p [In Russian: Sarkisov SV (2008) International logistics systems in the context of globalization. Doctoral dissertation, Moscow]

66. Beata Skowron-Grabowska. Development of logistics centers in Poland. Available at: http://www.oeconomica.uab.ro/upload/lucrari/920072/2.pdf/

67. Миротин ЛБ (2009) Эволюция логистической отрасли // Проблемы региональной экономики. 2009. Т.1, pp 110–118 [In Russian: Mirotin LB (2009) Evolution of the logistics industry. Probl Reg Econ 1:110–118]

68. Вагенер Н (2009) Опыт взаимодействия федеральных, региональных и муниципальных органов власти и частного капитала при создании и развитии транспортно-логистических центров в Федеративной Республике Германия. In: Транспортные коридоры в инновационном развитии экономики регионов: Матер. Междунар. научн. - практ. конф. http://www.council.gov.ru/media/files/41d44f243f73f2f51345.pdf [In Russian: Vagener N (2009) Experience of interaction between federal, regional and municipal authorities and private capital in the creation and development of transport and logistics centers of Germany. In: Transport corridors in the innovative development of the economy of the regions: Materials of the International Scientific-Practical Conference]

69. Копылова ОА, Рахмангулов АН (2011) Проблемы выбора места размещения логистических центров. Современные проблемы транспортного комплекса России. No. 1, pp 58–67 [In Russian: Kopylova OA, Rahmangulov AN (2011) Problems of choosing the location of logistics centers. Modern problems of Russian transport complex]

70. Транспортно-логистический комплекс "Южноуральский" (2017) http://ru.dev.ytlc.ru/investors [In Russian: Transport and logistics complex "Yuzhnouralsky"]

71. Поток грузов из Китая в ЕС проходит мимо транспортно-логистического комплекса "Южноуральский" (2017) https://www.nakanune.ru/articles/112777/ [In Russian: The flow

of goods from China to the European Union passes by the transport and logistics complex "Yuzhnouralskiy"]

72. Покровская ОД, Самуйлов ВМ, Воскресенская ТП (2013) Интеграция региональной терминально-логистической сети в международные транспортные коридоры. Инновационный транспорт. No. 1(7), pp 33–37 [In Russian: Pokrovskaya OD, Samuylov VM, Voskresenskaya TP (2013) Integration of regional terminal and logistics network into international transport corridors. Innovative Transport]

73. Проект закона Республики Беларусь "О логистической деятельности" (2011) Минск. 42 p [In Russian: Project of Law of Belarus "About Logistics Activity", Minsk]

74. Прокофьева ТА, Лопаткин ОМ (2003) Логистика транспортно-распределительных систем: региональный аспект. Москва: РКонсульт, 400 p. [In Russian: Prokofieva TA, Lopatkin OM (2003) Logistics of transport and distribution systems: a regional aspect. Moscow]

75. Higgins CD, Ferguson M, Kanaroglou PS (2012) Varieties of logistics centers: developing a standardized typology and hierarchy. Transportation Research Record. No. 2288, pp 9–18

76. Копылова ОА, Рахмангулов АН (2015) Размещение региональных логистических центров: монография. Магнитогорск: Изд-во Магнитогорск. гос. техн. ун-та Г.И. Носова. 172 p [In Russian: Kopylova OA, Rahmangulov AN (2015) Placement of regional logistics centers: monograph. Nosov Magnitogorsk State Technical University, Magnitogorsk]

77. Рахмангулов АН, Копылова ОА (2014) Оценка социально-экономического потенциала региона для размещения объектов логистической инфраструктуры. Экономика региона. No. 2(38), pp 254–263 [In Russian: Rahmangulov AN, Kopylova OA (2014) Assessment of socio-economic potential of regions for placement of the logistic infrastructure objects. Economy of Region]

78. Рахмангулов АН, Копылова ОА, Аутов ЕК (2012) Выбор мест для логистических мощностей. Мир транспорта. No. 1 (39), pp 84–91 [In Russian: Rahmangulov AN, Kopylova OA, Autov EK (2012) Logistics facilities distribution. The world of transport]

79. Петров МБ (2003) Региональная транспортная система: концепция исследования и модели организации. Екатеринбург: Институт экономики УрО РАН, Уральский государственный университет путей сообщения. 187 p [In Russian: Petrov MB (2003) Regional transport system: the concept of research and the model of organization. Ekaterinburg: Institute of Economics of the Ural Branch of the Russian Academy of Sciences, Ural State University of Railway Transport]

80. Рахмангулов АН, Копылова ОА (2011) Анализ спроса и предложения на рынке транспортно-логистических услуг России. Современные проблемы транспортного комплекса России. No. 1, pp 67–75 [In Russian: Kopylova OA, Rahmangulov AN (2011) Analysis of supply and demand in the market of transport and logistics services in Russia. Modern problems of Russian transport complex]

81. Кайгородцев АА, Рахмангулов АН (2012) Система методов выбора места размещения логистического распределительного центра. Современные проблемы транспортного комплекса России. No. 2, pp 23–37 [In Russian: Kaygorodtsev AA, Rahmangulov AN (2011) The system of methods for choosing the location of the logistics distribution center. Modern problems of Russian transport complex]

82. Рахмангулов АН, Багинова ВВ, Копылова ОА, Аутов ЕК (2012) Методика формирования энергоэффективной транспортно-логистической инфраструктуры. Бюллетень транспортной информации. No. 5, pp 26–30 [In Russian: Rahmangulov AN, Baginova VV, Kopylova OA, Autov EK (2012) Method of formation of an energy efficient transport and logistics infrastructure. The Bulletin of Transport Information]

83. Апатцев ВИ, Басыров ИМ (2017) Оценка факторов, влияющих на выбор оптимального месторасположения объектов логистической инфраструктуры. Наука и техника транспорта. No. 1, pp 33–37 [In Russian: Apattsev VI, Basyrov IM (2017) Factor score

on the choice of an optimum location of logistic infrastructure objects. Science and Technology in Transport]

84. Alam SA (2013) Evaluation of the potential locations for logistics hub: a case study for a logistics company. MSc thesis. KTH Royal Institute of Technology, Stockholm, 78 p

85. Lipscomb RT (2010) Strategic criteria for evaluating inland freight hub locations. 77 p. https://scholarsmine.mst.edu/cgi/viewcontent.cgi?article=5990&context=masters_ theses

86. Ларин ОН (2008) Теоретические и методологические основы развития транзитного потенциала автотранспортных систем регионов (на примере Челябинской области): автореф. дис. ... д-ра техн. наук: 05.22.01. Москва. 39 p [In Russian: Larin ON (2008) Theoretical and methodological bases of development of transit potential of automobile transport systems of regions (on an example of the Chelyabinsk Region). Doctoral Dissertation, Moscow]

87. Прокофьева ТА (2012) Развитие системы национальных и международных транспортных коридоров на основе формирования опорной сети логистических центров. Интегрированная логистика. No. 1, pp 20–23 [In Russian: Prokofieva TA (2012) Development of the system of national and international transport corridors on the basis of the formation of a network of logistics centers. Integrated logistics]

88. Панасейкина ВС (2010) Оценка инвестиционной привлекательности территориальных образований: основные концепции. Общество: Политика, Экономика, Право. No. 2, pp 27–32 [In Russian: Panaseikina VS (2010) Assessment of the investment attractiveness of territorial entities: basic concepts. Society: Politics, Economics, Law]

89. Коробов ВБ (2005) Сравнительный анализ методов определения весовых коэффициентов « влияющих факторов » . Социология: 4 М. No. 20, pp 54–73. [In Russian: Korobov VB (2005) Comparative analysis of methods for determining the weight coefficients of " influencing factors". Sociology 4 M]

90. Сай ВМ, Сизый СВ, Фомин ВК (2010) Интегральная оценка предприятий. Экономика железных дорог. No. 1, p 18 [In Russian: Saiy VM, Sizyi SV, Fomin VK (2010) Integral valuation of enterprises. Economics of railways]

91. Сай ВМ, Шутюк СВ (2005) Интегрированный коэффициент эффективности проектов при взаимодействии ОАО "РЖД" с региональными хозяйствующими субъектами. Транспорт Урала. No. 2(5), pp 4–11 [In Russian: Saiy VM, Shtyuck SV (2005) Integrated coefficient of project efficiency in JSC Russian Railways cooperation with regional economic management. Transport of the Urals]

92. Фомин ВК (2009) Пособие по определению итоговой оценки предприятий, взаимодействующих с железной дорогой: практическое пособие. Екатеринбург. 42 p [In Russian: Fomin VK (2009) Manual for determining the final assessment of enterprises interacting with the railway: a practical guide. Ekaterinburg]

93. Рахмангулов АН, Мишкуров ПН, Копылова ОА (2014) Железнодорожные транспортно-технологические системы: организация функционирования. Магнитогорск: Изд-во Магнитогорск. гос. техн. ун-та им. Г.И. Носова. 300 p. [In Russian: Rakhmangulov AN, Mishkurov PN, Kopylova OA (2014) Railway transport technological systems: organization of functioning. Nosov Magnitogorsk State Technical University, Magnitogorsk]

94. Сизый СВ (2011) Теория и методология формирования сетевого организационного взаимодействия на железнодорожном транспорте: дисс. ... д-ра техн. наук: 05.02.22. Екатеринбург. 385 c [In Russian: Sizyi SV (2011) Theory and methodology of formation of network organizational interaction in railway transport. Doctoral Dissertation, Ekaterinburg]

95. Копылова ОА (2013) Кластеризация региональных транспортно-логистических систем. Современные проблемы транспортного комплекса России. No. 4, pp 73–81 [In Russian: Kopylova OA (2013) Clustering of regional transport and logistics systems. Modern problems of Russian transport complex]

96. Официальный сайт Федеральной службы государственной статистики России (2017) http://www.gks.ru [In Russian: Official website of the Russian Federal Service of State Statistics]

97. Официальный сайт Федеральной таможенной службы РФ (2017) http://search.customs. ru/ [In Russian: Official website of the Federal Customs Service of the Russian Federation]

98. Фурсов ВА (2011) Формирование и функционирование региональных рынков транспортных услуг: теория, методология, практика: автореф. дис. … д-ра эк. наук: 08.00.05. Ставрополь. 45 р [In Russian: Fursov VA (2011) Formation and functioning of regional markets for transport services: theory, methodology, practice. Doctoral dissertation, Stavropol]

99. Гольская ЮН (2013) Оценка влияния транспортной инфраструктуры на социально-экономическое развитие региона: автореф. дис. … канд. эк. наук: 08.00.05. Екатеринбург. 24 р. [In Russian: Golskaya UN (2013) Assessment of the impact of transport infrastructure on the socio-economic development of the region. PhD thesis, Yekateriunburg]

100. Копылова ОА, Рахмангулов АН (2012) Применение метода системной динамики для исследования факторов размещения элементов транспортно-логистической инфраструктуры. Современные проблемы транспортного комплекса России. No. 2, pp 92–97 [In Russian: Kopylova OA, Rahmangulov AN (2011) Application of the system dynamics method for the study of placement elements factors of transport and logistics infrastructure. Modern problems of Russian transport complex]

101. Широков АА, Янтовский АА (2008) Опыт разработки инструментария долгосрочного макроэкономического прогнозирования. Российская академия наук. Научные труды: Институт народнохозяйственного прогнозирования РАН. vol 6, pp 96–110 [In Russian: Shirov AA, Yantovskii AA (2008) On the Development of Long-term Macroeconomic Forecasting Tools. Scientific works: Institute of Economic Forecasting of the Russian Academy of Sciences]

102. Рахмангулов АН, Осинцев НА, Мишкуров ПН, Копылова ОА (2014) Интеллектуализация транспортного обслуживания металлургических предприятий. Сталь. No. 4, pp 115–118 [In Russian: Rakhmangulov AN, Osintsev NA, Mishkurov PN, Kopylova OA (2014) Intellectualization of Transport Service of the Metallurgical Enterprises. Steel]

103. Zhao B (2017) China's "One Belt, One Road" Initiative: "A New Silk Road linking Asia, Africa and Europe". https://www.globalresearch.ca/chinas-one-belt-one-road-initiative-a-new-silk-road-linking-asia-africa-and-europe/5589427

104. Kaszańska D (2012) Advanced intermodal freight logistics centers. https://www.slideshare.net/Mimi0127/advanced-intermodal-freight-logistics-centers

105. Communication from the Commission—Freight Transport Logistics Action Plan (2007) http://eur-lex.europa.eu/legal-content/EN/TXT/?uri=celex:52007DC0607

Analysis and Development Perspective Scenarios of Transport Corridors Supporting Eurasian Trade

Aleksander Sładkowski and Maria Cieśla

Abstract The chapter deals with freight transport between Europe and Asia, supporting intercontinental exchange of goods. So first the impact of Eurasian trade on transport and logistics branch is presented. The analysis is focused on position of European Union in global trade, especially with China and its influence on the increase of services for logistics operators. In another part of the chapter analysis of transport corridors between Europe and Asia are presented. Then, the paper focuses on future development of Eurasian transport corridors. The research objective of this chapter was to prepare strategic analysis of future transport development in Eurasian region. Eurasian transport network SWOT analysis is shown and scenario building based on STEEPLE environmental analysis. The purpose of the scenario analysis was to acquire knowledge in order to rationalize future decisions in strategy building for the investments and research connected with the topic.

Keywords Eurasian trade · Eurasian transport network development
Environmental analysis · Environmental scenarios

1 Introduction

Analysing the trends in the development of trade and export and import, the importance of the flows of goods between Europe and Asia cannot be ignored. Globalization, which involves moving away from local markets to global ones in terms of consumer behaviour, implies the need to develop transport between the two continents. Many studies and research at the moment goes towards the

A. Sładkowski (✉) · M. Cieśla
Department of Logistics and Aviation Technologies, Faculty of Transport,
Silesian University of Technology, Krasinskiego 8, 40-019 Katowice, Poland
e-mail: aleksander.sladkowski@polsl.pl

M. Cieśla
e-mail: maria.ciesla@polsl.pl

© Springer International Publishing AG, part of Springer Nature 2018
A. Sładkowski (ed.), *Transport Systems and Delivery of Cargo
on East–West Routes*, Studies in Systems, Decision and Control 155,
https://doi.org/10.1007/978-3-319-78295-9_2

development of new transport corridors, which will prove to be not only economically efficient but also environmentally friendly and human-friendly.

A thorough analysis of the current transport corridors shows that the transport network must be modernized and expanded. The new initiative is the New Silk Road, which is to connect Europe's railways with Asia and will therefore enable the transport of goods not only to China, but also to Azerbaijan, India, Turkey and Iran. It is very important that companies will start importing and exporting goods to a higher level of quality customer service at a lower total cost of transportation. Experts note that some products are losing value in shipping, such as groceries. The transport of these products to China must be made by rail because the aggregates in special containers can keep the temperature constant for about two weeks. Roughly the same it takes for a train journey. By boat it takes too long for this type of cargo and air transport is too expensive for its value.

Some European countries, including Poland, even estimate the costs associated with the development of the broad-gauge railway, but the overall concept of rail transport development in the east-west relationship depends on the transport policy of the European Union.

Socio-economic changes in recent years have undoubtedly contributed to the change in the operating conditions of most activities. Today's development is very important. The modern market does not accept the indiscriminate and unprofessionally conducted undertakings. Every future decisions involving geographical, social, technical changes around different countries should have a development strategy that matches the opportunities and threats from the environment. Effective action should be the domain of project management.

The research objective of this chapter is to analyse and verify the applicability of methods of environmental scenario analysis to develop recommendations for the development of transport network strategy. In the theoretical analysis, the chapter focuses on the presentation of a multi-stage methodology of scenario building as one of the methods of analysis of the enterprise environment. In the research part of this chapter environmental scenarios of Eurasian future transport development is analysed. The purpose of the scenarios analysis was to gain knowledge of possible future events for rationalizing successive decisions in developing strategies for the research subject.

2   Impact of Eurasian Trade on Transport and Logistics Branch

2.1   European Union in Global Trade

The EU-28, China and the United States have been the three largest global players for international trade since 2004 when China passed Japan [4] causing big relationship between international trade and economic growth [32]. In 2015, the total level of trade in goods (exports and imports) recorded for the EU-28, China and the United States was almost identical, peaking at EUR 3633 billion in the United States, which was EUR 61 billion higher than for China and EUR 115 billion above the level recorded for the EU-28 (note the latter does not include intra-EU trade); Japan had the fourth highest level of trade in goods, at EUR 1127 billion.

Also, the global role of multinational and transnational companies is expending [5], as small and medium sized companies play a considerable role, although the 600 leading multinational firms which account almost 20% of global value-added in manufacturing and agriculture [22] are primarily shaping international economic relations.

Looking at the flows of exports and imports, the EU-28 had the second largest share of global exports (Fig. 1 [13]) and imports of goods (Fig. 2 [13]) in 2015: the EU-28's exports of goods were equivalent to 15.5% of the world total, and in 2014 were surpassed for the first time since the EU was founded by those of China (16.1% in 2014, rising to 17.8% in 2015), but still ahead of the United States [7] (13.4%); the United States had a larger share of world imports (17.4%) than either the EU-28 (14.5%) or China (12.7%).

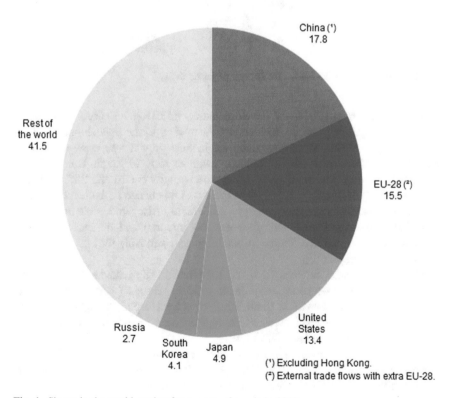

Fig. 1 Shares in the world market for exports of goods in 2015

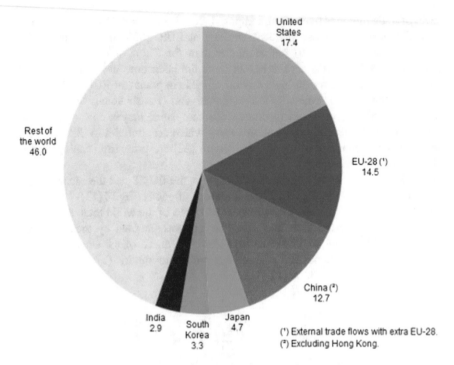

Fig. 2 Shares in the world market for imports of goods in 2015

Between 2006 and 2016, the development of the EU-28's exports of goods by major trading partner varied considerably. According to the data shown in Fig. 3 [14], among the main trading partners, the highest growth rate was recorded for exports to China which almost trebled, while exports to South Korea almost. Exports to Norway and Japan grew more slowly and were 26 and 30% higher in 2016 than they had been in 2006, while there was no change in the level of EU-28 exports to Russia over the period under consideration. Also, studies ale made on the links between international trade (exports and imports) and dimensions of firm performance (productivity, wages, profitability and survival) [36] as well as different trading analysis [2].

On the import side, between 2006 and 2016 the EU-28 notified a decrease in the value of its imports of goods from Japan (−15%), Russia (−17%) and Norway (−23%); for the latter two these changes reflect, at least in part, changes in the price of oil and gas. The greatest increases were registered for imports from China (76%), India (74%) and Switzerland (70%) [14].

The United States remained, by far, the most common destination for goods exported from the EU-28 in 2016 (Fig. 4 [14]), although the share of EU-28 exports destined for the United States fell from 28.0% of the total in 2002 to 16.7% in 2013 before recovering to 20.8% by 2016. China was the second most important destination market for EU-28 exports in 2016 (9.7% of the EU-28 total), followed by

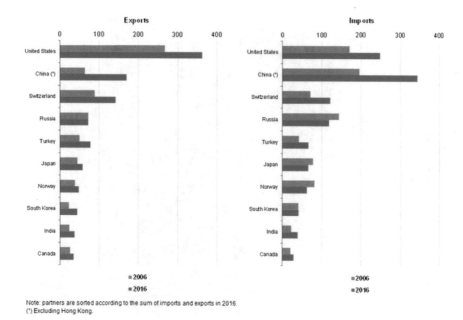

Fig. 3 Extra EU-28 trade in goods (billion EUR) by main trading partners in 2006 and 2016

Switzerland (8.2%). In 2015, Turkey overtook Russia to be the fourth largest destination for EU-28 exports of goods and this pattern continued in 2016 when Turkey accounted for 4.5% of EU-28 exports. The seven largest destination markets for EU-28 exports of goods—the United States, China, Switzerland, Turkey, Russia, Japan and Norway—accounted for more than half (53.4%) of all EU-28 exports of goods [14].

The seven largest suppliers of EU-28 imports of goods were the same countries as the seven largest destination markets for EU-28 exports, although their order was slightly different (comparing Fig. 4 to Fig. 5 [14]). These seven countries accounted for a larger share of the EU-28's imports of goods than their share of EU-28 exports of goods: just over three fifths (60.2%) of all imports of goods into the EU-28 came from these seven countries. China was the origin for more than one fifth (20.2%) of all imports into the EU-28 in 2016 and was the largest supplier of goods imported into the EU-28. The United States' share of EU-28 imports of goods (14.5%) was around 6 percentage points lower than that of China, while the shares of Switzerland (7.1%) and Russia (7.0%), which were the third and fourth largest suppliers of goods to the EU-28, were a further 7 percentage points smaller. Turkey was the fifth largest supplier of EU-28 imports of goods, followed closely by Japan and Norway [14].

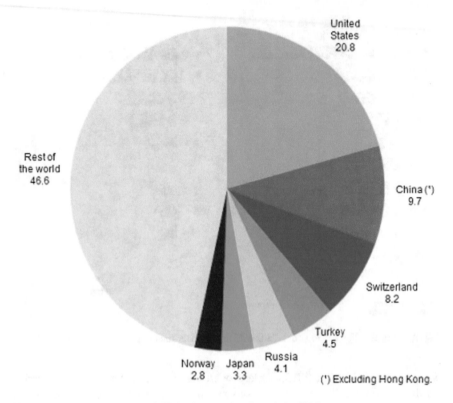

Fig. 4 Main trading partners of EU-28 for exports of goods in 2016

Between 2011 and 2016, the value of the EU-28's extra-EU exports increased for most product groups shown in Fig. 6 [14], although there were two exceptions: exports of raw materials (fell overall by 5.1%) and exports of mineral fuels and lubricant products (fell by 26.0%). The highest growth rate for exports was reported for food, drinks and tobacco for which an increase of 31.0% was observed, while there was also a relatively rapid increase in the level of extra-EU exports of chemicals and related products (up 23.1%), while double-digit growth rates were also recorded for machinery and transport equipment and for other manufactured products [14].

Theoretical research [6] examines how firms determine the range of products they will export and import or the breadth of countries they will export to or import from—or how any of these margins are influenced by globalization [19]. Even though for example [8] only a fraction of UK firms engages in international trade in services, that trade participation varies widely across industries and that services

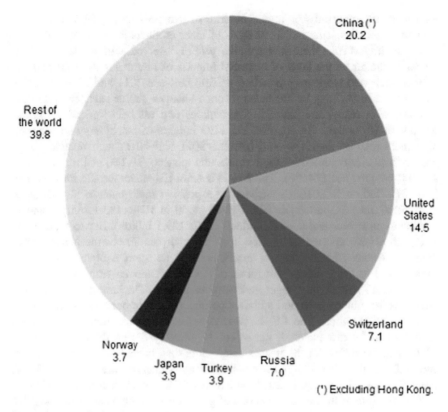

Fig. 5 Main trading partners of EU-28 for imports of goods in 2016

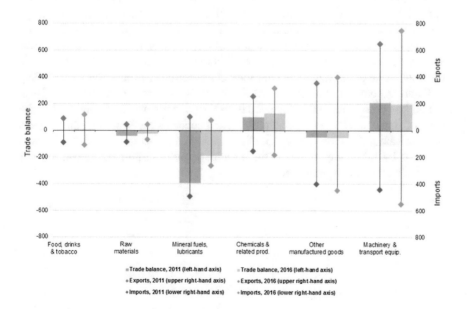

Fig. 6 External EU-28 trade by main products for 2011 and 2016

traders are different from non-traders in terms of size, productivity [9, 16] and other firm characteristics. That is why analysis of main products flows is rational.

On the import side, there was a similar pattern observed, with a relatively large overall reduction in the level of extra-EU imports of raw materials (−20.2%) and mineral fuels and lubricant products (−46.6%) between 2011 and 2016; note that some of the losses may be attributed to price changes and/or exchange rate fluctuations, with many raw material commodities and oil being priced on global markets in US dollars. By contrast, extra-EU imports of machinery and transport equipment rose by 24.9% overall between 2011 and 2016, with relatively high growth rates also recorded for food, drinks and tobacco (19.1%) and for chemicals and related products (18.9%). The EU-28's extra-EU trade surplus for goods of EUR 37.7 billion in 2016 was driven by a positive trade balance in relation to machinery and transport equipment, which stood at EUR 193.1 billion, and in relation to chemicals and related products (EUR 129.1 billion). Between 2011 and 2016, the EU-28 reported an increase in its trade surplus for chemicals and related products, whereas the surplus for machinery and transport equipment narrowed somewhat. For food, drinks and tobacco, the EU-28 moved from a small trade deficit in 2011 to a slightly larger trade surplus in 2016. The largest trade deficit in 2016 was for mineral fuels and lubricant products where imports exceeded exports by EUR 190.0 billion. The EU-28 trade deficits for mineral fuels and lubricant products and for raw materials narrowed considerably during the period 2011 to 2016, with the deficit for the former being more than halved during this five-year period. By contrast, the EU-28 trade deficit for other manufactured goods widened, reaching EUR 53.5 billion in 2016, which was 3.5% higher than in 2011 [14].

The structure of the EU-28's exports of goods changed between 2011 and 2016 most notably among the smaller product groups. The share of food, drinks and tobacco products increased from 5.7 to 6.6% between these years while the share of mineral fuels and lubricant products fell from 6.4 to 4.2%. The largest change between 2011 and 2016 in the structure of the EU-28's imports was for mineral fuels and lubricant products; whose share fell from 28.6 to 15.5%. By contrast, over the same period the share of other manufactured goods rose from 23.3 to 26.3%, while the share of machinery and transport equipment rose from 25.6 to 32.3% [13].

Figure 7 contrasts the structure of the EU-28's imports and exports in 2016: it should be borne in mind that the overall level of exports was 2.2% higher than the level of imports. The most notable difference concerns the share of mineral fuels and lubricant products which was 3.6 times as high for imports as exports. This was balanced by lower import shares for machinery and transport equipment and for chemicals and related products.

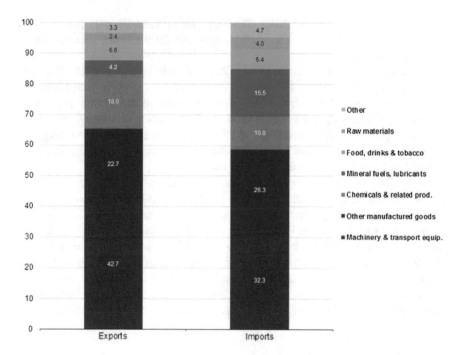

Fig. 7 Main percentage exports and imports of EU-28 by product in 2016

2.2 Trade Between EU and China

Trade and economic cooperation between Europe and Asia mainly concern import and export of goods between Europe and China. European Union and China are qualified as two of the biggest traders in the world. China is the EU's second largest trading partner. For China, the biggest is the Union. In 2015 the value of trade between them amounted to EUR 467 billion, which means that more than a billion euros of goods are transported daily by sea, but also by trains and airplanes. China is now the EU's second-biggest trading partner behind the United States and the EU is China's biggest trading partner. The EU is committed to open trading relations with China. However, the EU wants to ensure that China trades fairly, respects intellectual property rights and meets its obligations as a member of the World Trade Organization (WTO). In 2013 the EU and China launched negotiations for an Investment Agreement. The aim is to provide investors on both sides with predictable, long-term access to the EU and Chinese markets and to protect investors and their investments.

In 2015, Chinese GDP increased by 6.9%. Although this is impressive from a European perspective, this is the worst result in the Middle Kingdom for a quarter century. In March, Standard & Poor's downgraded China's rating outlook from stable to negative. As reported by the China National Development and Reform

Commission (NDRC) in June 2016, China's logistic revenue in 2015 reached \$ 34 trillion. This represents a 5.8% increase compared to 2014 and significantly slower than the GDP [24].

Exports from China to Europe in 2015 have fallen compared to 2014. The same trend is seen for 2016. In the first quarter, China's ocean exports recorded declines. It is clear from Fig. 8, which is based on the statistics of trade turnover between Europe and China [12], that in recent years the disproportion in between import and export is biggest in China. According to this data, European Union imported goods worth 344.6 €bn and services worth 26.4 €bn from China in 2016, while for export it was 170.1 €bn of goods and 37.3 €bn for services.

Comparing the statistics of import and export between European Union and China over the decade, a clear increasing trend line is visible in Fig. 9.

In recent years, however many changes in bilateral trade mainly in agri-food products between the EU and China were noticed. The analyses conducted in [30] indicate that in 2008–2015 there was a significant increase in the bilateral trade in agri-food products between the EU and China. As a result of a faster growth in the EU exports to China compared to the Chinese exports to the European markets, in 2013 the EU changed from a net importer into a net exporter of agri-food products, generating a trade surplus of EUR 2.7 bln in 2015. Despite the strengthening of

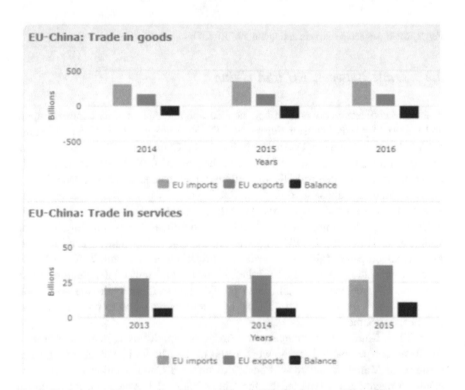

Fig. 8 Trade in goods and services between EU and China in the years 2014–2016

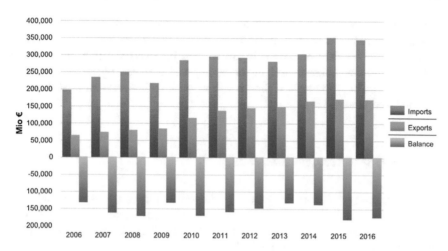

Fig. 9 Turnover between China and Europe in billion tons in the years 2006–2016

trade ties with China at the cost of other non-EU countries (e.g. USA and Japan), the European Union's trade remains concentrated in the Single European Market, and the Chinese share in the EU agri-food exports remains relatively small (below 2%).

According to European policy, when China joined the WTO in 2001 it agreed to reform and liberalise some important parts of its economy. EU's interactions with international institutions is shown in [28] in four modes (policy export, policy promotion, policy protection and policy import), establishes different rationales motivating EU actors to engage through a given mode and relates recent empirical research to this comprehensive typology.

While China has made, progress adapting to the requirements of EU, some problems remain:

- a lack of transparency,
- industrial policies and non-tariff measures that discriminate against foreign companies,
- strong government intervention in the economy, resulting in a dominant position of state-owned firms, unequal access to subsidies and cheap financing,
- poor protection and enforcement of intellectual property rights.

In 2016 the EU adopted a new strategy on China mapping out the European Union's relationship with China for the next five years [11]. The Strategy promotes reciprocity, a level playing field and fair competition across all areas of co-operation and is a part of globalization tendency [18]. What is even more interesting, research [31] show that importers and exporters trading with more distant countries appear to be the most skill-intensive and to pay the highest wages.

The strategy also includes a trade agenda with a strong focus on improving market access opportunities—including negotiations on a Comprehensive

Agreement on Investment. It also deals with overcapacity and calling on China to engage with ambition at multilateral level.

The natural element of economic growth and the exchange is to increase the potential by-sector transport freight forwarding and logistics, operating cash flow between partners. The result is the growth of outsourcing services related to the transport and storage [35].

2.3 Logistics Services Increase

One of the biggest factors for growth in import and export exchanges between countries is the globalization and the exchange of goods within e-commerce. The value of the e-commerce market in Europe this year will exceed half a trillion euros (in 2015 it was 455 billion euros) [24].

Analysing Table 1 in terms of the share of e-commerce in GDP based on Eurostat, Asia-Pacific is the clear frontrunner. Its eGDP rate of 4.48% is significantly above the global average of 3.11%. With an eGDP of 0.77 and 0.71%, Latin American and the Middle East and Northern Africa are at the bottom of the list. Still, these figures grew significantly as well compared to 2014, when they amounted to 0.51 and 0.54%, respectively.

Internet commerce is a challenge for every major transport, forwarding and logistics industry, opportunity and area to develop. Demand for e-commerce transportation services among some businesses has grown by several hundred percent. This is due to the avalanche increase in the number of packages larger than typical parcels, i.e. parcels weighing over 50 kg, but delivered to individual customers, in which this segment counts the highest competencies. In the e-commerce marketplace, not only the Internet sales force like Amazon, EBay and Alibaba, but also companies that distribute the flow of goods, because their quality of service has an impact on the perceived quality of customer service. Figure 10 [17] presents diagram of relationships between e-business, e-commerce and e-logistic.

Logistic operators look at ecommerce as a whole process for financial gain: from order acceptance to web-based service, preparation of sales documents, shipment of goods, co-operation with other couriers within the last mile service, to handling

Table 1 GDP at market prices and share of e-commerce in GDP, 2015

Region	GDP at market prices (bn $)	Share of e-commerce in GDP
Global	73,106	3.11
Asia-Pacific	23,564	4.48
North America	20,642	3.12
Europe	19,518	2.59
Latin America	4295	0.77
MENA	3606	0.71

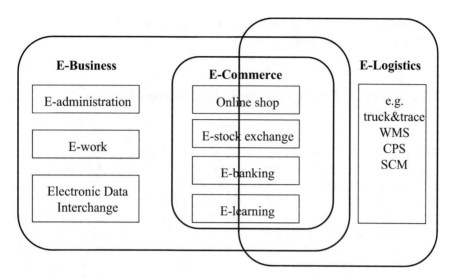

Fig. 10 Diagram of relationships between e-business, e-commerce and e-logistic

possible returns. In e-commerce deliveries, time plays an increasingly important role. Only those companies will win that will provide fast service, because they expect customers who do not want to wait too long for the goods. In terms of delivery times, Europe is at a higher level than the United States. The standard delivery time in America is 72 h (48 h shorter). In Poland, the standard becomes 24 h and it is not only the distance differences [24]. In Europe, including Poland, there are growing expectations about deadlines and flexibility of supply, for example, customers expect them on Saturdays and Sundays, afternoons and at different addresses, which creates a complex matriculation of terms with little possibility of optimization. Shorter waiting times will continue and supply chains will be built in such a way that these requirements are met at least for the part of the product. However, the customer should be able to choose not only the mode of transport but also the different pricing options within the proposed solution. Flexibility in the customer approach is the key issue to success.

In order to prevent conflicts of interests between customers, online shops and shipping companies, a drop-ship logistics solution is implemented and tested. It is a model of cooperation between a manufacturer or an importer and an online shop and a logistics operator for direct delivery. Instead of driving from the manufacturer to the e-commerce store and from there to the final supplier, you can pick up the goods directly from the manufacturer and go to the final destination.

Another logistics solution is fulfilment. Before goods are sent to the customer, the vendor has to order them from the supplier. Next, they are taken into the warehouse and stored, and after the order has been placed, they are packaged and sent to the customer. These processes are defined in e-commerce as fulfilment. Due to their complexity and time and cost pressure of e-customers, they are more and more often conducted by external operators. Despite the dynamic development of the fulfilment service in e-commerce practice, this still remains a relatively unknown issue in management theory [20].

3   Analysis of Transport Corridors Between Europe and Asia

During the last decade of the 20th century and at the beginning of the 21st century, Asian countries appeared in Europe as cheap places for the production of export oriented products. Nearly two thirds of all global freight traffic is in developing countries, more than three quarters of which are in three countries: China, India and Russia. However, this situation is changing and Asian countries diversify their production while focusing on domestic absorptive markets. In addition, more and more prosperous Asians are interested in increasing demand for imported goods from Europe, and this in turn significantly contributes to a change in the structure of trade and international trade. These changes also have implications for logistics.

Use of large-capacity container ships on routes Europe—Asia are economically justified. It allows to reduce the cost of delivery of containers. As it is known, the sea tariff is inversely proportional to capacity of a vessel: bigger capacity—lower transportation cost. On the other hand, it causes additional problems. Delivery of containers in ports of the Baltic or Black seas of transport means such as container ships are impossible because channels or water in ports are too shallow. That is why ocean container ships deliver containers to the ports equipped with hubs (e.g. Rotterdam, Hamburg and Antwerp).

One of the most important questions in delivery of containers from Asia to Europe concerns the time of such delivery. It usually takes from 30 to 45 days from East Asia to the countries of Central and Eastern Europe. This time includes additional delay necessary for processing and transfers of containers, for example, in Rotterdam or Hamburg. Also from 2 to 5 days it is necessary to transit from the port - hub to the port of destination. In the characteristics specified above, considerable time leaves not on transportation, and on performance of port procedures on reception and departure of ships during concrete time.

Interesting illustration of the length of the routes and the time needed to overcome them is presented in Fig. 11 [25]. The main focus here is in the context of the New Silk Route, where over 11,000 km can be overcome in several days of travel. Compared to other travel variants, it is a great time saver. The new Silk Route, which is a freight rail linking China with Europe, is on the list of transport priorities of the Chinese government. The shortest route of the land route is from China to Germany—through Poland. Thanks to the transhipment terminal in Malaszewice in the area of the Polish-Belarusian border, our country has a chance to become a hub at the western end of the Silesian Route.

There are some problems, arising on the routes of railway container transportations. The first one is the various width of a railway track in different countries.

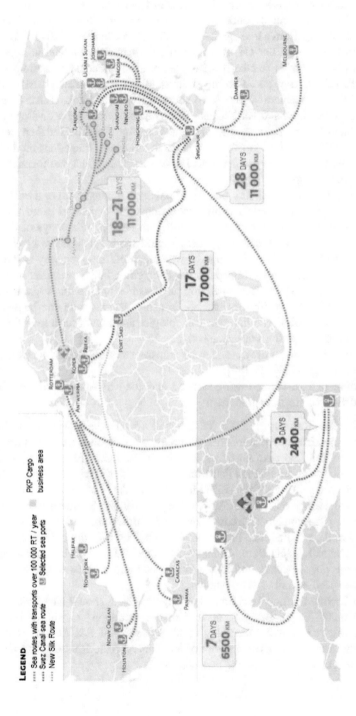

Fig. 11 Duration of cargo carriage from China to Poland by different modes and transport routes

And if to consider transportation of containers from China to the Central or the Western Europe without dependence from a choice of a transport corridor it is necessary to change from a track gauge of 1435 mm in China to 1520 (1524) mm in the countries of the former USSR, Mongolia, Afghanistan, Finland and again to 1435 mm (the European standard). Examples of such boundary transitions on the ways of containers transportation are stations Manchzhurija/Zabaykalsk between China and Russia, Mamonovo/Braniewo between Russia and Poland, Ala Shankou/ Dostyk between China and Kazakhstan, Brest/Małaszewicze between Belarus and Poland, Chop/Zahony between Ukraine and Hungary, Sarakhs/Tedzhen between Turkmenia and Iran. It is obvious, that it is much more such transitions on the ways of containers, however the problems arising at such transitions, are identical. It is caused by technology of a reload.

According to on the boundary transitions having different width of a track, three basic technologies are used. The first is a reload of containers from one cars (platforms) on others. In this case, it is enough to have parallel ways of different width and container reach stackers, which should work between the given ways. The second technology is a exchange of wheel pairs. The given way is the most expedient for application by transportation dangerous, bulk, oversized and other cargoes demanding care. This operation is possible only with wagons of the European standard. The third technology is using wagons with the bogies with expandable wheel pairs. The first two technologies are known for a long time, but the third one is interesting with perspective.

The most important railway and road transport corridors linking China and Europe are shown in Fig. 12 [29].

According to Fig. 12, following railway and road transport corridors linking China and Europe are most important:

Trans-Siberian Railway (TSR): Vladivostok (Nakhodka)–Khabarovsk–Chita–Irkutsk–Krasnoyarsk–Novosibirsk–Ekaterinburg. Further on a map the three possible exits on Pan-European corridors are shown: Northern (Ekaterinburg–Kirov–ports of Baltic sea or on the First Pan-European corridor of Helsinki–Tallinn–Riga–Kaliningrad–Gdansk); Central (Ekaterinburg–Yaroslavl or Nizhni Novgorod–Moscow–and further on Second Pan-European corridor); Southern (the Kurgan–Chelyabinsk–Ufa–Samara–Kharkov–Kiev—and further on Third Pan-European corridor). On the resulted map connections of the TSR with the railway system of Mongolia (Naushki–Ulan Bator–Erenhot) and China (Zabaykalsk–Harbin–Beijing or Seoul) are not allocated, and also a site Baikal-Amur Railway (BAR), nevertheless, the specified railways are shown.

Northern Trans-Asiatic Corridor: Lianyungang–Zhengzhou–Lanzhou–Urumqi–Dostyk–Almaty–Astana–Kurgan—is further transportation possibility on three directions resulted above). According to classification of ESCAP this route has received the name "Northern corridor" of the Trans-Asiatic Railway, and on classification of OSZhD (OSJD)—the first corridor.

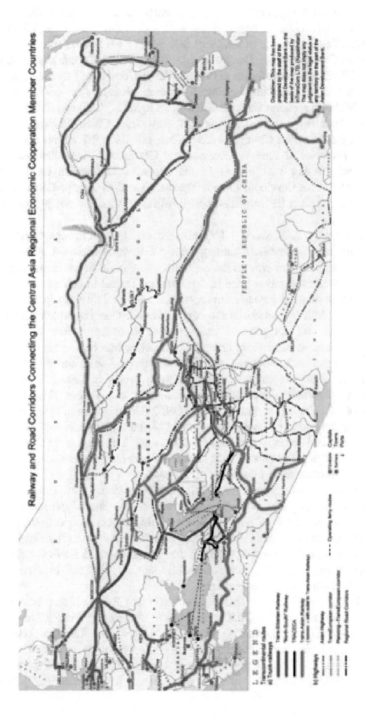

Fig. 12 Transcontinental railway and road routes connecting central Asia and Europe

Southern Trans-Asiatic Corridor: Beijing–Urumqi–Almaty–Tashkent–Chardzhou–Sarakhs–Mashhad–Tehran–Istanbul in the extent of 12 thousand the kilometres, almost coinciding with a line of the Great Silk Way of the Middle Ages.

Corridor   TRASECA: Dostyk–Tashkent–Ashkhabad–Turkmenbashy–Baku–Tbilisi–Poti, further through ferries to Odessa, Varna, Constance, Istanbul. In May of 1993 in Bruxelles the working meeting of representatives of Azerbaijan, Armenia, Georgia, Moldova, Mongolia, Kazakhstan, Kirghizia, Tajikistan, Turkmenia, Uzbekistan, Ukraine, Romania, Bulgaria, Turkey, Iran, China, Pakistan and the countries of the European Union has taken place, on which the various routes connecting the countries of Europe, Caucasus and Central Asia were discussed. The result of this meeting was signing of the Bruxelles Declaration on Technical Assistance of EU for development of a Transport Corridor Europe-Caucasus-Asia (TRASECA), including the combined system of railways, highways, pipelines, airlines and sea ways.

European Transport Network (ETN) provides joining with transport systems of the East European countries, including with the Russian transport system. The choice of a transportation way of the container more often depends on economic factors, and the transportation price in this case has crucial importance.

One more problem of container transportations is instability of a political situation in region. Military actions at the conflict Russia–Georgia, which have mentioned port Poti, did not promote trust increase to transportations on corridor TRASECA. Straining in relations around Afghanistan, Iran or Iraq also influences on development of transportations on a Southern Trans-Asiatic corridor.

Due to the changing structure of trade, new means of transport are being developed at the international level between Asia and Europe. The Eurasian Railway Corridor Project, supported by the OSJD (Organization of Cooperation of Railways), is the most representative example of the initiative to increase operational efficiency and economic importance of railways. This project aims to increase the efficiency of passenger and freight services between Asia and Europe with the use of tens of thousands of kilometres of railways.

The idea of activating international rail freight is nothing new. However, it is appropriate in recent years to return to the idea of the 50s of the last century, through the development of railway infrastructure, to stimulate economic links between the countries of Asia and Europe. It is the idea of transcontinental rail corridors, developed by the Organization for Cooperation of Railways (OSJD), established on June 28, 1956, emits so much emotion today and involves many governments of the beneficiary countries.

OSJD brings together 27 countries from Asia and Europe along the planned corridors. The members of the OSJD are 14 Asian countries: Vietnam, China, North Korea, Kyrgyzstan, Kazakhstan, Turkmenistan, Uzbekistan, Tajikistan, Georgia, Azerbaijan, Armenia, Iran, Mongolia, Asian parts of Russia and 13 European countries: Lithuania, Latvia, Estonia, Belarus, Ukraine, Czech Republic, Slovakia, Hungary, Bulgaria, Romania, Moldova and Albania. Many countries, including

Germany, France, Greece, Finland and Serbia, which are not directly involved in the project, have adopted observer status.

Freight and passenger transport takes place along following thirteen transport corridors shown in Fig. 13 [37]:

Corridor no 1—the longest land transport corridor in the world—runs through 11 countries: Estonia, Latvia, Lithuania, Poland, Belarus, Russia, Kazakhstan, Uzbekistan, China, Mongolia and North Korea. Its total length, including all its branches, is 24,800 km. The corridor, which runs a large part of its course through the territory of the Russian Federation, uses the Transsiberian Railway.

Corridor no 2—runs through the territory of the Russian Federation, Kazakhstan, China and Vietnam. The total length of the corridor along with numerous branches is 15,212 km.

Corridor no 3—has its course through the territory of Poland, Belarus and the Russian Federation. The length of this corridor is 2209 km.

Corridor no 4—is located in the Czech Republic, Slovakia, Poland, Hungary and Ukraine. The total length of the corridor is 2711 km.

Corridor no 5—runs through Hungary, Slovakia, Ukraine, Moldova, the Russian Federation, Azerbaijan, Kazakhstan, Kyrgyzstan and China. Length: 11,486 km.

Corridor no 6—is located in the Czech Republic, Slovakia, Hungary, Romania, Serbia, Bulgaria, Greece, Turkey, Iran and Turkmenistan. The length of this corridor is 12,442 km.

Corridor no 7—runs through the territory of Poland and Ukraine. Length— 1520 km.

Corridor no 8—is located in the territory of Ukraine, the Russian Federation, Kazakhstan, Uzbekistan and Turkmenistan. The length of this corridor with its branches is 5115 km.

Corridor no 9—runs through the territory of Lithuania, Belarus and the Russian Federation. The length of this corridor is 845 km.

Corridor no 10—links the countries of the former Soviet Union (Georgia, Azerbaijan, Turkmenistan, Uzbekistan, Kyrgyzstan, Kazakhstan and Tajikistan) to the European states (Bulgaria, Romania and Ukraine). On the route of this corridor there are ferry crossings through the Black Sea (1100 km) and the Caspian Sea (310 km). The total length of the corridor is 7437 km and the length of the part of the corridor is 8847 km.

Corridor no 11—is located in the territory of the Russian Federation, Azerbaijan and Iran. The main part of the corridor has a length of 5576 km, and with its branches its overall length rises to 7891 km.

Corridor no 12—runs through the territory of Moldova, Romania and Bulgaria. The length of this corridor is 1461 km.

Corridor no 13—trail of this corridor connects Poland, Lithuania, Latvia, Estonia and the Russian Federation, with a length of 1497 km.

On the global scale, the public railway network is almost one million kilometers. Approximately 56% of this network (512 thousand km) is located in developing

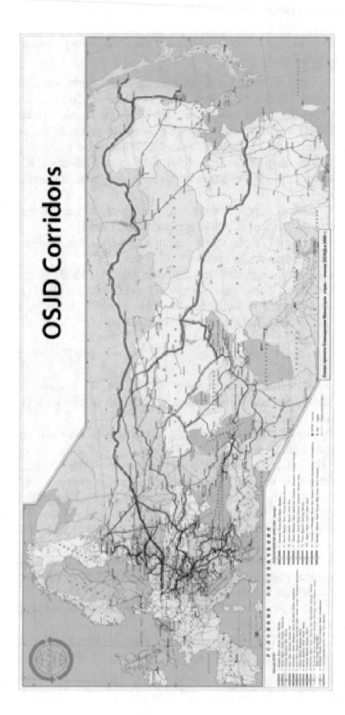

Fig. 13 OSJD transport corridors

countries. Over the past twenty years, the total mass of goods transported by rail in developing countries (as measured by net tonne-kilometres) increased by almost 50%, although this increase has shown a large regional variation. The highest rates were observed in East Asia, Pacific Ocean, South and Central Asia and Eastern Europe. The main reason for this growth was economic growth in China, India and Russia.

Although at the end of the 20th century and early XXI century, road transport clearly dominated international transport of goods, both in terms of mass and transport performance, far ahead of rail transport, this does not, however, mean total the fall of the railway as one of the basic branches of transport. Land is still in fierce competition and none of the parties has yet said the last word. This can be substantiated by statistical studies, which clearly show that rail transport still plays an important role in international freight traffic (e.g. in the Russian Federation it is around 90%, in Kazakhstan over 72%, in Belarus 87% and in Ukraine about 70%) [38]. As it might be noticed a particularly important role it fulfils in countries characterized by low density of traditional roads designed for cars with large tracts of wilderness areas and prevent effective linear expansion of road infrastructure.

In previous years, the objectives of the OSJD were to focus on the reconstruction and modernization of infrastructure in designated transport corridors and management structures, to adapt them to the current market environment, to increase the mass of the freight carried and to meet the trend of rapid technological progress.

The achieved results of establishing international transport corridors were possible thanks to coordinated and deliberate work of all OSJD member countries.

In 2012–2014, the activities of the OSJD focused on the further development of international rail traffic between Europe and Asia, including, above all, the improvement of the regulatory and regulatory framework, the enhancement of rail transport competitiveness, the demand for passenger transport in the Member States and passengers in the vast territories of Eurasia. The cost of developing all TAR infrastructure projects is estimated by ESCAP professionals at nearly $ 500 billion a year over the next 10 years [38]. A very interesting concept that should also be considered when discussing transport routes is the **New Silk Route**.

The design of the New Silk Route refers to the trade route connecting China to Europe and the Middle East. That route was over 12 thousand km and was used since the 3rd century BC to the 17th century AD. Then he ceased to be of importance due to the discovery sea route to China about 1650.

The new concept was presented by China's President Xi Jinping in September 2013 during a visit to Kazakhstan. That's when he proposed the creation of Silk Road Economic Belt, i.e. land transport corridors linking China with Europe. Later on this concept was added 21st Century Maritime Silk Road, as an extension of sea routes. Both elements over time began to function under a common name One Belt and One Road.

It is not only a great logistics project, but also a way to strengthen the so-called. Chinese soft power, or the influence of the region. The Silk Road is a counter-proposal for the other economic and political projects of the region: the American (Transpacific Partnership (TPP)) and the Russian (Eurasian Economic

Union). In order to realize the associated investments, China has established a Silk Road Fund, which has a $ 40 billion capital. The second source of funding is the Asian Infrastructure Investment Bank (AIIB), which was also joined by Poland.

The silk route of the XXI century will go through several countries as shown in Fig. 14 [26]. The beginning will be in China, the end in Germany. Poland is also included in the project. Russia is also involved in the project, which has to go through a large part of the route for geographical reasons. Kazakhstan and Azerbaijan are also important participating countries. These are located at the junction of the main transport corridors and connect the two Eurasian regions. Land and rail are cheaper and faster than maritime transport, causing so much interest in so many countries on two continents.

The new silk route is a logistic megaproject that will revolutionize trade between China and Europe—including Poland. The importers we have confirmed confirm that the short-term delivery time is 2 weeks, which is two times shorter than the ship's time (if the vessel is flying around Cape of Good Hope). However, they complain about the high costs that cause the pulling of goods from Chinese factories are often unprofitable. Importing electronic products from China to Poland costs between 1.5 and 2 thousand. dollars. for a container. Meanwhile, their transport by rail is at least 5–6 thousand. dollars. Most governments in both Europe and Central Asia are seeking to encourage the development of rail freight, mainly due to low levels of negative environmental impacts and low transport costs.

However, it must be acknowledged that the main disadvantages of this type of transport are, on the one hand, the lack of flexibility and, on the other, the presence of a number of physical and non-physical barriers hindering freight traffic on international routes. To overcome these obstacles, it is necessary to remove these obstacles, at least at the administrative level, while emphasizing the advantages of long-distance intercontinental rail routes as compared to sea lanes, for example by demonstrating the economic benefits of significantly reducing transit times.

The domain of transport as a branch of transport is the transport of bulk goods and containers. For this reason, it should be seen and treated as the most effective branch of land transport. But its proper use and further development largely depend on the quality of management. In most countries railways are still state-owned companies that do not operate on a private sector basis, where "to be or not to be" depends on factors such as performance and competitiveness. Despite the potentially large potential, the railroads in Eurasia have, despite a marked increase, relatively low utilization of their potential. This is primarily due to the low number of specialist rolling stock, obsolete line and point infrastructure (not adapted to today's tasks and requirements), and low labor productivity. All of this results in negative consequences for national allocation of resources and significantly limits transport capacity.

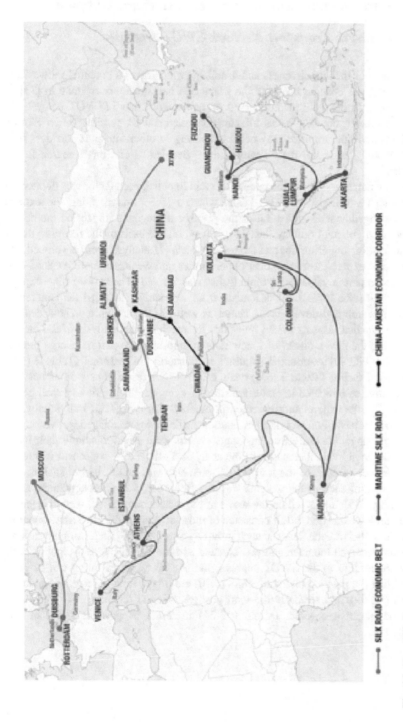

Fig. 14 New Silk Route

4   Future Development of Eurasian Transport Corridors

4.1   Eurasian Transport Network SWOT Analysis

Assessment of the attractiveness and usability evaluation of a potential railway line Connecting the Caspian Sea with the Black Sea has already been done in [33]. It was done with heuristic method based on expert knowledge (SWOT analysis) in order to present the key issues affecting development of possible route. Similar problems and opportunities may occur according to other transport corridors possibilities and that is the reason why authors decided to cite this research in the chapter.

For the determination of the external factors final list, each factor was defined by evaluating its validity in a 0–5 scale, which is the average rating the experts opinions together with indication of the positive or negative impact on the development of a potential railway line. In this way, on the basis of the resulting list of external factors opportunities and threats in the SWOT analysis were developed. On the basis of expert knowledge then opportunities and weaknesses were established and after experts evaluation a ranking list was formed. There were nine experts familiar with the topic both: scientists from two universities, participant representing research and development center as well as practical entrepreneur dealing with the practical aspects of rail transport. Research on the analysis of the factors influencing the development of a new railway line was one part of a large project connected with future-oriented solutions in transport. All external (Table 2 [33]) and internal factors (Table 3 [33]) evaluated from the research as well as positive and negative as indicated in tables are sorted by the weighted rating in each group indicated by the experts and presented in brackets for each factor individually.

The study basing on weighted evaluation method was conducted in two stages. In the first stage, the brainstorming stage, experts were asked to submit their ideas related to the individual elements of the analysis. Followed by consensus hierarchy of importance of each of the four SWOT elements was established in such a way that the sum of each of the weights was 1. After a predetermined agreement as to the validity of elements within the scope of the factors considered, second stage of the research was carried out. It consisted of individual and subjective assessment in the scale of 0–5 of each factor in each group by every expert. Then from the experts scorecard received the averaged was counted and multiplied by the weight of each factor. According to the SWOT analysis, the overall weighted rating can be concluded that the positive internal factors (3.40) are bigger than the negative internal factors (−3.09), and similarly the total positive external factors (3.74) are higher than the negative external factors (−3.24) in terms of possible Eurasian rail connection.

Table 2 External factors of new railway connection SWOT analysis

Opportunities (total: 3.74)	Threats (total: −3.24)
1. Increase in expenditure on rail transport (0.72)	1. Capital intensity of new transport technologies (−0.50)
2. Development of new technologies in the European and global transport (e.g. handling automation) (0.68)	2. Differences in gauge, infrastructure, access to technological progress of countries (−0.45)
3. Economic growth of the region and incomes of the population (0.4)	3. Insufficient number of platforms and wagons in relation to the transport potential (0.40)
4. International conventions in transport, harmonizing the conditions of carriage and documentation (0.4)	4. Military conflicts politically or religiously motivated (−0.36)
5. High density of railways including those belonging to the TEN-T (0.25)	5. Necessity of introducing a system of waste utilization derived from transport activities (−0.21)
6. Emphasis on the development of intermodal infrastructure points (0.2)	6. Diversity of countries in terms of tax rates and duties (−0.15)
7. Standardization of intermodal transport units on a global scale (0.2)	7. Contraband smuggling (−0.12)
8. Uniform rules of international trade (0.15)	8. High level of corruption (−0.12)
9. Existing market structures taking into account the liberalization of the transport market (0.12)	9. Different conditions of the time, speed, axle loads (−0.10)
10. Increase in the use of environmentally friendly technologies in transport (0.12)	10. Further development of the road transportation with a decrease in intermodal transport (−0.08)
11. Possibility of using private capital and public funds (especially from EU) (0.12)	11. Cultural linguistic and religious differences between countries along the railway connection (−0.06)
12. Availability of land for the construction of a modern transport infrastructure (0.09)	12. Diversity of countries' economic policies (−0.06)
13. Increasing environmental awareness in the society (0.09)	13. Ageing of the population (−0.03)
14. Demographic potential: a large number of residents in the region, age structure, level of education (0.08)	
15. Growing importance of intermodal transport in policy and strategy development in the countries (0.06)	
16. Stimulating innovation in transport through political lobbying (0.04)	
17. Conscientiousness and the liability of carriers in transport (0.02)	

Further, basing on the SWOT analysis, further matrix can be elaborated with actionable strategies with multiple solutions among:

– SO (Strengths-Opportunities)—maxi-maxi strategy: strategies that use strengths to maximize opportunities,
– ST (Strengths-Threats)—maxi-mini strategy: strategies that use strengths to minimize threats,
– WO (Weaknesses-Opportunities)—mini-maxi strategy: strategies that minimize weaknesses by taking advantage of opportunities,

Table 3 Internal factors of new railway connection SWOT analysis

Strengths (total: 3.40)	Weaknesses (total: −3.09)
1. Research potential in the field of intermodal transport and transhipment (0.75)	1. Decapitalisation of infrastructure fragments (−0.60)
2. Educated human resources in the field of rail and intermodal transport (0.40)	2. Weak position of intermodal transport in acquiring the EU and national funds (−0.50)
3. Enormous potential for transport from China to Europe (0.30)	3. Little interest in business enterprises for cooperation and research funding (−0.40)
4. Financing the development of transport from the national and the European Union funds (0.30)	4. Demand for workers with appropriate qualifications (−0.30)
5. Possibility of using good practices for national and international promotion of a new railway line (0.25)	5. Property decapitalisation of transport operators (−0.30)
6. Enterprise interest of new technologies in transport implementation (0.20)	6. Small initiative in the development of public-private partnerships for the construction of new transhipment points (−0.20)
7. Research development of Automatic Track Gauge Changeover Systems (0.20)	7. Weak activity associated with the promotion of intermodal transport (−0.20)
8. Formation of an integrated transport structure of the transcontinental scale (0.16)	8. Unfavourable cost structure of intermodal transport (−0.24)
9. Research development in the field of energy-efficient technologies in transport solutions (0.16)	9. Small number of innovative transport projects (−0.15)
10. Active manufacturing capabilities of modern intermodal transport technologies (0.15)	10. Research funding from different sources, different decision-making centres, lack of concentration of resources in the implementation of projects (−0.10)
11. Higher education development related to railway and intermodal transport (0.12)	11. Low level of potential use of the R&D and academic institutions dealing with transport (−0.10)
12. Large volume of trade in the transport sector (0.12)	
13. Demand for transport services, resulting from the import and export volume between Europe and Asia (0.10)	
14. Gradual launch of innovative potential in the rail transport sector (0.09)	
15. Large number of scientific conferences with scope of transport container for popularizing issues of container transportation (0.09)	
16. Popularization of container for the transport of goods (0.09)	

– WT (Weaknesses-Threats)—mini-mini strategy: strategies that minimize weaknesses and avoid threats.

Basing on the results gained with the weighted evaluation method based on expert opinions several strategies can be developed, which are presented in Table 4 [33]. The table shows only the most important suggestions and possibilities of strategies because of the enormous number of factors generated during research on the SWOT analysis.

Table 4 TOWS strategic alternatives matrix of the Eurasian railway connection

	External opportunities (O)	External threats(T)
Internal Strengths (S)	**SO** 1. Increasing number of research projects associated with rail transport (O_1, S_1, S_6, S_4) 2. Developing research projects on new handling possibilities (O_2, S_1, S_4, S_6, S_7, S_9) 3. Developing TEN-T railway corridors especially on East - West routes (O_5, O_3, S_3) 4. Integrating transport structure by further standardization (O_7, S_8) 5. Developing good practice policy among partners involved in transport chains (O_8, S_5) 6. Developing environmental awareness among transport systems (O_{10}, O_{13}, S_{11}, S_{15}) 7. Increasing higher education potential among intermodal transport (O_{14}, S_{11}) 8. Popularization of containers for carriage harmonization O_4, S_{16})	**ST** 1. Increasing number of research projects associated with cheap transport methods (T_1, S_1) 2. Coping with integrated transport systems according to rail gauge etc. (T_2, S_8, S_{10}) 3. Building interest potential for rail platforms manufacturing (T_3, S_6) 4. Building integrated wastes utilization for transport chains (T_5, S_8)
Internal Weaknesses (W)	**WO** 1. Developing model of prices for different price levels according to infrastructure owner (O_1, W_1, W_5, W_8) 2. Introduction of business enterprises in research projects seeking to uniform transport (O_7, O_8, W_3) 3. Increasing expenditures on public-private initiatives among transhipment points of railway corridors (O_1, W_6, W_8) 4. Increasing different funding sources for innovative transport projects (O_1, W_9, W_{10}) Increasing usage of academicals potential for innovations in transport (O_2, O_{13}, W_{11})	**WT** 1. Strengthening position of intermodal transport by minimalizing differences in gauge and infrastructure (T_2, W_2) 2. Employees and users education on corruption and contraband smuggling (T_7, T_8, W_4) 3. Strengthening activities associated with intermodal transport promotion and development (T_{11}, W_7)

4.2 Scenarios of Eurasian Corridors Future Development

There are different environmental analysis techniques [15]. Macro-analysis methods include, but are not limited to: extrapolation of trends, analysis of strategic gaps, heuristic methods based on expert opinions (Delphi method), or environmental scenario analysis.

There are methods for analysing discontinuous changes that are not extrapolative continuation of processes that take place in the future in a given time. Discontinuous change means the transition between the past and the present and between the present and the future. Scenario methods are used to analyse discontinuous changes.

Scenario analysis is a method of creating many different possible events in the future with a probability of occurrence with improving quality and effectiveness [1]. This is to reduce the uncertainty associated with the planned project. Scenarios are a long-term forecasting method, and the need to develop them is due to the high uncertainty surrounding the environment, which requires appropriate tools to anticipate the future and adapt the organization's activities.

The current research based on scenario planning reach years over 2020 [21] and even until 2050 [10]. Scenarios surrounding states are qualitative. Scenario analysis is a procedure that allows scanning the future in an organized way and internalizing human choice into sustainability science [34] which is the biggest advantage of this method. Answers to how qualitative and quantitative scenario methods differ and what are the advantages and disadvantages may also be found in [3].

Scenario methods are a strategic management tool and are used to analyse strategic planning in changing environment. They do not derive an accurate picture of the future, but they stimulate management to anticipate the different phenomena and to study their impact on the enterprise or problem. An important feature of scenario scenarios is the inclusion of variants. The need for their preparation is due to the very high uncertainty surrounding the environment requiring managers to have adequate tools to anticipate the future and adapt the organization's activities.

Scenarios are considered as a long-term forecasting method, and can be divided into four basic groups:

(1) scenarios of possible events-based on intuitive logic, the essence of which is to create lists of possible future events that are important to the organization. The scenario method of possible events is an instrument useful for assessing the degree of anticipation risk, fundamental change, and the selection of competing strategic objectives of an enterprise;

(2) simulation scenarios serve to make a forward-looking assessment of the value of the organization's strategic choices based on the impact of the environment;

(3) scenarios of the environment—they are assumed to be of a qualitative nature. This means that evaluating the potential impact of individual processes occurring in the environment on the organization and estimating the probability of occurring these processes in a specific future is based on the knowledge of the scenario creators. We distinguish four basic scenarios: optimistic,

pessimistic, surprise and most likely. This are the scenarios which will be further discussed and elaborated in this chapter;

(4) environmental scenarios—give a generalized view of the environment, determine the impact of individual processes on the organization, show opportunities and threats and possible surprises. On the other hand, scenarios of processes in the environment focus on the most important processes with potentially longer impact on organizations.

Creating environment scenarios is based on several stages. The first is to identify the organization's environment (both macro- and microenvironmental factors).

In the second stage, basing on trend analysis of individual factors the environment script for all states of separate spheres is built as well as suitable environment assessment of factors identified. Then one of the three potential trends—growth, stabilization or decline—is identified for each of them.

The formulation of scenarios of the environment consists in ordering the trends according to the different scenarios of the surrounding environment. The optimistic scenario is created based on the trends of the factors of the individual spheres, which have the most positive influence on the organization. The pessimistic scenario is built on the basis of trends that have the biggest negative impact on the organization. Surprise scenario—is based on the trends of which the probability of occurrence is the smallest and independent of the potential positive or negative influence, its analysis is the starting point for early warning systems. On the other hand, the most likely scenario—consisting of trends that are most likely to occur, regardless of potential positive or negative impact.

In the fourth stage, evaluation and graphical presentation of the scenarios surrounding states is being prepared. Graphical presentation of scenarios allows you to visualize and quickly identify which sphere of environment is the source of the opportunity, and which threats, and how strong is the impact of each sphere of the environment.

Scenario building is a significant tool and support of environmental decision-making [23] and will be presented in practise.

In the first stage of scenario building, the company environment should be identified in order to determine which macroeconomic factors have the strongest impact on the company and which of the competitive environment are decisive. The most commonly used method of analyzing the impact of macro-business on a company is the PEST analysis. The name was created from the first letters of the political factors—P; economic—E; social—S and technological—T. Authors, however, decided to use more detailed method with brainstorming which may bring breadth of factors identified within categories, and the number of factors identified in the most distant future time-period [27].

PEST analysis is a simple tool for preliminary analysis of the macro-environment of the organization. Compared to the version based on the four basic spheres of the environment, there are many modifications to extend the scope of analysis to additional areas of the environment. Among them following can be distinguished:

SLEPT—political sphere is added (Social, Legal, Economic, Political, Technological);

STEEP—environmental sphere is added (Social, Technological, Economic, Environmental, Political);

PESTEL—legal sphere is added (Political, Economic, Social, Technological, Environmental, Legal);

STEEPLE—legal, ecological and ethical spheres are added (Social, Technological, Economic, Ecological, Political, Legal, Ethical);

STEEPLED—legal, ecological, ethical and demographic spheres are added (Social, Technological, Economic, Ecological, Political, Legal, Ethical, Demographic);

STEER—ecological sphere is added and political sphere is replaced with regulatory sphere (Social, Technological, Economic, Ecological, Regulatory).

The scope of the analysis is adjusted according to the specific needs and the adopted division of the environment. Contrary to appearances, the environment is not unique to all businesses, technologies or processes analyzed. In fact, every research problem should be analyzed under appropriate conditions, depending on, among others, from location, sector type, industry, problem size etc.

Methodology proceed, irrespective of the division made, can be hereinafter the same. After isolating from the environment, the areas that will be analyzed, we identify factors in the individual areas that have or can affect the problem being analyzed.

The environmental factors of Eurasian transport connections were identified according to STEEPLE analysis, which is an acronym of following factors: social, technological, economic, ecological, political, legal and ethical.

The next step is to evaluate the identified processes in the environment. This assessment is made in terms of the strength and direction of the impact of the developing transport connections between Europe and Asia, on a scale from −5 (most negative impact) to +5 points (the most positive impact) according to Table 5, and the probability of a given process in three variants.

In the scenario surrounding states, it is assumed that in the future, for which formulated the organization's strategy, each with a distinct trends or processes in the environment can be characterized by regression, stagnation or growth. Accordingly, three potential trends may be identified for each of the phenomena or processes analyzed:

Table 5 Scale of trend evaluations in points and semantic groups

Type of influence	Negative					Positive				
Points	−5	−4	−3	−2	−1	+1	+2	+3	+4	+5
Signification	Very high	High	Medium	Small	Very small	Very small	Small	Medium	High	Very high

- growth—upward tendency of the process in the future; potential negative or positive force of trend influence and probability of occurrence,
- stabilization—future stability of the process; potential positive or negative strength of trend impact and probability of occurrence,
- recession—downward trend of the process in the future; Positive or negative force of trend impact and probability of occurrence.

On the basis of the STEEPLE analysis, which is an acronym of following factors: social, technological, economic, ecological, political, legal and ethical, forty-one individual environmental factors were identified together with trend and probability of occurrence evaluation. In the first, social sphere there are taken into account such factors affecting the problem analysed, such as values, attitudes, beliefs, patterns of consumption, lifestyle and cultural traditions, labour standards and attitudes towards work, the level of education and training, ethics, business, health care, openness for international products in other technologies. Table 6 presents list of six environmental trends in the social sphere of Eurasian transport connection development.

Next, a very important issue was identified and evaluated, connected with technological and technical aspects of transport connecting the two continents. In this area following elements can be taken into account: quality of research facilities, rate of change in production processes, ecological technologies, information technology and communication systems, transport infrastructure, production technologies, design technologies in products, level of engineering and technical staff etc.

Table 6 Analysis of environmental trends in the social sphere of Eurasian transport connection development

Index	Social factor	Trend	Influence strength of trends +5 to −5	Probability of trend 0−1
S1	Awareness and communication patterns of people behavior	Growth	+5	0.6
		Stabilization	+3	0.3
		Recession	−4	0.1
S2	Difference in lifestyle and cultural behavior of Eurasian population	Growth	−5	0.3
		Stabilization	−2	0.6
		Recession	+3	0.1
S3	Inequality of income distribution of Eurasian population	Growth	−3	0.3
		Stabilization	+1	0.4
		Recession	+2	0.3
S4	Level of education and training of people	Growth	+4	0.5
		Stabilization	+2	0.4
		Recession	−2	0.1
S5	Openness to innovative transport solutions	Growth	+5	0.4
		Stabilization	+2	0.3
		Recession	−2	0.3
S6	Language barriers and time zones that affect working time	Growth	−4	0.3
		Stabilization	−1	0.5
		Recession	+3	0.2

Table 7 Analysis of environmental trends in the technological sphere of Eurasian transport connection development

Index	Technological factor	Trend	Influence strength of trends +5 to −5	Probability of trend 0–1
T1	Pace of changes in transport processes	Growth	+5	0.7
		Stabilization	+3	0.2
		Recession	−3	0.1
T2	Development of linear and point transport infrastructure	Growth	+5	0.6
		Stabilization	+4	0.3
		Recession	−4	0.1
T3	Development of pro-ecological transport technologies	Growth	+3	0.4
		Stabilization	+1	0.3
		Recession	−2	0.3
T4	Co-financing of investment in new transport technologies by the involved Eurasian countries	Growth	+5	0.3
		Stabilization	+3	0.5
		Recession	−3	0.2
T5	Development of new channels of communication and information on developing transport lines	Growth	+5	0.5
		Stabilization	+2	0.3
		Recession	−4	0.2
T6	Realization of joint research projects in the field of innovative transport solutions	Growth	+3	0.7
		Stabilization	+2	0.2
		Recession	−2	0.1

Table 7 presents list of six environmental trends in the technological sphere of Eurasian transport connection development.

In the next step issues connected with important sphere of economic issues were identified among problems connected with: economic systems, size of the economy of a given country by GDP, employment and unemployment, economic structures and direction of change, cyclical changes: recession, economic climate, stagnation, economic growth rates, income level and distribution, energy, transport, materials, official price range, currency value and fluctuations in exchange rates, level of development of investment and capital markets etc. Table 8 presents list of six environmental trends in the economic sphere of Eurasian transport connection development.

Following group of factors is called ecological or environmental and usually following issues are taken into consideration: global warming and climate change, carbon footprint, recycling, environmental regulation, attitude towards sustainability etc. Table 9 presents list of six environmental trends in the ecological sphere of Eurasian transport connection development.

Further analysis was based on regulatory aspects of the environment surrounding transport development. In the political trends issues connected with the economic and social program of the ruling party and the largest opposition party, foreign policy priorities were discussed. Other topics related were: political system in a given country, parties and national groups, political stability, membership of a country to international groups and trade blocs, availability of government

Table 8 Analysis of environmental trends in the economic sphere of Eurasian transport connection development

Index	Economic factor	Trend	Influence strength of trends +5 to −5	Probability of trend 0–1
E1	Varied level of unemployment	Growth	−3	0.2
		Stabilization	−1	0.3
		Recession	+4	0.5
E2	Level of interest rates	Growth	−4	0.3
		Stabilization	−2	0.5
		Recession	+3	0.2
E3	Varied level of the GDP in the countries of Europe and Asia	Growth	−4	0.6
		Stabilization	−1	0.3
		Recession	+4	0.1
E4	The development of the world economy related to the exchange of goods between Europe and Asia	Growth	+4	0.5
		Stabilization	+2	0.4
		Recession	−3	0.1
E5	Level of development of an economy based on innovation in the field of transport	Growth	+3	0.3
		Stabilization	+1	0.5
		Recession	−2	0.2
E6	Inflation levels in the countries of the continents analysed	Growth	−3	0.5
		Stabilization	−2	0.3
		Recession	+3	0.2

Table 9 Analysis of environmental trends in the ecological sphere of Eurasian transport connection development

Index	Ecological factor	Trend	Influence strength of trends +5 to −5	Probability of trend 0–1
C1	Impact of transport emissions on global warming and climate change	Growth	−4\5	0.2
		Stabilization	−2	0.4
		Recession	+5	0.4
C2	Development of pro-ecological branches and means of transport	Growth	+4	0.2
		Stabilization	+1	0.5
		Recession	−3	0.3
C3	Common environmental policy of the Eurasian countries	Growth	+3	0.3
		Stabilization	+1	0.4
		Recession	−2	0.3
C4	The attitude of countries to the concept of sustainable development	Growth	+3	0.3
		Stabilization	+1	0.5
		Recession	−3	0.2
C5	Obtaining energy for transport from renewable sources	Growth	+4	0.1
		Stabilization	+1	0.7
		Recession	−3	0.2
C6	Protection of nature and landscape for designed transport investments	Growth	+3	0.6
		Stabilization	+1	0.3
		Recession	−4	0.1

subsidies, level of commercial protection. List of six environmental trends in the political sphere of Eurasian transport connection development are shown in Table 10.

Another step was to identify legal (regulatory) aspects, such as: financial and banking system, scope of state interventionism, contract law, employment regulations, trade union law, monopoly law and restrictive practices, consumer protection legislation, tax law, anti-corruption law, transport and customs. Six legal trends in the political sphere of Eurasian transport connection development are listed in Table 11.

While considering ethical trends, following aspects were analysed: ethical behaviour in business, trust and reputation in business, commercial secrecy, protection of personal data, protection and protection of customer data. Table 12 presents list of five ethical factors with the influence and probability estimation.

The next step is to sort the trends according to the different scenarios. On the basis of the table containing trend analysis, environmental scenarios can be created. In the methodology for building development scenarios, the third step is to organize the trends and create exemplary scenarios:

- optimistic scenario consists of choosing in the different spheres the trend, which have most positive influence on the problem,
- pessimistic scenario is the trend that has the most negative impact,
- most probable scenario includes those trends that, regardless of the potential positive or negative impact, are most likely to occur,

Table 10 Analysis of environmental trends in the political sphere of Eurasian transport connection development

Index	Political factor	Trend	Influence strength of trends +5 to −5	Probability of trend 0–1
P1	Compatibility of the policies of the ruling parties and the opposition countries of Eurasia	Growth Stabilization Recession	+2 +1 −1	0.3 0.4 0.3
P2	Pro-development and open state policy	Growth Stabilization Recession	+4 +2 −1	0.1 0.4 0.5
P3	Peaceful policy, which prevents wars and armed conflicts	Growth Stabilization Recession	+5 +2 −5	0.2 0.5 0.3
P4	Common economic and transport policy	Growth Stabilization Recession	+4 +2 −4	0.4 0.3 0.3
P5	Prioritization of foreign policy	Growth Stabilization Recession	+4 +2 −3	0.3 0.5 0.2
P6	Membership of the countries of Eurasia in international political-military unions	Growth Stabilization Recession	+4 +2 −4	0.3 0.5 0.2

Table 11 Analysis of environmental trends in the legal sphere of Eurasian transport connection development

Index	Legal factor	Trend	Influence strength of trends +5 to −5	Probability of trend 0–1
L1	Differentiation of the legal systems of the countries of Eurasia	Growth	−4	0.3
		Stabilization	−2	0.6
		Recession	+3	0.1
L2	Corporate governance	Growth	+3	0.2
		Stabilization	+1	0.6
		Recession	−1	0.2
L3	National and international regulations on trade between continents	Growth	+5	0.4
		Stabilization	+3	0.5
		Recession	−4	0.1
L4	Competition regulation in the market	Growth	+3	0.3
		Stabilization	+1	0.4
		Recession	−2	0.3
L5	Attitude of society and entrepreneurs to regulation and legal procedures	Growth	+3	0.4
		Stabilization	+2	0.4
		Recession	−2	0.2
L6	Legal regulations on the introduction of transport infrastructure innovations	Growth	+3	0.3
		Stabilization	+1	0.5
		Recession	−4	0.2

Table 12 Analysis of environmental trends in the ethical sphere of Eurasian transport connection development

Index	Ethical factor	Trend	Influence strength of trends +5 to −5	Probability of trend 0–1
H1	Ethical behaviour in business and prevention of bribery	Growth	+3	0.3
		Stabilization	+2	0.5
		Recession	−2	0.2
H2	Trust and reputation in business	Growth	+2	0.2
		Stabilization	+1	0.7
		Recession	−2	0.1
H3	Preservation of business secrets	Growth	+4	0.2
		Stabilization	+1	0.6
		Recession	−3	0.2
H4	Protection and safety of customer data	Growth	+4	0.3
		Stabilization	+1	0.4
		Recession	−5	0.3
H5	Respect for cultural and religious differences	Growth	+4	0.7
		Stabilization	+2	0.2
		Recession	−5	0.1

– surprising scenario—here trends with the least probability of occurrence are taken into account.

The first evaluated are optimistic and pessimistic scenarios compared together in Table 13. The optimistic scenario indicates which undertakings may have the greatest impact on future development of transport system between Europe and Asia. The optimistic scenario is created in such a way that the individual areas for each process selected this trend, which has the greatest positive impact on the issue. The pessimistic scenario emphasizes these factors that have the greatest negative and destructive impact on the network extension. The pessimistic scenario creates these trends which, in relation to a given factor, have the greatest negative impact on the organization.

The bigger spread between positive and negative impact of the factor in the scenario, the stronger is the dependence of the problem analysed on the environment in this area. The biggest gap (from +5 in the optimistic scenario to −5 in the negative) is only observed in ecological sphere: impact of transport emissions on global warming and climate change (C1) and in political sphere: peaceful policy, which prevents wars and armed conflicts (P3). Development of transport between continents can also be less sensitive to other factors:

– in social sphere: awareness and communication patterns of people behaviour (S1),
– in technological sphere: development of linear and point transport infrastructure (T2) and development of new channels of communication and information on developing transport lines (T5),
– in legal sphere: national and international regulations on trade between continents (L3),
– in ethical sphere: protection and safety of customer data (H4) and respect for cultural and religious differences (H5).

The smallest dependence of successful transport network's development seems to depend on three scenario elements: compatibility of the policies of the ruling parties and the opposition countries of Eurasia (P1) from the political sphere, corporate governance (L2) from the legal sphere as well as trust and reputation in business (H2) from ethical sphere.

Figure 15 presents the spread between optimistic and pessimistic scenarios in social, technological, economic, ecological, political, legal and ethical spheres. The greater the spread between these scenarios on an axis, the stronger is the dependence of the analysed problem on the environment in these spheres. Looking at the picture we may deduct that further development of transport network between Europe and Asia is most strongly and dependent on the environment in the technological sphere, probably because of the technical structure of problem analysed. On the other hand, the weakest dependency is observed in the legal sphere and it may be caused by already existing laws and conventions in transport.

Analysing the gap between the optimistic scenario and the pessimistic scenario in the individual spheres greatest dependence to environment can be observed in the

Table 13 Optimistic and pessimistic scenarios for Eurasian transport system development

Sphere	Scenario element (index)	Optimistic scenario			Pessimistic scenario		
		Trend	Influence strength +5 to −5	Influence strength sphere average	Trend	Influence strength +5 to −5	Influence strength sphere average
Social	S1	G	+5	+3.67	R	−4	−3.33
	S2	R	+3		G	−5	
	S3	R	+2		G	−3	
	S4	G	+4		R	−2	
	S5	G	+5		R	−2	
	S6	R	+3		G	−4	
Technological	T1	G	+5	+4.33	R	−3	−3.00
	T2	G	+5		R	−4	
	T3	G	+3		R	−2	
	T4	G	+5		R	−3	
	T5	G	+5		R	−4	
	T6	G	+3		R	−2	
Economic	E1	R	+4	+3.50	G	−3	−3.17
	E2	R	+3		G	−4	
	E3	R	+4		G	−4	
	E4	G	+4		R	−3	
	E5	G	+3		R	−2	
	E6	R	+3		G	−3	
Ecological	C1	R	+5	+3.67	G	−5	−3.33
	C2	G	+4		R	−3	
	C3	G	+3		R	−2	
	C4	G	+3		R	−3	
	C5	G	+4		R	−3	
	C6	G	+3		R	−4	

(continued)

Table 13 (continued)

Sphere	Scenario element (index)	Optimistic scenario			Pessimistic scenario		
		Trend	Influence strength +5 to −5	Influence strength sphere average	Trend	Influence strength +5 to −5	Influence strength sphere average
Political	P1	G	+2	+3.83	R	−1	−3.00
	P2	G	+4		R	−1	
	P3	G	+5		R	−5	
	P4	G	+4		R	−4	
	P5	G	+4		R	−3	
	P6	G	+4		R	−4	
Legal	L1	R	+3	+3.33	G	−4	−2.83
	L2	G	+3		R	−1	
	L3	G	+5		R	−4	
	L4	G	+3		R	−2	
	L5	G	+3		R	−2	
	L6	G	+3		R	−4	
Ethical	H1	G	+3	+3.67	R	−2	−3.40
	H2	G	+2		R	−2	
	H3	G	+4		R	−3	
	H4	G	+4		R	−5	
	H5	G	+4		R	−5	

G growth, *S* stabilization, *R* recession

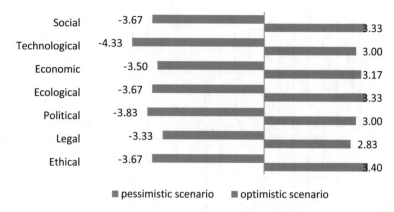

Fig. 15 The spread of the analysed external environment spheres in the optimistic and pessimistic scenario

technological sphere (3.67 + 3.33 = 7.33) and slightly smaller in ethical sphere (3.67 + 3.4 = 7.07) and equal in the social (4.33 + 3.00 = 7) and ecological sphere (3.67 + 3.33 = 7). In the political sphere (3.83 + 3.0 = 6.83) the dependence is smaller, than economic sphere (3.5 + 3.17 = 6.67) and the weakest dependence in legal area (3.33 + 2.83 = 6.16). The technological sphere is the most turbulent for transport development and should be given the most attention in the process of formulating the strategy of its future development. The degree of the dependence on this sphere of environment is biggest.

Third scenario is the most probable one, based on trends that are most likely to occur (Table 14). In the case of the Eurasian transport network development, experts have classified as the most probable the following factors: pace of changes in transport processes (T1), obtaining energy for transport from renewable sources (C5), trust and reputation occurring in business (H2) and respect for cultural and religious differences (H5).

The last surprising scenario is also seen in the same table and presents factors that should not occur in the future (or have a small probability of occurrence). The least likely to happen according to the research are much more factors indicated with 0.1 probability:

- S1—awareness and communication patterns of people behaviour,
- S2—difference in lifestyle and cultural behaviour of Eurasian population,
- S4—level of education and training of people,
- T1—pace of changes in transport processes,
- T2—development of linear and point transport infrastructure,
- T6—realization of joint research projects in the field of innovative transport solutions,
- E3—varied level of the GDP in the countries of Europe and Asia,
- E4—the development of the world economy related to the exchange of goods between Europe and Asia,

Table 14 Most probable and surprising scenarios for Eurasian transport system development

Sphere	Scenario element (index)	Probable scenario				Surprising scenario			
		Trend	Probability	Influence strength		Trend	Probability	Influence strength	
				Negative	Positive			Negative	Positive
Social	S1	G	0.6		+5	R	0.1	-4	
	S2	S	0.6	-2		R	0.1		+3
	S3	S	0.4		+1	G	0.3	-3	
	S4	G	0.5		+4	R	0.1	-2	
	S5	G	0.4		+5	R	0.3	-2	
	S6	S	0.5	-1		R	0.2		+3
Technological	T1	G	0.7		+5	R	0.1	-3	
	T2	G	0.6		+5	R	0.1	-4	
	T3	G	0.4		+3	S	0.3		+1
	T4	S	0.5		+3	R	0.2	-3	
	T5	G	0.5		+5	R	0.2	-4	
	T6	G	0.7		+3	R	0.1	-2	
Economic	E1	R	0.5		+4	G	0.2	-3	
	E2	S	0.5	-2		R	0.2		+3
	E3	G	0.6	-4		R	0.1		+4
	E4	G	0.5		+4	R	0.1	-3	
	E5	S	0.5		+1	R	0.2	-2	
	E6	G	0.5	-3		R	0.2		+3
Ecological	C1	S	0.4	-2		G	0.2	-5	
	C2	S	0.5		+1	G	0.2		+4
	C3	S	0.4		+1	R	0.3	-2	
	C4	S	0.5		+1	R	0.2	-3	
	C5	S	0.7		+1	G	0.1		+4
	C6	G	0.6		+3	R	0.1	-4	

(continued)

Table 14 (continued)

Sphere	Scenario element (index)	Probable scenario					Surprising scenario				
		Trend	Probability	Influence strength			Trend	Probability	Influence strength		
				Negative	Positive				Negative	Positive	
Political	P1	S	0.4		+1		G	0.3		+2	
	P2	R	0.5	−1			G	0.1		+4	
	P3	S	0.5		+2		G	0.2		+5	
	P4	G	0.4		+4		R	0.3	−4		
	P5	S	0.5		+2		R	0.2	−3		
	P6	S	0.5		+2		R	0.2	−4		
Legal	L1	S	0.6	−2			R	0.1		+3	
	L2	S	0.6		+1		G	0.2		+3	
	L3	S	0.5		+3		R	0.1	−4		
	L4	S	0.4		+1		G	0.3		+3	
	L5	G	0.4		+3		R	0.2	−2		
	L6	S	0.5		+1		R	0.2	−4		
Ethical	H1	S	0.5		+2		R	0.2	−2		
	H2	S	0.7		+1		R	0.1	−2		
	H3	S	0.6		+1		G	0.2		+4	
	H4	S	0.4		+1		G	0.3		+4	
	H5	G	0.7		+4		R	0.1	−5		

G growth, *S* stabilization, *R* recession

- C5—obtaining energy for transport from renewable sources,
- C6 - protection of nature and landscape for designed transport investments,
- P2—pro-development and open state policy,
- L1—corporate governance,
- L3—national and international regulations on trade between continents,
- H2—trust and reputation in business,
- H5—respect for cultural and religious differences.

Verification of elements in the most probable and surprising scenarios allow to observe that there is one factor in technological sphere (T1), one in ecological sphere (C5) and two in ethical sphere (H2 and H5), all from the probable scenario that are also divided in the surprising one. The reason of that is the sudden change of the trend that may occur, taken into account in both possibilities. This also indicates vulnerability of the issue analysed in these areas.

Figure 16 shows graphical presentation of most probable scenario in each sphere of the STEEPLE environmental analysis. The larger the span, the weaker and the more heterogeneous environment around the analysed problem. The economic sphere is the most important to be focused on. The most likely scenario consists of trends that are most likely to occur, regardless of the potential positive or negative impact.

Analysis of the most likely scenario in the individual areas shows that the highest range is in the field of economy (3 + 3 = 6) and social (1.5 + 3.75 = 5.25), smaller in the case of technological domain (0 + 4 = 4), slightly smaller in the legal (2 + 1.8 = 3.8), ecological (2 + 1.4 = 3.4) and political (1 + 2.2 = 3.2) spheres. The smallest in the ethical sphere (0 + 1.8 = 1.8). It means that the most hetero-geneous and poorly structured are the economic and social spheres in the

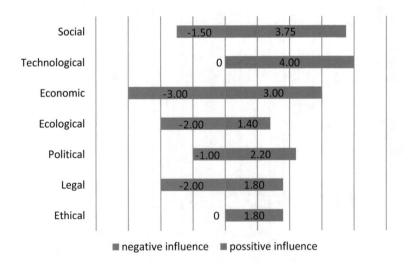

Fig. 16 The spread of the external environment spheres negative and positive influence in the most probable scenario

environment and thus it is stated that it is important to focus first of all on them and to make further research in this area. This is due to the differences in wealth and worldviews of residents in the different regions of Europe and Asia, and the barriers resulting from this.

Analysis of the least likely scenario, also called a surprise scenario, is the starting point for warning system in case of the Eurasian transport network development. Surprise scenario includes those trends that, regardless of the potential force of positive or negative influence, are least likely to occur. Figure 17 shows graphical presentation of surprising scenario in each sphere of the STEEPLE environmental analysis. In this case the biggest spread between negative and positive influence is observed in case of ecological (3.5 + 4 = 7.5), political (3.67 + 3.67 = 7.33) and ethical (3 + 4 = 7) spheres. Smaller dependency is noticed in legal (3.33 + 3=6.33) and economic (2.67 + 3.33 = 6) spheres. Then there is social sphere (2.75 + 3 = 5.75) in the ranking and the lowest spread in technological sphere (3.2 + 1 = 4.2).

A summary of the results of the scenarios of the surrounding events for the analysed problem of transport network development between Europe and Asia, with particular regard to rail transport is presented in Fig. 18.

By analysing Fig. 18, it can be seen that the problem of intercontinental transport network development in Eurasia is characterized by an unstable environment, as indicated by large differences in scenarios: optimistic and pessimistic in all analysed STEEPLE areas. The spread of the most probable scenario is that the most turbulent environment is characterized by economic sphere, which makes it the least homogeneous. In turn, the most structured is the ethic sphere, which, unlike the aforementioned, does not require so much involvement. Other spheres are moderately homogeneous and structured.

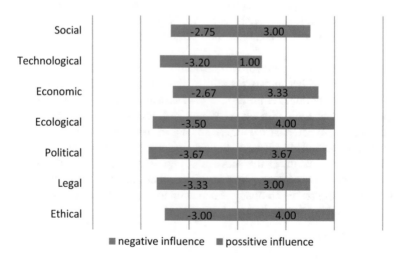

Fig. 17 The spread of the external environment spheres negative and positive influence in the surprising scenario

In the technological sphere the greatest advantage over threats can be seen, but it can also be noted for the social, political and ethical sphere. In the other cases there is an inverse relationship. Among the opportunities listed in the technology and technology segment following can be distinguished: development of transport infrastructure and pro-ecological transport technologies as well as new channels of communication and information on developing transport lines.

Analysing the leading processes that have the greatest impact on the development of transcontinental transport and the most likely to occur, the technological factor should be emphasized: T1 - pace (rate) of changes in transport processes with 3.5 rating for growing trend.

The cost-effectiveness and future development of transport can be diminished by various factors, among which the greatest risk according to the research results is the E3 economic factor: varied level of the GDP in the countries of Europe and Asia (with −2.4 rating for growing trend). This external factor of the analysed problem should be considered as the main barrier and must be analysed more closely when carrying out research and development work on the transport network of the geographic areas mentioned.

To summarize, he development of scenarios surrounding the environment sued to assess the problem's environment according to the criteria of turbulence, stability and degree of structuralisation. There are five ways of inference:

(1) An analysis of the turbulence of the environment and the degree of dependence of the analysed problem on changes occurring in it. The greater the spread between the optimistic scenario and the pessimistic scenario in the individual spheres, the stronger the dependence of the issue on the environment. This means that such an area of the environment, referred to as turbulent, must pay special attention to the processes of strategy formulation.

(2) In analysing the span of the most likely scenario, the greater the span in the individual spheres, the more heterogeneous and less structured the environment will be. In this case, special attention should be paid to the events that concern this sphere.

(3) The environment is defined by the spheres in which the opportunities prevail, and those in which threats prevail. In these segments of the environment in which the opportunities prevail, organizational strategies should include projects aimed at their utilization. Where risk prevails, the strategy should be targeted at projects that reduce the impact of their impact.

(4) From the most likely scenario, the leading processes should be exposed in the environment, i.e. those that have a strong impact on the issue and a high probability of occurrence. Once identified, these trends are evaluated in terms of their impact on the company. Adaptation to opportunities and threats must be a policy-making principle.

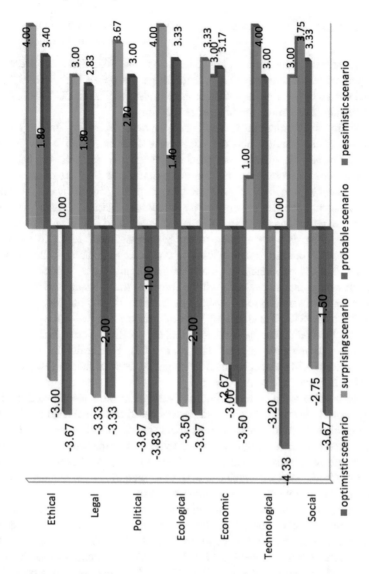

Fig. 18 Environmental scenario layout

(5) You can also estimate the potential force of phenomena that may unexpectedly affect the execution of the strategy by creating a surprise scenario. There are trends that are characterized by low probability of occurrence but with strong potential impact on the organization—positive or negative. Surprise scenario analysis is the starting point for early warning systems.

5 Summary

Trade and economic cooperation between Europe and Asia mainly concern import and export of goods between Europe and China. European Union and China are qualified as two of the biggest traders in the world. China is the EU's second largest trading partner. For China, the biggest is the Union.

The share of e-commerce in retail trade varies, depending on the degree of digitization of the country and the habits of consumers, from a few to several percent. However, its growth rate reaches several percent year on year, which means a systematic increase in the share of e-commerce in retail sales.

When analyzing the level of customer service quality, special attention is paid to the improvement activities undertaken by companies that are designed to meet the customer's expectations and even offer more value than the customer expects. In e-commerce, there is a clear trend in the way consumers adjust to the pace of change and force them to adapt quickly to companies operating in this area. As a result, if this trend continues, all actors in the supply chain, especially transport companies and logistics operators, will be forced to make changes to their strategic decisions (business models) and operations (route optimization, transport, etc.).

Basing on the SWOT analysis of Eurasian transport network development, the overall weighted rating can be concluded that the positive internal factors (3.40) are bigger than the negative internal factors (−3.09). Similarly, the total positive external factors (3.74) are higher than the negative external factors (−3.24) in terms of possible Eurasian rail connection.

Scenarios of Eurasian corridors future development were developed according to seven groups of external environmental factors based on STEEPLE analysis (social, technological, economic, ecological, political, legal, ethical areas). Evaluation of forty-one factors in those groups according to possible trend growth, stabilization or recession and possibility of occurrence. Then, the evaluated factors' list was the basis for environmental scenarios creation. Four scenarios were created: optimistic, pessimistic, most probable and surprising scenario.

Analysing the gap between the optimistic and the pessimistic scenario of Eurasian transport development in the individual spheres greatest dependence to environment can be observed in the technological sphere (3.67 + 3.33 = 7.33) and slightly smaller in ethical sphere (3.67 + 3.4 = 7.07) and equal in the social (4.33 + 3.00 = 7) and ecological sphere (3.67 + 3.33 = 7). In the political sphere (3.83 + 3.0 = 6.83) the dependence is smaller, than economic sphere (3.5 + 3.17 = 6.67) and the weakest dependence in legal area (3.33 + 2.83 = 6.16). The technological sphere is the most turbulent for transport development and should be given the most attention in the process of formulating the strategy of its future development. The degree of the dependence on this sphere of environment is biggest.

Analysis of most probable scenario showed that the most heterogeneous and poorly structured are the economic and social spheres in the environment and thus it is stated that it is important to focus first of all on them and to make further research in this area. In case of surprising scenario, case the biggest spread between negative and positive influence is observed in case of ecological (3.5 + 4 = 7.5), political (3.67 + 3.67 = 7.33) and ethical (3 + 4 = 7) spheres.

The final figure of all scenarios evaluated in the research allowed to draw conclusion that the problem of intercontinental transport network development in Europe and Asia is characterized by an unstable environment, as indicated by large differences in the scenarios presented. Also, a leading process could be selected which was a technological T1 factor: pace of changes in transport processes. In fact, the fastest rate of changes in according to transport processes, the fastest the development will be. Also, the weakest economic E3 parameter: varied level of the GDP in the countries of Europe and Asia should be specially considered when planning Eurasian transport network because it may prove to be the biggest barrier for its' development.

Bibliography

1. Alcamo J (ed) (2008) Environmental futures: the practice of environmental scenario analysis, vol 2. Elsevier, Amsterdam
2. Altomonte C, Békés G (2009) Trade complexity and productivity (No. MT-DP-2009/14). IEHAS Discussion Papers. https://www.econstor.eu/bitstream/10419/108157/1/ MTDP0914. pdf
3. Amer M, Daim TU, Jetter A (2013) A review of scenario planning. Futures 46:23–40
4. Andersson M, Lööf H, Johansson S (2008) Productivity and international trade: firm level evidence from a small open economy. Rev World Econ Weltwirtschaftliches Archiv 144 (4):774–801
5. Barba Navaretti G, Venables AJ (2004) Multinational firms in the world economy. Princeton University Press, Princeton
6. Bernard AB, Jensen JB, Redding SJ, Schott PK (2007) Firms in international trade. J Econ Perspect 21(3):105–130
7. Brambilla I, Lederman D, Porto G (2012) Exports, export destinations, and skills. Am Econ Rev 102(7):3406–3438
8. Breinlich H, Criscuolo C (2011) International trade in services: a portrait of importers and exporters. J Int Econ 84(2):188–206
9. De Loecker J (2007) Do exports generate higher productivity? Evidence from Slovenia. J Int Econ 73(1):69–98
10. Ercin AE, Hoekstra AY (2014) Water footprint scenarios for 2050: a global analysis. Environ Int 64:71–82
11. European Commission (2016) Joint communication to the European Parliament and the Council. Elements for a new EU strategy on China. Brussels, 22.6.2016, JOIN (2016) 30 final. http://eeas.europa.eu/archives/docs/china/docs/joint_communication_to_the_european_ parliament_and_the_council_-_elements_for_a_new_eu_strategy_on_china.pdf
12. European Commission (2017) Trade. Countries and regions. http://ec.europa.eu/trade/policy/ countries-and-regions/
13. Eurostat (2017) Share of EU in the World Trade [ext_lt_introle]

14. Eurostat (2017) Statistics Explained. http://ec.europa.eu/eurostat/statistics-explained/pdfscache/1188.pdf
15. Fleisher CS, Bensoussan BE (2003) Strategic and competitive analysis: methods and techniques for analyzing business competition. Prentice Hall, Upper Saddle River, 457 p
16. Foster L, Haltiwanger J, Syverson C (2008) Reallocation, firm turnover, and efficiency: selection on productivity or profitability? Am Econ Rev 98(1):394–425
17. Frąś J, Scholz S, Olsztyńska I (2017) Modern information technologies in the logistics of e-business. Res Log Product 7:185–197. doi:https://doi.org/10.21008/j.2083-4950.2017.7.4.1
18. Greenaway D, Gullstrand J, Kneller R (2008) Surviving globalization. J Int Econ 74(2):264–277
19. Hayakawa K, Machikita T, Kimura F (2011) Globalization and productivity: a survey of firm-level analysis. J Econ Surv 25(1):19–68
20. Kawa A (2017) Fulfilment service in e-commerce logistics. LogForum 13(4):429–438
21. Kirchgeorg M, Jung K, Klante O (2010) The future of trade shows: insights from a scenario analysis. J Bus Ind Mark 25(4):301–312
22. Klein MW, Welfens PJ (2012) Multinationals in the New Europe and global trade. Springer Science & Business Media
23. Mahmoud M, Liu Y, Hartmann H, Stewart S, Wagener T, Semmens D, Hulse D (2009) A formal framework for scenario development in support of environmental decision-making. Environ Model Softw 24(7):798–808
24. Majszyk K (2016) Branża logistyczna spogląda w stronę Chin, które dostały zadyszki, Dziennik Gazeta Prawna Nr 122 (4269) 27.06.2016 [In Polish: The logistics sector is looking towards China, which got short of breath]
25. Majszyk K, Otto P (2015) Czas to pieniądz. Firmy liczą koszty transportu z Chin do Europy. 29.12.2015. http://biznes.gazetaprawna.pl/ [In Polish: Time is money. Companies charge transport costs from China to Europe]
26. McBride J (2015) Building the New Silk Road. https://www.cfr.org/backgrounder/building-new-silk-road
27. More E, Probert D, Phaal R (2015) Improving long-term strategic planning: an analysis of STEEPLE factors identified in environmental scanning brainstorms. In: 2015 Portland international conference on management of engineering and technology (PICMET). IEEE, New York, pp 381–394
28. Müller P, Kudrna Z, Falkner G (2014) EU–global interactions: policy export, import, promotion and protection. J Eur Public Policy 21(8):1102–1119. http://dx.doi.org/10.1080/13501763.2014.914237
29. O'Reilly B (2014) Transport at centre of China's drive for regional supremacy. http://chinaoutlook.com/transport-centre-chinas-drive-regional-supremacy/
30. Pawlak K, Kołodziejczak M, Xie Y (2016) Changes in foreign trade in agri-food products between the EU and China. J Agribus Rural Dev 4(42):607–618. https://doi.org/10.17306/jard.2016.87
31. Serti F, Tomasi C, Zanfei A (2010) Who trades with whom? Exploring the links between firms' international activities, skills, and wages. Rev Int Econ 18(5):951–971
32. Singh T (2010) Does international trade cause economic growth? A survey. World Econ 33 (11):1517–1564
33. Sładkowski A, Cieśla M (2015) Influence of a potential railway line connecting the Caspian Sea with the Black Sea on the development of Eurasian trade. Naše more 62(4):264–271
34. Swart RJ, Raskin P, Robinson J (2004) The problem of the future: sustainability science and scenario analysis. Glob Environ Change 14(2):137–146
35. Wagner J (2011) Offshoring and firm performance: self-selection, effects on performance, or both? Rev World Econ 147(2):217–247

36. Wagner J (2012) International trade and firm performance: a survey of empirical studies since 2006. Rev World Econ 148(2):235–267. https://doi.org/10.1007/s10290-011-0116-8
37. With small steps…the establishment of freight corridors takes shape (2013) URL: http://www.railwaypro.com/wp/with-small-stepsthe-establishment-of-freight-corridors-takes-shape/
38. Wojcieszak A (2015) Eurazjatyckie korytarze transportowe. https://www.log24.pl/artykuly/eurazjatyckie-korytarze-transportowe,5987 [In Polish: Eurasian transport corridors]

The Role of Railway Transport in East-West Traffic Flow Conditions of the Slovak Republic

Anna Dolinayova, Lenka Cerna and Vladislav Zitricky

Abstract Transport is a significant part of international trade in the economic globalization. Continual growth of goods flows between Europe and Asia forces the business companies to find new possibilities of transport routes by the maintenance of speed, reliability and economic efficiency of transport chains. These facts represent opportunities for railway transport to participate on the transport flows of Eurasian continent. The Slovak Republic can be part of a new transport flows between Europe and Asia due to its geographical location. Presented study shows possibilities connected to railway transport on the Eurasian traffic flows in conditions of Slovak republic. The first part of study defines possibilities of Slovak railway infrastructure to the connection to Eurasian transport routes and study also contains international transport law rules valid on the Europe and Asia continent. Study comprises calculation of transport costs of the potential use of railway transport on European railway infrastructure. Economical part of study is processed alternately—for intermodal transport units and for rail transport in conventions wagons.

Keywords International trade · Railway transport · Transport flows
Transport costs

A. Dolinayova (✉) · L. Cerna · V. Zitricky
Department of Railway Transport, Faculty of Operation and Economics
of Transport and Communications, University of Žilina, Žilina, Slovakia
e-mail: anna.dolinayova@fpedas.uniza.sk

L. Cerna
e-mail: lenka.cerna@fpedas.uniza.sk

V. Zitricky
e-mail: vladislav.zitricky@fpedas.uniza.sk

© Springer International Publishing AG, part of Springer Nature 2018
A. Sładkowski (ed.), *Transport Systems and Delivery of Cargo
on East–West Routes*, Studies in Systems, Decision and Control 155,
https://doi.org/10.1007/978-3-319-78295-9_3

1   Introduction

Economic development of East Asian countries has influenced the volume of international trade between Europe and Asia in recent years. Many multinational corporations moved their factories to Asia because of cheaper manpower and lower taxes in China and other East Asian countries. This situation is especially advantageous for employment and economic growth in East Asian economies. The situation in international trade between Europe and Asia has had an impact on the trade of goods. Currently, the majority of goods are transported by sea in intermodal transport units [1–3].

The largest container ships have a capacity of several thousand twenty-foot equivalent units (TEUs). For example, the world's largest, the Maersk Mc-Kinney Moller, can hold 18,270 containers. The main driver of the development of maritime transport is its low cost, which results in building new and larger ships. This reduces the average price per transport unit, but costs could start to rise in the future, since the price of oil, the price of which is now at its minimum level, is likely to rise again [4, 5].

The biggest European ports (Rotterdam, Antwerp, Hamburg etc.), located in Western Europe and through which imports flow to Central Europe, are overloaded. The current infrastructure of sea ports in Europe cannot handle an increase of transported goods from Asia to Europe [1, 4, 6].

These problems require the search for new solutions to integrate the transport network between regions and seek solutions that would lead to lower transport costs and higher transport speed, while preserving shipment quality, safety and consistency. These solutions should encourage the use of adequate transport infrastructure in each of the countries concerned.

2   Analysis of International Trade in Goods in the EU and Third Countries

International trade is the oldest and most frequently used form of economic relations between countries. The rapid development of international trade is a decisive indicator of an effective economy and strengthens economic competition. As a result, a number of international entities have emerged. Because of the consolidation of national economies, these entities have focused their goals on international trade. Globalization of companies has created a need for unified international trade legislation.

International trade delivery clauses are common rules that explain the most commonly used clauses in a sales contract. Delivery clauses were produced by the International Chamber of Commerce in Paris. Those clauses are intended for entities that prefer the reliability of common rules, as opposed to uncertainty arising from a misunderstanding of the delivery clauses between contracting parties.

Each country can ensure the free circulation of goods by introducing a set of measures (customs, import/export/transit premiums, limits) that will ensure the protection of the domestic market. By its membership to the European Union, the Slovak Republic has entered the free circulation zone, which is implemented on the basis of the common customs policy of EU members.

Not all countries have sufficient capacity to meet the needs of their populations—availability of factors of production (labour, technological equipment, raw materials, capital, etc.). The development of science and technology has brought with it a surplus of goods that are not in demand on the domestic market [7].

At present, international trade is of great importance, especially in the economic, political and cultural spheres. The consolidation of relations between countries and the identification of a way of life and culture feature in the individual countries involved in international trade.

International trade includes several areas, namely the export of goods, the importation of goods, transit as a special form of relocation of goods, the exchange of services and the movement of populations (labour forces).

In order for a country to engage in the international division of labour and develop international trade, it should have the following conditions for its development, which are divided:

- Internal conditions of international trade—natural conditions, climatic conditions, raw material wealth, size of the internal market, industrial potential, population, land wealth and historical conditions
- External conditions of international trade—geographical location, industrial potential of the country, maturity of neighboring countries, relations with neighboring countries, economic situation (crisis) and political situation (war) [8].

Currently among the world's most industrially developed states are the G8 (Canada, Germany, France, Japan, the United Kingdom, Italy, Russia and the US). However, China's and India's hidden industrial potential may lead these countries to become the world's largest economic powers in 2017 [9].

2.1 Development of Foreign Trade in the Slovak Republic

By 2019 the leaders of world economic growth will, according to diplomacy, be developing countries and the emerging markets of Africa and Asia.

In 2016 the focus of Slovakia's export and economic diplomacy remained in the trade sector, where 82.81% of exports are directed, and where 62.5% of its imports come from.

A trade priority area is with the EU's neighbouring territories, namely the Balkans, the Commonwealth of Independent States (CIS), the Eastern Neighbourhood and select countries of the Southern Neighbourhood.

The focus is on cooperation with countries with strong scientific research and innovation potential such as the US, Turkey, Japan, Israel, South Korea, Switzerland, Germany, the Netherlands, Scandinavia and China, which opened cooperation in this area with 16 Central and Eastern European countries.

The area of investment cooperation remains a priority with strong offerings in foreign direct investment coming from the US, UK, France, Germany, the Netherlands and other EU countries, Japan, South Korea, Israel, the oil countries of the Middle East and the Gulf, Singapore, Hong Kong and China.

Priority countries in the export of Slovak capital will be the Western Balkans, Turkey, Vietnam, and prospectively selected African countries. Exploitation of technological units (power plants, production technologies) to create export opportunities in these countries for other Slovak SMEs as subcontractors or complementary suppliers (infrastructure, housing, services) will be used. In 2016 the balance of foreign trade was active in the volume of EUR 3672.1 million. (EUR 352.9 million higher than the same period in 2015.)

The Slovak Republic saw the highest active balance with Germany (EUR 4204.8 million), the United Kingdom (EUR 2940.5 million), France (EUR 2262.1 million), Austria (EUR 2057.2 million), Poland (EUR 1885.3 million), the Czech Republic (EUR 1260.5 million), Italy (EUR 1236.4 million), Spain (EUR 1126.6 million), the Netherlands (EUR 1062.7 million) and the United States (EUR 1007.6 million).

It saw the largest passive balance in foreign trade with China (EUR 4426.8 million), Korea (EUR 3805.4 million), the Russian Federation (EUR 1281.7 million), Taiwan (EUR 508.3 million), Japan (EUR 496.3 million), Malaysia (EUR 366.8 million), India (EUR 212.5 million), and Ireland (EUR 91.5 million).

Export

Goods worth 70,073.9 million EUR were exported from the Slovak Republic in 2016. Total exports increased by 3.5% compared to the same period in 2015.

In the automobile industry the SR exported motor cars and other vehicles worth 1503.3 million EUR, components and accessories of motor vehicles worth 460.2 million EUR, monitors, projectors and television receivers worth 366.2 million EUR, oil and other gaseous hydrocarbons worth 269.1 million EUR, and flat-rolled products of other alloy steel valued at 137.3 million EUR.

Exports of petroleum and other oils obtained from bituminous minerals, dropped by 406.9 million EUR, car bodies 370.9 million EUR, telephone sets including cellular network phones 211.4 million EUR, flat-rolled products of iron or non-alloy steel, clad 168.5 million EUR, and rev counters, product calculators, taximeters, mileage meters 110.3 million EUR.

From the most significant trade partners, exports to Germany increased 1%, France 13.5%, UK 13.1%, Hungary 2.8%, Italy 10.6%, Spain 11.2%, the Netherlands 22.5%, the United States 16.3%, and Romania 6.1%. Exports decreased to the Czech Republic 1.2%, Poland 5.2%, Austria 1.5% and the Russian Federation 4.4%.

In terms of the main economic clusters, exports to EU countries increased 3.5% (85.2% of total Slovak exports) and to OECD countries 3.9% (88.6% of total Slovak exports).

Import

The Slovak Republic imported goods worth 66,401.8 million EUR, with year-on-year growth of 3.2%.

Imports of motor vehicle parts and accessories increased the most, worth 416.4 million EUR. Passenger cars and other motor vehicles worth 374.8 million EUR, seats, also convertible to beds, 286.5 million EUR, machinery and electric, laser or ultrasonic equipment 213 million EUR, and piston engines 204.4 million EUR.

Imports fell for petroleum oils and oils obtained from bituminous minerals, crude by 541.7 million EUR, parts and components of broadcasting or television broadcasting equipment by 213.6 million EUR, railway and tramway passenger and freight wagons by 173.8 million EUR, oil and other gaseous hydrocarbons by 101.5 million. EUR, and monitors, projectors and television receivers by 94.9 million EUR.

From the most significant trade partners, imports from Germany increased by 10%, Poland 4%, Austria 13% and the UK by 17%. Imports from the Czech Republic fell by 2.1%, China 0.9%, Korea 7.6%, Hungary 3.2%, the Russian Federation 23.2%, Italy 1.7%, and France by 4.3%.

From the point of view of the main economic groupings, compared to the same period in 2015, imports from EU countries increased by 6% (67.3% of total imports) and from OECD countries imports increased by 3.6% (66.1% of total imports).

The largest exporters of Slovakia (exports of goods) in 2016:

- Volkswagen Slovakia, a.s., Bratislava
- Kia Motors Slovakia, s.r.o., Teplička nad Váhom
- Samsung Electronics Slovakia, s.r.o., Galanta
- PCA Slovakia, s.r.o., Trnava
- Slovnaft, a.s., Bratislava
- Foxconn Slovakia, s.r.o., Nitra
- Slovenské elektrárne, a.s., Bratislava
- Schaeffler Slovensko, s.r.o., Kysucké Nové Mesto
- Continental Matador Rubber, s.r.o., Púchov
- Continental Matador Truck Tires, s.r.o., Púchov
- IKEA Components, s.r.o., Malacky
- Mondi SCP, a.s., Ružomberok
- Duslo, a.s., Šaľa
- Železiarne Podbrezová, a.s., Podbrezová

2.2   Development of Major International Trade Actors

The next chapter is a description of the development of international trade in goods majors. The total level of trade in goods in 2016 (exports and imports) recorded for the EU-28, China and the United States, was almost similar, peaking at 3455 billion EUR in the EU-28 (note this does not include intra-EU trade), which was 109 billion EUR higher than for China and 125 billion EUR above the level recorded for the US. Japan had the fourth highest level of trade in goods, at 1131 billion EUR.

Figure 1 shows the export and import values of the largest players in international trade for 2016 [10].

In terms of flows of exports and imports should the EU-28 in 2016, the second largest share of world exports and imports of goods (see Figs. 4 and 5) the export of goods from the EU-28's exports of goods were equivalent to 15.6% of the world total. In 2014 they were surpassed for the first time since the EU was founded, by those of China (16.1% in 2014, rising to 17.0% in 2016), but still ahead of the United States (11.8%). The United States had a larger share of world imports (17.6%) than either the EU-28 (14.8%) or China (12.4%).

2.3   Analysis of the Main Import and Export Product Groups

Between 2011 and 2016, EU-28 exports increased in most of the product groups shown in Fig. 13 with two exceptions: exports of raw materials (total decrease by 5.1%) and exports of mineral fuels and lubricants (a decrease of 26.0%). The highest growth rate of exports was recorded for food, beverages and tobacco, which saw an increase of 31.0%, while the EU export level for chemicals and related products (23.1% increase) and double digit values growth in machinery and transport equipment as well as other industrial products were recorded.

On the import side a similar trend was observed with a relatively large overall reduction in the level of imports from non-EU countries during 2011–2016, in terms of raw materials (−20.2%) and mineral fuels and lubricants (−46.6%). This

Fig. 1 The main players in international trade in goods (%). *Source* [10]

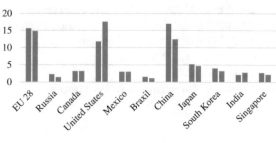

decline can be partly attributed to price changes and/or fluctuations in the exchange rate due to prices of raw materials and oil being set in US dollars. However, exports of machinery and transport equipment increased by 24.9% between 2011 and 2016 and relatively high growth rates were also recorded for food, beverages and tobacco (19.1%) as well as chemicals and related products (18.1%).

The EU's largest surplus in trade with non-EU countries amounted to 37.7 billion EUR. In 2016 surplus was also achieved due to a positive trade balance for machinery and transport equipment worth 193.1 billion EUR, and chemicals and related products worth 129.1 billion EUR). From 2011 to 2016, the EU-28 saw an increase in its trade surplus in chemicals and related products, while surplus in machinery and transport equipment fell somewhat. In the case of food, beverages and tobacco, the EU-28 reached a slightly higher trade surplus in 2016, compared to the small trade deficit in 2011. The largest trade deficit in 2016 was recorded for mineral fuels and lubricants, where imports exceeded exports by 190.0 billion EUR.

The EU-28 trade deficit in the case of mineral fuels, lubricants and raw materials has declined significantly over 2011–2016. In terms of mineral fuels and lubricants, there has been more than half the reduction over this five-year period. On the other hand, deficit increased for other industrial products, reaching 53.5 billion EUR in 2016 representing an increase of 3.5% compared to 2011.

From 2011 to 2016, the structure of the EU-28's exports changed, especially for smaller product groups (see Fig. 2). The share of food, beverages and tobacco products increased from 5.7 to 6.6% in the period, while the share of mineral fuels and lubricants declined from 6.4 to 4.2%.

The largest change in the structure of imports into the EU-28 was in the case of non-fossil fuels and lubricants, which fell from 28.6 to 15.5% between 2011 and 2016 (see Fig. 15). The share of other industrial products, on the other hand, rose from 23.3 to 26.3%, while the share of machinery and transport equipment increased from 25.6 to 32.3%.

Figure 2 compares the structure of EU-28 imports and exports in 2016. Keep in mind that the total export volume was 2.2% higher than the import level. The most significant difference concerns the share of mineral fuels and lubricants, which was 3.6

Fig. 2 Main items of imports and exports by products, EU-28, 2016. *Source* [10]

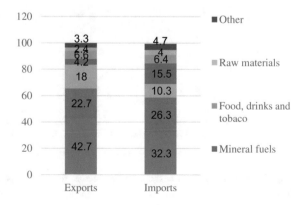

times higher when imported than exported. However, it was offset by lower shares of
machinery and transport equipment, chemicals and related products for imports.

2.4 Analysis of Freight Transport Development
in the Slovak Republic

The development of freight transport in the Slovak Republic in the monitored
period from 2012 to 2015 was slightly declining [11]. The main reason for this is
the structural changes in the Slovak economy, including the construction of new
industrial plants in the electro technical and automotive industries, whose products
have a higher added value.

The transport system in the Slovak Republic also affects the cultural, social and
educational development. The rise should match these needs and focus on the future
demand for transport services. The basis for decision-making on future developments is the financial possibilities and technical backgrounds of transport engineering, safety, development concept, traffic and other [12].

The following Fig. 3 shows the comparison of transport performance in individual transport modes in the Slovak Republic in the monitored period 2002–2016
[13].

In the case of the transport performance of road freight transport in the period
2000–2015, an increase in output was recorded by approximately 46.5%, with
year-on-year decreases and increases. The performance of water freight transport
experienced fluctuations, in 2002 (decrease) and in 2010 (sharp increase). The
water transport performance after 2010 recorded a significant decline, which lasted
until 2014. The rail transport performance was stable and did not show any significant differences between 2005 and 2014.

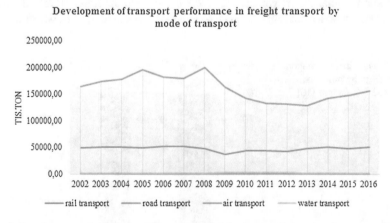

Fig. 3 Development of transport performance in freight transport by mode of transport. *Source*
[13]

The development of freight transport in the Slovak Republic is predominantly conditional upon the adoption of new measures. The most important are liberalization of freight transport, customs and trade restrictions, transport costs, fuel prices, charging infrastructure and others [12].

In general, it can be said that the transport of goods by road haulage has a dominant position not only within the Slovak Republic but also across the European Union. It constitutes almost 70% of total shipments. The Slovak Republic, as well as the European Union, are trying to limit road transport performance and transfer it to rail transport through strategic documents.

These objectives, however, are problematic to ensure only legislative requirements. Customers choose the type of transport that is more tailored to meet demands and can meet all of its transport needs [12].

2.5 Analysis of Freight Transport Development in the EU

The main transport problems in the European Union (EU) include the unbalanced development of transport modes, congestion of road infrastructure and airspace, as well as negative impacts of transport on the environment [12].

Efforts to limit the performance of road haulage and transport it to rail have been developed by the EU in a number of materials and national documents, specifically the White Paper and the Lisbon Strategy.

Based on analysis of transport it has been concluded that Slovakia and other EU countries (EU-28) have road haulage representing the largest share. The reason is that road transport is cheaper, more attractive, and more adaptable to the specific requirements of customers. Due to high competition, carriers are willing to carry out some under-priced transport (dumping price), which is not realistic in the long run.

One of the most important criteria is transport speed [14]. Figure 4 shows that Germany accounts for 36% of the largest percentage of goods transported by road haulage. France is second with 22% share, and third is the UK with 18% of transported goods [12].

For rail freight, the freight rate is longer, especially when it is not a logistic train, and the freight cost is more complex. It also includes other charges related to the

Fig. 4 Percentage of freight traffic by road in EU countries. *Source* [13]

Percentage of freight traffic by road in EU countries

Fig. 5 Graph 2. Percentage share of rail freight traffic in EU countries. *Source* [13]

Share of rail freight transport in EU countries

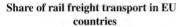

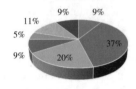

- Czech Republic
- Germany
- Austria
- France
- Hungary
- Great Britain
- Poland

shipment itself, such as weighing, displacement, sealing, or wagon loading and landing fees. For this reason, rail transport may be less attractive for a customer.

Figure 5 shows that the largest share of goods transported by rail freight is led by Germany with 37%, followed by Austria with 20%, and in third place, Great Britain with 11% of goods transported by rail.

Where there is untapped potential, water transport may play an increasing role in the transport of goods inland. European nodes from which multimodal logistics flows can be managed have as their primary objective to minimize road congestion and energy consumption by rail and inland waterway transport. Germany has the largest share of inland waterway transport. Its transport share is up to 70% and it is due to the Rhine-Gaelic delta, which is the busiest in Europe. Behind Germany, France ranks second with 20% and the Czech Republic with a 3% share [12].

2.6 Analysis of Global Freight Transport Development

Transport is a key part of economic growth, globalization and the development of societies. Globally, the transport sector is one of the fastest growing major economic sectors of a country, but it has a major impact on the environment.

In the last ten years, freight transport has grown by about 15%. The increase in freight transport has been more balanced across sectors, with greater regional differences.

Freight transport has the largest share of water transport (around 3/5 of transport capacity). The share of road and rail transport in overall transport performance is approximately the same.

Significant transport of road freight transport, as recorded in the EU, is relatively unique in the world. In the US, paradoxically, freight is carried by rail, at 43% of transport performance, whereas road transport accounts for only 31%, and shares of other freight transport are smaller. Rail transport also has the largest share of land transport in Russia, China, India and Kazakhstan. These four countries are among

35 countries in Asia, where rail transport is most dominant. Conversely, in the countries of Southeast Asia, goods transport is largely provided by water transport.

Air freight transport is used in the world to transport valuable goods with high value and low volume. Worldwide, only about 1% goods are transported by air, but its cost is about 40% of the total cost of transported goods worldwide. Most of this type of transport is for electronics and branded clothing.

Since the 1950s, freight transport has been affected mainly by containerization. Standardized ISO containers have revolutionized international transport and nationalized countries and have made significant savings in the transhipment of goods.

Transport in developing countries is characterized in particular by low freight transport performance. Developing countries are referred to as very poor countries in Africa and Southeast Asia. In most of these countries, the bulk of transport is provided by road transport, largely under conditions that do not meet the conditions. In the more developed countries of Asia and Africa, but which cannot be considered developed (Egypt, Thailand, South Africa, Malaysia) and large cities of the two continents (Cairo, Bangkok, Hanoi etc.) the traffic situation is closer to European standards.

3 Transport Routes Between Europe and Asia

A transport route connecting west European countries with eastern countries in Asia is the sea. In the past, this sea route was around of Cape of Good Hope, but after the Suez Canal was built, cruise time between the Indian Ocean and the North Atlantic was reduced by about 10,000 km. Today, this is the route used to transport most of the goods in the world [1, 2, 15].

Figures 6 and 7 show the volume of transport on the main sea routes in the world in TEUs [16]. We can see that the Europe–Asia route has had a share of sea transport performance since 2009. The main reason for this has been the great world economic crisis [17].

However, it is important to note that the total volume of shipping on sea routes directed from east to west is dominated by imports more than exports. The import and export imbalance is due to the additional costs associated with the shipping of empty containers, as can be seen in Fig. 2.

The Asia-Europe route reached an annual output of about 22.4 million TEU, with the difference between the currents from east to west (15.4 million TEU) and from Europe to the east (7 million TEUs), representing almost a 50% difference. The situation in Asia/North America is similar. Imports to the USA was 14.7 million TEU and in the opposite direction exports to Asia were only 7.5 million TEU. Here it is possible to see the volume shifts from Asia to Europe, and from Asia to North America they are approximately two times larger. This also results in a constant imbalance in container flows and the movement of empty containers is still one of the main concerns of transporters. The year-on-year

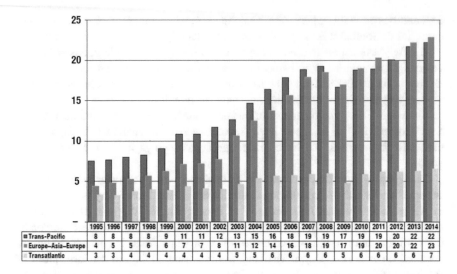

	1995	1996	1997	1998	1999	2000	2001	2002	2003	2004	2005	2006	2007	2008	2009	2010	2011	2012	2013	2014
▣ Trans-Pacific	8	8	8	8	9	11	11	12	13	15	16	18	19	19	17	19	19	20	22	22
▥ Europe–Asia–Europe	4	5	5	6	6	7	7	8	11	12	14	16	18	19	17	19	20	20	22	23
▨ Transatlantic	3	3	4	4	4	4	4	4	5	5	6	6	6	6	5	6	6	6	6	7

Fig. 6 Estimated containerized cargo flows on major East–West container trade routes (million TEUs), 1995–2014. *Source* [16]

increase in imports from Asia reached up to 7.5%, demonstrating the importance of addressing infrastructure capacity and the performance of intermodal terminals for such a continuous increase. On the Transatlantic route, which is the smallest East-West route, the annual output was about 6.6 million TEU, which is only a slight increase. As can be seen in Fig. 10, there was a year-on-year increase ranging from 1.3 to 8.3% on all routes, except one. The only route whose annual decline was 4.5% was the North America–Asia route.

On the main sea routes, the issue of infrastructure solutions and transport services configuration (loading, unloading, etc.) is currently very important. High-capacity vessels are not always fully reloaded, and for the sake of efficient management, they also move part of the transport to smaller vessels. This is mainly because of the reuse of capacities, because shipping is often unloaded in Asia and the cost of shipping increases. For example, waste is shipped to Asia and especially to China [17].

Low fuel prices result in increased demand for cheaper transportation. Profitability is one of the carriers' priorities, and there is an attempt to maintain dominance in the market, as evidenced by the Group of Major Container Ship Operators in the Mega Alliance. The four key alliances include 2M, Ocean Three, G6, and CKYHE Container volumes are growing. It is therefore necessary to look for new solutions, either in intermodal logistics or on technical solutions on individual routes. Maritime transport has grown enormously and the profitability on these trade routes is very high. Therefore, boat owners also seek to increase the speed and load capacity of vessels. Proof of this is the order of 11 vessels of the second generation Triple-E container ships, each with a capacity up to 19,630 TEU [15].

	Transpacific		Europe Asia		Transatlantic	
	Asia–North America	North America–Asia	Asia–Europe	Europe–Asia	Europe–North America	North America–Europe
2009	10.6	6.1	11.5	5.5	2.8	2.5
2010	12.3	6.5	13.3	5.7	3.2	2.7
2011	12.4	6.6	14.1	6.2	3.4	2.8
2012	13.1	6.9	13.7	6.3	3.6	2.7
2013	13.8	7.9	14.3	6.9	3.6	2.7
2014	14.7	7.5	15.4	7.0	3.9	2.7
Percentage change 2013–2014	6.3	-4.5	7.5	1.3	8.3	0.0

Fig. 7 Estimated containerized cargo flows on major East–West container trade routes, 2009–2014 (million TEUs and percentage annual change). *Source* [16]

The world's largest shipping operator is Maersk Line. The Denmark-based company has up to 607 container ships with a total capacity of 3.1 million TEU. The following figure shows the Maersk Line transport route to the ports of Rotterdam, Bremen and Felixstowe, as well as to those of the UK, except those of Great Britain in Central and Western Europe. Today, Western Europe's ports are the centres of trade with Asia and are likely to maintain their dominant position in the years to come. The main transport line of Maersk is shown in Fig. 8 [19].

The main business routes between Europe and Asia include two major canals that were built to shorten routes, thus accelerating transport time. In our case, the most important canal is the Suez Canal [17].

The Suez Canal connects the Red Sea with the Mediterranean Sea and is one of the world's most important waterways. It shortens the length of sea routes between Europe and India by more than 10,000 km. Before it was officially opened in 1869, seagoing vessels transporting cargo between Europe and Asia had to encircle Africa or land cargo on the Mediterranean coast and have it transported to the Red Sea coast overland [17].

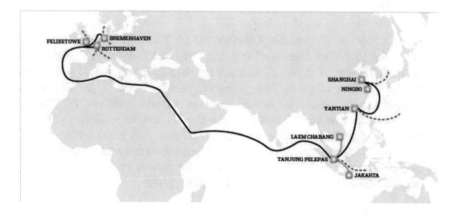

Fig. 8 Maersk Line shipping route for Asia-Europe. *Source* [19]

Not all vessels, however, meet the required technical parameters such as the permissible dimensions. Vessels have only a certain depth and therefore ships that exceed the maximum permissible draft cannot pass through these passages. Other limitations include, for example, lights, the underpass height under bridges and the length and breadth of chambers. The maximum allowable draft for seagoing ships is 20.7 m, a clear bridge height of 68 m and a maximum deadweight of 210,000 tonnes [17].

Due to ship weight and tonnage, today's large container ships have to flood the African continent, but such transport is still profitable in terms of the cost of one shipped container [17].

In 2015 Egypt opened a new branch of the Suez Canal. This second parallel branch allows simultaneous traffic in both directions and shorts the track by about half. This expansion, which required approximately 7.5 billion EUR of investment in Egypt, will help reduce waiting times by six or seven hours, increasing the number of day-to-day ships. The waterway is 193 km long and is crucial for trade between Asia and Europe. Even if the canal is not available for all container ships, it can save a carrier up to 16 days of cruise. If vessels have to sail around South Africa, their journey takes up to 32 days, depending on the size of the vessel and the tonnage of the cargo [15].

By 2023, Egypt expects an increase in the daily rate of ships through the Suez Canal from the current 49 vessels to 97 vessels. The generated profit should reach approximately 1.7 billion EUR. Today, revenue from the construction of the channel amounts to approximately 4.7 billion EUR. However, these forecasts do not have to be fulfilled. This is because the current trend is to reduce the cost of shipping as much as possible and no one is interested in accelerating it. The tendency is to reduce ship speeds to reduce fuel consumption [15].

4 Possible Railway Transport Connections to East–West Traffic Flows

Based on Chapters "Sustainable Development of Transport Systems for Cargo Flows on the East-West Direction" and "Analysis and Development Perspective Scenarios of Transport Corridors", there is potential for rail transport to be involved in the east-west flow of goods. The geographic position of the Slovak Republic is advantageous with regard to the existence of supplies between the Russian Federation, Ukraine and the Slovak Republic, there is also the possibility of linking the transport of goods flows between Europe and Asia with railway transport.

In the following chapters, rail transport is characterized from the point of view of the Slovak transport market, and the analysis of the railway transport infrastructure for international freight transport in the Slovak Republic.

4.1 Characteristics of Railway Transport in the Slovak Republic

Transport volume in the Slovak Republic is growing continuously every year. This rising trend is influenced by the development of the automotive industry and its suppliers. The Slovak Republic also has a geographically strategic position in middle Europe regarding transport corridors (east-west and north-south). The development of freight transport volume depends on the transport and business processes between the European Union and China and it is an opportunity for the Slovak Republic to obtain transport flows [20].

In the Slovak Republic road transport dominates the transport market. The volume of road transport has gradually increased during recent years. The increase of road transport is reflected by the highways and speed roads in regions which have higher economic potential. The increase in rail transport can be seen on the main rail corridors, but is not as significant as the increase in road transport use. Trade globalization has influenced the increase in transport volume and intermodal transport. Multiple increases based on the prediction of transport volume for this transport mode are from 2.3 million tonnes per year, currently, to 8 million tonnes in 2020 [20]. Table 1 shows the development of transport volume of goods in Slovakia. [13].

The dominant position of road transport stems from its advantages and the character of the Slovak economy (industry). After the planned economy, a Slovak business market based on services was created and some factories in engineering closed. Then railway transport lost its customers, resulting in a loss position on the transport market. The Slovak trade market is currently more oriented to customer satisfaction and companies oriented to the cost side of economics are faced with the transportation problems. Road transport is faster and more efficient for short distances, so road transport has a dominant position [20].

The strength of railway freight transport is Slovakia's geographic position in the middle of Europe. Railways are important transport crossings connecting transport flows in east-west and north-south directions. Other advantages are the higher density of the Slovak rail network and the possibility to move bulk substrates for heavy industry. Disadvantages include the poor technical condition of rail

Table 1 Transport volume of goods in Slovakia (thousand tons)

Indicator	2010	2011	2012	2013	2014	2015
Rail transport	44,327	43,711	42,599	48,401	50,997	47,358
Road transport	143,071	132,568	132,074	128,855	142,622	147,275
Inland waterway transport	3109	2454	2472	1920	1838	1683
Air transport	11	1	4	7	9	24
Together	190,507	178,733	177,145	179,176	195,466	193,340

infrastructure, the low speed of moving consignments from place to place and the low response rate to customer requirements [21].

The opportunity for railway transport is the development of intermodal transport. This is due to the globalization of the world economy. For this transport opportunity to succeed, the right conditions need to be created, such as responsible railways, an optimized network of intermodal terminals and carriers flexible to customer requirements. When railway transport connects the intermodal transport mode effectively, then it is possible to shift the volume of goods being transported from roads to railways [21].

4.2 Transport Corridors in Slovak Rail Freight Transport

The conception of rail freight corridors is one of the elements supporting the aims of the EU White Paper published in 2011—Roadmap to a Single European Transport Area—towards a competitive and resource efficient transport system. The creation of a competitive rail market in the European area may strengthen the position of rail transport [22].

Currently, there is a noticeable growing trend to use road transportation. Road transport is particularly effective for its flexibility but it lacks the capacity of rail transport. Negative impacts on the environment, included the increasing frequency of traffic congestion, noise, construction of road networks and land occupation. Transport safety is closely related to the damage caused during transport. Road safety advances are decreasing dramatically due to increased traffic accidents [20].

Development and support of the automobile industry in Slovakia has had a synergistic effect on the using and development of combined transport. The inclusion of three-track lines in Slovak railways has an important impact on the speed and quality of rail connections between European ports (Koper, Rijeka, Hamburg and Bremerhaven) and intermodal terminals in Slovakia (Žilina, Bratislava, Dunajská Streda, Košice). Common rail operation (passenger and freight) on the rail freight corridors can cause several problems: construction of traffic diagrams (train paths for passenger trains and freight trains), and consumption capacity of the track lines or operation problems [22].

Opening the rail freight market allows new rail operators to enter the rail network. To optimise the use of the network and ensure its reliability, it is useful to introduce additional procedures to strengthen cooperation on allocation of international train paths for freight trains between infrastructure managers [23–25].

The implementation of international rail freight corridors, forming a European rail network for competitive freight, should be conducted in a manner consistent with the trans-European Transport Network (TEN-T) and/or the European Railway Traffic Management System (ERTMS) corridors. The coordinated development of the networks is necessary, in particular as regards the integration of the international corridors for rail freight into the TEN-T and the ERTMS corridors [23].

The planned measures to improve the performance of rail freight transport should have little impact for users of rail. Therefore, it is necessary to include all the planned measures to the implementation plan. Also, all planned measures, which can impact lower-capacity infrastructure, must be regularly published or discussed with infrastructure managers [23].

The management of freight corridors should also include procedures for the allocation of the infrastructure capacity for international freight trains running on such corridors. Those procedures should recognise the need for capacity of other types of transport, including passenger transport [23]. Through the territory of the Slovak Republic are routed three rail freight corridors:

- rail freight corridor—RFC 5,
- rail freight corridor—RFC 7,
- rail freight corridor—RFC 9.

Characteristics of Rail Freight Corridor 5

Rail freight corridor (RFC) 5 (the Baltic–Adriatic corridor) connects the northern part of Europe (Baltic Sea) to the southern part (port of Khoper in Slovenia). The corridor crosses Poland, Slovakia, Austria, Italy and Slovenia. On the network of Slovak railways the entry point of the corridor is Skalité station, the cross-border station with the Polish railway network. The exit point is Devínska Nová Ves (alternative station is Petržalka), the cross-border station with the Austrian railway network. Figure 9 and Table 2, show the RFC 5 in Slovakia [26].

In Fig. 10, the speed profile of RFC 5 on the network of Slovakia railways are illustrated. The dynamic profile shows that the greatest limitation of this corridor is the track speed from Skalité to Čadca. In this track section, speed is limited and there are poor track and slope conditions. The maximum length of train in the track section Skalité–Zwardoň (PKP) is 300 m. It is therefore necessary to divide the intermodal trains running from Žilina to the port of Hamburg (or Bremerhaven or the Kaliningrad area). This causes technological problems on the rail freight corridor and then railway undertakings use other rail routes for their intermodal trains [27, 28].

Figures 11 and 12 show the analyses of a number of trains RFC 5 in Slovakia where it is a double line with a short line section at Čadca–Skalité (single line). Operation on RFC 5 is mixed and consists of passenger and freight transport. It is not ideally situated, based on the conditions of directive EU 913/2010. Common operation (passenger and freight) on the corridor could consume much of the railway infrastructure capacity.

Characteristics of Rail Freight Corridor 7

RFC 7 connects middle Europe (the city of Prague) with the east part of Europe (the Black Sea). The corridor crosses the Czech Republic, Slovakia, Hungary, and Romania with a branch via Bulgaria and Greece.

The entry point of RFC 7 on the railway network in Slovakia is the cross-border station Kúty Gr. The exit point is the cross-border station Štúrovo Gr. (alternative

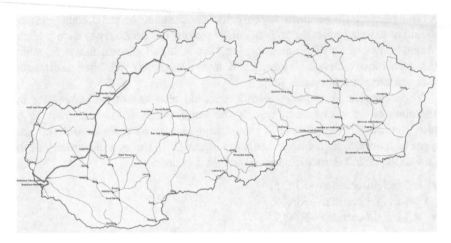

Fig. 9 Rail Freight Corridor 5 on the Slovakian railways network. *Source* [26]

Table 2 Rail Freight Corridor 5 on the Slovakian railways network

Corridor	Route	Distance (km)
RFC 5	Skalité št. hr.–Čadca–Žilina–Púchov–Trnava–Bratislava hl. st.–Devínska Nová Ves št. hr.	264
RFC 5	Skalité št. hr.–Čadca–Žilina–Púchov–Trnava–Bratislavské spojky–Petržalka št. hr.	252

Fig. 10 Dynamic profile of Rail Freight Corridor 5 on the Slovakian railways network (NM). *Source* [26]

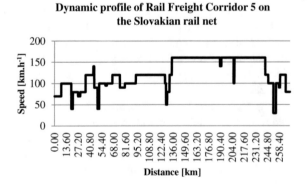

exit point are stations Rusovce and Komárno, respectively). Table 3 shows the stations in the corridor. [28]. Figure 13 shows the corridor map.

The speed profile of RFC 7 is shown in Fig. 14. The maximum track speed on the rail corridor is 140 km h^{-1}, but a significant part of the rail line has a maximum speed of 120 km h^{-1}. At some points the maximum speed is less as speed is restricted through railway stations due to old technical bases (tracks and switches) that need

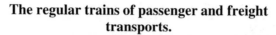

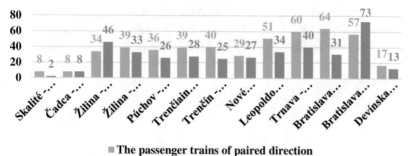

Fig. 11 The regular trains of passenger and freight transports of paired direction in the RFC 5. *Source* [22]

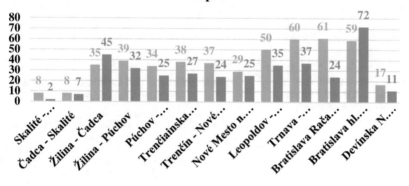

Fig. 12 The regular trains of passenger and freight transports of unpaired direction in the RFC 5. *Source* [22]

reconstruction. Nevertheless, the technical base of the Slovak portion of rail freight corridor 7 has relatively good technical conditions for rail operations [23].

On RFC 7 in Slovakia, there is common rail operation—passenger and freight— and a double-track line. Analysis of transport volume on the corridor is in Figs. 15 and 16. Analysis of transport volume on the rail freight corridor has been processed for even and odd directions.

Table 3 Rail Freight Corridor 7 on the Slovakian railways network

Corridor	Route	Distance (km)
RFC 7	Kúty št. hr.–Bratislava–Nové Zámky–Štúrovo št. hr.	300
RFC 7	Kúty št. hr.–Bratislavské spojky–Rusovce št. hr.	176
RFC 7	Kúty št. hr.–Bratislava–Nové Zámky–Komárno št. hr.	280

Fig. 13 Rail Freight Corridor 7 on the Slovakian railways network. *Source* [28]

Fig. 14 Dynamic profile of Rail Freight Corridor 7 on the Slovakian railways network (Source: [28])

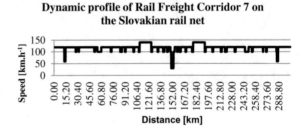

Dynamic profile of Rail Freight Corridor 7 on the Slovakian rail net

4.2.1 Characteristics of Rail Freight Corridor 9

RFC9 connects Czech Republic and Slovakia. The entry point is the cross-border station Čierna nad Tisou and the exit point is the cross-border station Lúky pod Makytou Gr. Table 4 and Fig. 17 show a preview of this corridor network in Slovakia [28].

The speed profile (Fig. 18) of rail freight corridor 9 is varied. This corridor is also the main rail line in Slovakia, but its technical conditions are poor. Future repairs will be based on the conditions of the AGTC agreement, then, it can be assumed that technical specifications will be improved [28].

The regular trains of passenger and freight transports.

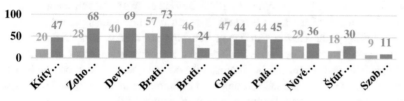

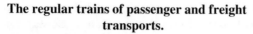

■ **The passenger trains of paired direction**

■ **The freight trains of paired direction**

Fig. 15 The regular trains of passenger and freight transports of paired direction in the RFC 7. *Source* [22]

The regular trains of passenger and freight transports.

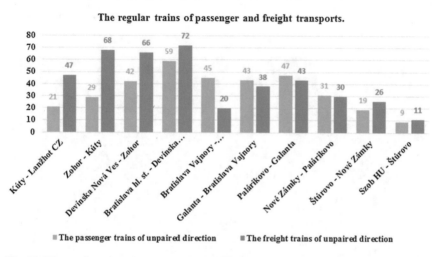

■ The passenger trains of unpaired direction   ■ The freight trains of unpaired direction

Fig. 16 The regular trains of passenger and freight transports of unpaired direction in the RFC 7. *Source* [22]

Rail operation on RFC 9 is common (passenger and freight) are RFC 5 and RFC 7. Transport volume on RFC 9 is showed in Figs. 19 and 20. This corridor is a double-track line and analysis of transport volume has been made separately for each direction.

4.3 The East Slovak Transhipment Area in Railway Transport

The existence of a different track gauge between the Slovak Republic and Ukraine is a technical limitation for goods in the east–west direction and vice versa. This fact led to the construction of the first transhipment site in 1946 in the village of Čierna (now the town of Čierna nad Tisou), which was first built as a one-way transfer site from the Soviet Union and Asia. Already in 1947 the first railway transhipment station was built. As a result of the great drought that year, the first transhipment of grain was made. Initially, the Čierna nad Tisou transhipment station consisted only of four normal-gauge tracks, three wide-track gauges, and one transhipment where they were hand transhipped. From the beginning, oil was transported, later on machines and equipment, chemical fertilizers, along with building supplies, food, and other special goods. The most important commodities were iron ore [29].

Table 4 Rail Freight Corridor 9 on the Slovakian railways network

Corridor	Route	Distance (km)
RFC 9	Lúky pod Makytou št. hr.–Žilina–Košice–Čierna nad T isou št. hr.	408

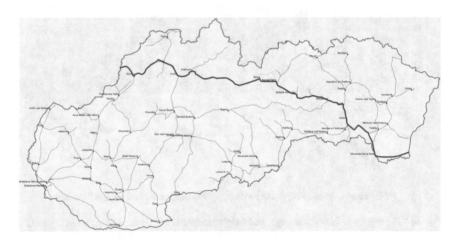

Fig. 17 Rail Freight Corridor 7 on the Slovakian railways network. *Source* [28]

Fig. 18 Dynamic profile of Rail Freight Corridor 9 on the Slovakian railways network. *Source* [28]

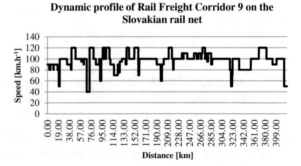

Dynamic profile of Rail Freight Corridor 9 on the Slovakian rail net

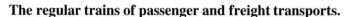

The regular trains of passenger and freight transports.

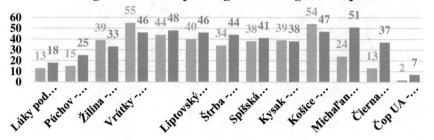

■ The passenger trains of paired direction

■ The freight trains of paired direction

Fig. 19 The regular trains of passenger and freight transports of paired direction in the RFC 9.
Source [22]

The regular trains of passenger and freight transports.

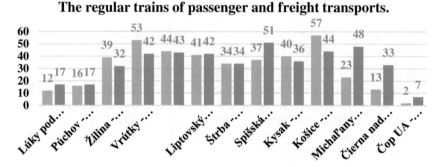

■ The passenger trains of unpaired direction

■ The freight trains of unpaired direction

Fig. 20 The regular trains of passenger and freight transports of unpaired direction in the RFC 9.
Source [22]

An increase in volume of transported goods was seen at the border crossing station of Maťovce. It was here that, in 1988, that construction of wagons greatly simplified the passage of goods from Ukraine to Slovakia [29].

In 2015 the East Slovakian Transhipment Section of Železničná spoločnosť Cargo Slovakia, a.s. (ESTS) was formed. This section is part of Slovak national rail cargo operator. It united two transhipment cross-border stations (Fig. 21): Čierna nad Tisou and Maťovce [30].

In 1996, in accordance with the European Agreement on the most important routes of international combined transport (AGTC), a combined transport terminal in Dobra near Čierna nad Tisou was built. It was designed for the transhipment of

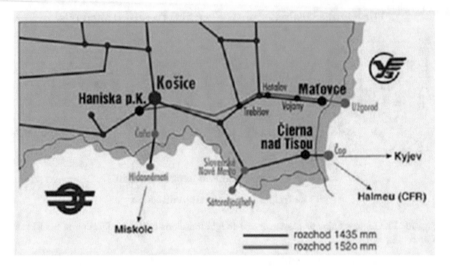

Fig. 21 Scheme of the East Slovakian transhipment area Čierna nad Tisou (ESTS). *Source* [29]

passenger cars, which contributed to further development and the possibility of connecting road transport to Ukraine [31].

Strategic Position of the ESTS in International Transport

The East Slovakian Transhipment Section of ZSSK CARGO has a significant position due to its position as the largest transfer area with a complete range of services between wide and normal-gauge tracks in Slovakia [30].

In the case of transhipment, overfilling and wagon transport, the following transport services are provided:

- loading, transloading, fixing and marking of consignments,
- negotiation of shipment of items or consignments that are unusual or have exceeded the loading rate,
- determining the weight of the consignment by weighing on the rail weight,
- palletizing, packing and banding goods,
- detecting damage to wagon consignments,
- supplying transport equipment and means of transport which belong to the carrier for carriage,
- distribute consignments under collective or partial orders and, where appropriate, ensure unloading,
- detecting the passage of the consignment across the Slovak state border,
- informing customers of the entry, transhipment and the departure of wagons from Čierna nad Tisov and Maťovce,
- issuing a transit declaration without guarantee,
- issuing and presenting a single customs document with annexes,
- representing the customs agent in the delivery of the customs declaration,

- delivering transport documents to an inspector to perform a health check on the transport of plants or products which may be carriers of harmful organisms,
- sending bills of lading and other documents to transporters providing written reports on the crossing of consignments [29].

At the same time, the ESTS also provides a change of the SMGS to CIM mode and vice versa. Since 1st August 2004, there is no need to change the transport regime for CIM consignments from all COTIF Member States with a destination station in Ukraine on the 1435 mm gauge–Esen– Batevo–Barkasovo–Stabičevo–Mukachevo [29].

In 2012, the Ukrainian Railways transported 15 million tonnes of goods to Slovakia in the export and transit mode via the Čierna nad Tisou and Maťovce railway crossings. For transports from Ukraine, ZSSK CARGO is the most important partner of Ukrainian Railways. The advantage of the Slovak national operator is that it is able to tranship and overpass almost all the raw materials and goods that customers want to transport at Čierna nad Tisou [29].

East Slovak transhipment stations are a gateway to Central Europe, as evidenced by the fact that almost half of the goods coming from the Ukrainian and Lviv Railways are transported across the Ukrainian Railways (Poland–PKP, Hungary–MÁV, Romania–CFR) compared to other railways (UZ) towards Central and Western Europe. According to 2012 data 31.769 million tonnes were transported from the UZ railway line to ŽSR, PKP, MÁV and CFR; 14.99 million tonnes (47.2%) were transported to ŽSR–ZSSK CARGO (through Čop-Čierna nad Tisou and Užhorod–Maťovce); 12.21 million tonnes (38.5%) towards the PKP (Mostiska II, Izov, Jagodin, Rava-Russia); 3.32 million tonnes (10.5%) towards MÁV (Batevo–Eperjeske); and 1.23 million tonnes (3.8%) towards CFR (via Halmeu and Vadul Siret) [29].

Although transhipment and overpassing constitute a significant part of ESTS performance (Fig. 22), they are not the only activity providing transhipment services. There are also comprehensive transhipment, overdraft, shipment and splitting services, liaison with state administration authorities, and a customer-centred pricing policy using a common price in cooperation with ČD Cargo.

Slovakia is a good alternative compared to other competitive shipping routes to Western Europe countries, such as Poland.

Table 5 shows the structure of manipulated goods at the workplace of the ESTS during 2012–2015. [30]. Iron ore dominates the transported goods in the ESTS.

The assumption of a significant increase the ESTS is linked to the relocation of production capacities to the eastern part of Europe, by increasing the importance of trans-Siberian railways in Euro-Asian transport and the dependence of metallurgical enterprises in Central and Western Europe on mineral raw materials from Eastern Europe [30].

The departments of the ESTS are of great importance for the transport market, not only in the SR, but also as a gateway to the European Union. Another

Fig. 22 Parts of goods transported from Ukrainian railways to neighbouring countries. *Source* [29]

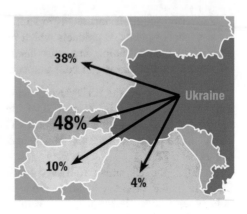

Table 5 Transport performance in the Čierna nad Tisou by the structure of goods

Type of Goods	Transport Performance by the year (thousand ton)			
	2012	2013	2014	2015
Chemistry	234	226	252	226
Wood	218.7	270.9	232.6	146.8
Foodstuff	1.4	1	2.2	1.6
Building	36.6	96.6	82.8	130.5
Metal	515	596	713	316
Petroleum products	132	81	120	42
Coal	187	107	209	144
Iron ore	6029	6257	6018	6244
Not specify	1.5	3,5	1	2.4
Total	7355.2	7639	7630.6	7253.3

possibility for the development of the ESTS sections is the use of transport potential from the People's Republic of China. In view of the continuous growth in international trade between the European Union and China and at the same time the marginal capacity of seaports in Europe, there is an opportunity to use rail transport for freight between Europe and Asia.

Therefore, it is essential for ZSSK Cargo to take account of the power of transit traffic and pay attention to the modernization of technology and the efficiency of its transhipment mechanisms. ESTS Čierna nad Tisou and Maťovce is a place where ZSSK Cargo provides complete services for transhipment. ESTS is the only provider of services that comply with European phytosanitary and veterinary standards. Goods are transported exclusively in one place, iron ore can be transported throughout the year due to de-icing reconstruction.

4.3.1 Intermodal Terminal Dobra

Intermodal terminal Dobra, whose operator is TransContainer—Slovakia, a.s., is located near the cross-border station Čierna nad Tisou, between the villages of Biel and Pribeník. The entire area of the terminal is a customs space. The intermodal terminal extends to an area of 180,000 m² [31].

The terminal provides the following basic services: load operations, short- and long-term storage of loaded intermodal loading units, and storage of empty intermodal loading units. Dobra terminal also provides its own complementary services, including cleaning of intermodal units, accommodation, customs, forwarding, logistics, and repair and maintenance of rail wagons.

Additional services provided by the terminal include the cooling-down of isothermal intermodal units, unit repairs, destruction of dangerous goods in damaged intermodal loading units, as well as banking services. The terminalis a bimodal terminal where transhipment is carried out between railway transport and road freight. It also has both rail gauges: normal—1435 mm and wide—1540 mm [32].

4.3.2 Technical Equipment of the Terminal

Intermodal terminal Dobra has two gantry cranes types of PKZ Brezno, whose load capacity is 50 tonnes. There is one reach stacker with a load capacity of 45 tonnes. There are also road and rail weights, and front and side ramps. The terminal has eight rail tracks. The length of the normal-gauge tracks are 570, 595, 684 and 735 m. The wide-gauge track lengths are 588, 593, 802, and 812 m [32].

4.3.3 Space Configuration of the Terminal

The storage area is designed for short and long-term storage of intermodal loading units. Other features of the terminal are the loading tracks that are connected to the loading area. The entire area of the intermodal terminal is equipped with roads for services to loading operations [32].

The basic parameters of the terminal are shown in Table 6 [32].

Table 6 Basic parameters of intermodal terminal Dobra

The storage area under the cranes	8700 m²
Transhipping capacity TEU	31,000 pcs
Time of one load capacity	4 min
The whole length of cranes track	750 m
Number of layers for stacking	3 layers

5 Carriage and Rules in International Railway Transport

International rail transport means transport that:

- commences in the territory of one state and ends in the territory of another state; takes place at least in the territory of two states,
- is carried out by rail transport operations of at least two states,
- has the same transport-legal conditions as agreed in the international agreement binding on railway users and on all the railways involved in the carriage,
- is carried out on the basis of a single contract of carriage, the agreement of which is documented in a single transport document, which corresponds in form and content to internationally agreed conditions [33].

International rail transport can be organized and qualitatively and quantitatively developed only if two conditions are met:

- if the railway networks of the individual states are technically interconnected in a way that allows the passage of wagons and, together with them, the passage of wagon and rail passengers from one state to the railroad of another state, or in the case of a different track gauge of individual railway lines at least, allows the passage of passengers and passengers from one country's rail to another country's railways,
- if between States and national rail administrations there are agreed conditions under which the technical capabilities of railway administrations are used between countries [33].

The creation of technical conditions for international rail transport is of prime importance. Equally important, however, is the creation of conditions for this international transport as well as in relations between individual states and between their rail administrations [34].

International rail transport can be carried out in principle in two basic forms, namely:

- indirect international rail traffic; indirect international rail transport,
- direct international rail traffic; direct international rail transport.

The distinction between the above forms of international rail transport is based on the differences in the relationship between the purpose of the shipment and the manner in which this purpose is met, both in terms of transport and the transport technology itself. In international rail transport, the distinction between forms of direct and indirect transport is particularly important in the transport of goods, which mainly serves the international circulation of goods and products.

The form of direct international rail freight traffic shall be carried out in accordance with the terms and conditions of carriage agreed in general terms generally applicable to their entire course, i.e., they are subject to internationally agreed transport conditions binding the carrier, or the passenger and all participating rail. Direct international transport is carried out on the basis of a single contract of carriage and is documented by a single transport document which corresponds to the form and content of internationally agreed transport standards and conditions [33].

The form of indirect international rail transport has its application:

- in the case of international transport from border to border, international carriage for which the different countries' transport conditions apply for their individual national rail transport,
- in the case of international carriage for which the various conditions of carriage for direct international carriage apply, taking into account the differences in their territorial validity,
- for international carriage for which the carrier is deliberately negotiated,
- the successive validity of the various transport schedules, the gradual validity of the different transport rules serving to achieve advantages.

In the case of border to border international transport, it is transport that is divided into two or more national transports that are linked from the point of view of the traffic-law regime. The national transport conditions of the participating States on whose territory the carriage is progressing shall apply within the framework of this carriage.

This transport requires a liaison person on the border between the two neighbouring countries' railways, which will provide for the new posting of the shipment to be followed by the further interconnection of the individual domestic transports to which this international transport is composed. At present, the use of international transport from border to border is not used, considering that this transport is only available under the conditions:

- if direct international rail freight is not at all regulated by agreement between participating states
- in the case of extraordinary events between certain states where the validity of the international transport conditions is abolished and the shipment is agreed on a case-by-case basis

Currently, more important is indirect international rail traffic, the realization of which is ensured by two modes of transport of direct international rail freight traffic under the Warsaw Agreements and under the Bern Conventions. The existence of these two direct orders for direct international rail freight, differ to a certain extent in their material content. They are especially different regarding to their respective areas of territorial validity. This indirect international rail transport operation is to ensure the interconnection of international shipments according to SMGS rules and CIM rules in conditions of participation of some states, both SMGS and CIM, namely:

- SMGS participating countries to countries that are both SMGS and CIM
- SMGS participating countries, passing through countries that are both SMGS and CIM parties.

International transport is divided into four modes of transport:

- SMGS shipment from the dispatching station in the country of dispatch to the entry border of the last transit country that is a member of SMGS and CIM
- for transport by CIM from this station to a destination station in the destination country
- for CIM transport from the dispatching station in the country of dispatch to the exit border station of the last transit country that is party to both CIM and SMGS
- SMGS shipment from this station to the destination station in the destination country.

5.1 Basic Characteristic of Transport Rules by International Rail Cargo Transport

The transport of goods in international rail transport is based on the EU's common transport policy and is based on agreements underlying the right of transport:

- Convention on the International Carriage of Goods by Rail–COTIF (Bern Convention)
- International SMGS and SMPS Passenger Transport Agreements (Warsaw Agreement).

The Bern Convention
The "Vilnius Protocol" of 3 June 1999, as amended by the COTIF Convention of 9 May 1980, was adopted in the light of the general belief that it was necessary and expedient to further develop the provisions of the Bern Convention, in particular its JSP CIV and CIP JPCs, to adapt to the new needs of international rail transport. The right of transport under the Bern Convention is so flexible that it can be used in the conditions of classical rail transport and in the conditions of separation from the operation of the infrastructure and the transport activity. The important provisions of the Convention for the creation of the conditions of carriage of goods are the Uniform Rules applicable to International Carriage by Rail, namely:

- Uniform legislation for the International Carriage of Goods (CIV) Convention, Appendix A to the Convention,
- Uniform legislation on the contract for the international carriage of goods by rail (CIM), Appendix B to the Convention,
- Convention on the International Carriage of Dangerous Goods by Rail (RID), Appendix C to the Convention,
- Uniform legislation on international carriage of goods by rail (CUV), Appendix D to the Convention,
- Uniform legislation for the contract for the use of railway infrastructure in international rail transport (CUI), Appendix E to the Convention,

- Uniform legislation for the binding disclosure of technical standards and the adoption of uniform technical prescriptions for APTU, Appendix F to the Convention,
- Uniform legislation on the technical approval of railway material used in international traffic (ATMF), Appendix G to the Convention.

Warsaw Agreements

The Warsaw Agreements consist of the Agreement on the International Carriage of Passengers by Rail–SMPS and the Agreement on the International Carriage of Goods by Rail. SMGS thus constitutes a counterpart to the CIM transport regime. The territorial validity of this agreement covers the territories of the former Eastern Bloc countries. The agreement applies to the carriage of goods on public railway lines in the direct international rail transport of goods with one transport document— the SMGS consignment note from the place of dispatch to the destination with the participation of at least two member countries.

5.2 Differences Between Transport Conditions for Bern Convention and Warsaw Agreements

International rail traffic begins in the territory of one country and ends in the territory of another one, i.e. it spans the territory of at least two countries. International agreements and the resulting regulations related to international carriage by rail include the Convention concerning International Carriage by Rail (COTIF) and the Warsaw agreements (SMGS, SMPS).

The transport of goods in international rail freight transport is regulated by the Convention on International Carriage by Rail (COTIF) dated 1 January 2001, COTIF 1999, Appendix B—Uniform Rules concerning the Contract of International Carriage of Goods by rail (CIM) and implementing international railway regulations, pursuant to which users of transport and railway workers or railway employees alone are to act. This contract defines the rules for transport of goods between rail carriers and users. The purpose of the contract is to establish responsibility for the organisation of transport, establish obligations to be met by the carrier and the user, create a unified tariff system to prevent monopoly, and ensure a transparent railway market is equally available to all. The CIM is in accordance with competition requirements under Articles 100 to 108 of the EU Consolidated Treaty. The contract of carriage shall be determined by the consignment note that is signed and confirmed by the carrier and the consignee. Acceptance of goods is confirmed in two copies of the consignment note, with the recipient receiving a duplicate of the consignment note as proof of accepting the goods. A consignment note is issued for each shipment. Table 7 shows sheets names and users of each sheet of CIM consignment note [35]. The consignor and

Table 7 Consignment note CIM

Sheet no.	Title	Retention of the sheet
1	Original of the consignment note	Consignee
2	Invoice	Carrier at the destination
3	Arrival notes/customs	Customs or carrier at the destination
4	Duplicate consignment note	Consignor
5	Duplicate invoice	Forwarding carrier

the carrier are to agree the transit period. The maximum transit period is as follows [35, 36]:

- for wagon-load consignments

  - period for consignment = 12 h,
  - period for carriage, for each 400 km = 24 h;

- for less than wagon-load consignments

  - period for consignment = 24 h,
  - period for carriage, for each 400 km = 24 h.

According to the SMGS agreement on carriage of goods, having an SMGS consignment note gives a railway administration party to the SMGS Agreement a direct transport passage through all the stations to the destination station. Table 8 shows sheets names and users of each sheet of SMGS consignment note [36].

The delivery times in SMGS are:

- period for consignment = 1 day;
- period for carriage:
- wagon loads for each 200 tariff kilometres = 1 day,
- large container loads for each 150 tariff kilometres = 1 day.

Clearly there is a difference between the number of sheets between CIM and SMGS consignment notes and a difference between the delivery times. There are other differences between CIM and SMGS in terms of transport law. For example, there are differences in the responsibility for loss or damage of goods, and in compensation if the delivery time is exceeded. Based on the transport conditions of

Table 8 Consignment note SMGS

Sheet No.	Title	Retention of the sheet
1	Original of the consignment note	Consignee
2	Card/(invoice)	Carrier at the destination
3	Delivery note	Carrier at the destination
4	Duplicate consignment note	Consignor
5	Note of acceptance of goods	Forwarding carrier
6	Arrival note	Consignee

CIM, the maximum compensation is four times the carriage charge (article 33 of CIM). SMGS transport rules grade compensation according to delay time (article 45 of SMGS):

- 6% carriage charge for exceeding the delivery times by up to 1/10
- 18% carriage charge for exceeding the delivery times by 1/10 up to 3/10
- 30% carriage charge for exceeding the delivery times by more than 3/10

Disparity also exists in the application of the Bern Convention (CIM) and the Warsaw Pact (SMGS) in the case of the use of wagons. CIM makes use of the AVV agreement and SMGS makes use of the PGV agreement. These agreements apply different rules. Lastly, there is also a difference in the case of payment of fees for the use of railway infrastructure in countries which use CIM or SMGS [35, 36].

The transport rules of CIM and SMGS help to better organise transport processes in international rail freight transport and they harmonise conditions for safety, speed and quality of international rail freight transport. Establishing the advantages and disadvantages of the transport rules of CIM and SMGS is a complicated process which is affected by several factors. For example, Belarus, the Russian Federation and Kazakhstan have common customs areas and for these countries and their railway companies SMGS transport rules are preferable. In the Slovak Republic it is more convenient to use CIM transport rules based on EU membership. One of the advantages of CIM rules is that the consignment note can be used as a customs paper in rail transport. When a carrier uses a CIM consignment note for import or export to/from the EU there is no need to make an additional customs document. With the SMGS consignment note this option does not exist and the carrier (or shipper) must create a transit declaration based on the Union Customs Code [37].

The International Committee of the Railway and the Organisation for Cooperation of Railways is working on the implementation of legal frameworks to enable simpler, more acceptable and faster transport of goods by rail. A unique CIM/ SMGS consignment note is to be created. For example, transport on Slovak Railways is to be regulated by the rules of CIM until the border station where the rules concerning SMGS for the further transport path in Ukraine will be applied, and vice versa. Benefits of a CIM/SMGS consignment note include the following: both contracts of carriage can be displayed on a single sheet of paper that can be issued in electronic form, with greater legal certainty for all participants in the process of transportation, reduced transportation costs, reduced transport time, improved quality, and greater competitiveness of railways in international freight [37].

5.3 Characteristics of Common Consignment Note CIM/SMGS

The CIM/SMGS consignment note is a product of the International Transport Committee (CIT) and the Organization for Co-operation between Railways (OSJD).

Why this transport document was created was to make international transport easier, faster and more efficient. The common consignment note reflects transport conditions which are contained in CIM rules and SMGS agreements [37].

Standardized consignment notes are established by international federations of carriers with the consent of the international federations of customers, the competent customs authorities of member states, and any intergovernmental organization existing in the regional economic community, which has legislative powers in customs [37].

A common CIM/SMGS consignment note can be used by the countries that apply CIM rules and SMGS agreements. This document contains a contract of carriage, therefore it is no longer necessary to conclude further contracts when the consignment is crossing the border of international transport law. Territorial validation of transport conditions is solved geographically, in countries using CIM rules, the common consignment note is applied as consignment note CIM. In countries where the SMGS agreement is used, the common consignment note is applied as consignment note SMGS [38].

Instructions for using the CIM/SMGS consignment note is described in the document: "CIM/SMGS Consignment Note Manual (GLV CIM/SMGS) which is published by the International Transport Committee. This manual describes an alternative to the classic system of consignment with retranscription of the SMGS consignment note to the CIM consignment note or from the CIM consignment note to the SMGS consignment note at the reconsignment point [38].

In Table 9 is a specimen of a common consignment note and it is described using the sheets of consignment note. Paper form of the CIM/SMGS consignment note has six sheets in A4 format [38].

The common CIM/SMGS consignment note has some specific conditions for its use. For example, when consignments come from states that apply the SMGS, the

Table 9 Consignment note CIM/SMGS sheet and their used

Sheet		Retention of the sheet
No.	Description	
1	Original of the consignment note	Consignee
2	Invoice	Carrier who delivers the goods to the consignee
CIM 5 SMGS 3	Duplicate of the consignment note	Consignor
4	Delivery note	CIM → SMGS traffic: carrier who delivers the goods to the consignee SMGS → CIM traffic: not used
CIM 3 SMGS 5	Arrival note/Customs	CIM → SMGS traffic: consignee/customs SMGS → CIM traffic: destination carrier/customs
6	Duplicate invoice	CIM → SMGS traffic: forwarding carrier SMGS → CIM traffic: not used

consignor is to make out additional copies of the invoice: two copies are to be made out for the contractual SMGS carrier and one copy for each successive SMGS carrier. When consignments come from states that apply CIM rules, these additional copies of the invoice are to be supplied by the SMGS carrier at the transhipment/gauge change point in the form of photocopies of the invoice which are to be authenticated by the date stamp [38].

When the consignment note is produced as a computer printout, the following conditions must be observed:

- content: no departure from the specimen, layout: as little departure from the specimen as possible.
- the back of CIM/SMGS consignment notes may be printed on special sheets (supplementary sheets) [38].

Use of the common CIM/SMGS consignment note has several advantages in international railway transport:

- both contracts of carriage are showed on one paper,
- increased legal security for all transport participants,
- can be used in electronic form,
- harmonization of both transport regimes,
- lower transport costs,
- shorter delivery times, and improved competitiveness of rail freight transport [38].

6 Case Study—Costs of Freight Rail Operators in RFC9 Corridor

Price calculation for transport is an important factor in a customer choosing a transport mode. Competitive and efficient price calculation in railway transport is conditional by the exact calculation of real actual costs on the realized transport.

The different conditions in international transport rules (CIM an SMGS) also includes the transport price calculation. SMGS agreement mostly uses tariff obligation in the price calculation. This is opposite to the CIOTIF convention tariff obligation does not exist, and a rail operator then calculates the transport price directly by the cost side.

The methodology of the total cost calculation on selected transports, which could be subject to east–west traffic flows, is in the following case study. International corridor RFC 9 (chapter 4.2) was chosen for case study due to its direct connection to the East Slovakian Transhipments Section and its potential to participate in the transport between Europe and Asia. Figure 23 shows the transport route that was chosen for case study [39].

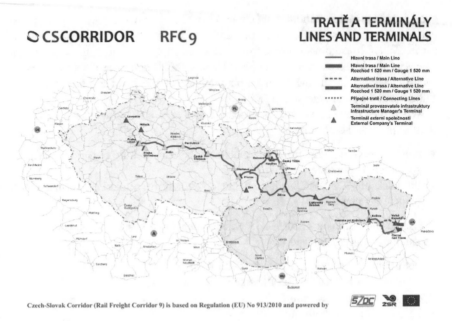

Fig. 23 Lines and terminals of RFC9 corridor. *Source* [39]

The case study deals with the comparison of costs in intermodal transport and bulk substrate transportation at different carrying capacities of a train in the RFC9 corridor. The frontier exchange station for intermodal transport between Ukraine and Slovakiais Dobrá and for other commodities is Čierna nad Tisou.

6.1 Technical Parameters of RFC9 Railway Lines

RFC9 corridors comprises railway lines in Slovak Republic:

- Track No. 101 A: Čop UA–Čierna nad Tisou–Košice, track length is 99 km
- Track No. 105 A: Košice–Kraľovany, track length is 203 km
- Track No. 106 A: Kraľovany–Púchov, track length is 83 km
- Track No. 106 F: Púchov–Horní Lideč CZ, track length is 21 km [40].

The total length of the corridor in the Slovak Republic is 406 km.

Table 10 shows the basic characteristics of the RFC 9 railway network in the Slovak republic [40]. Track Spišská Nová Ves–Poprad–Vrútky is the most critical line, where the gradient is 16%. The maximum gross tonne of train is listed for the Siemens ES 64 F4/189 locomotive in the traffic direction of Čierna nad Tisou št. hr.–Lovosice.

Table 10 Characteristic of track section RFC9 in the Slovak Republic

Track section	Length of a track section (km)	Number of track	Max. length of train (m)	Max. gross tonne of a train (t)	Load on track	Max. gradient (‰)
Čierna nad Tisou št. hranica–Čierna nad Tisou	4	1	700	2000	D4	↑3 ↓3
Čierna nad Tisou–Košice	95	2	670	1200	D4	↑15 ↓15
Košice–Kysak	16	2	650	2000	D4	↑7 ↓1
Kysak–Spišská Nová Ves	59	2	650	2000	D4	↑7 ↓0
Spišská Nová Ves–Poprad	26	2	650	1200	D4	↑15 ↓7
Poprad–Vrútky	120	2	650	1200	D4	↑15 ↓15
Vrútky–Žilina	21	2	650	2000	D4	↑1 ↓5
Žilina–Púchov	45	2	650	2000	D4	↑3 ↓7
Púchov–Lúky pod Makytou št. hr.	20	2	645	1000	D4	↑18 ↓15

The electric traction system is DC 3000 V in all RFC9 track sections in the Slovak Republic. The maximum speed of freight train is 100 km h^{-1}. If the gradient on the track section is higher than 8% we calculate a lower speed of freight train.

The RFC9 corridor comprises the following railway lines in the Czech Republic:

- Track No.308: Horní Lideč št. hr.–Hranice na Moravě, track length is 76 km
- Track No. 305B: Hranice na Moravě–Prosenice, track length is 16 km
- Track No. 309E: Prosenice–výh. Dluhonice, track length is 9 km
- Track No. 309A: výh. Dluhonice–Česká Třebová, track length is 107 km
- Track No. 501A: Česká Třebová–Praha, track length is 164 km
- Track No. 527A: Praha–Lovosice, track length is 84 km

The total length of the Czech portion of the corridor is 456 km.

Table 11 shows the basic characteristics of the RFC 9 railway network in the Czech Republic [41]. Track Horní Lideč št. hr.–Hranice na Moravě is the most critical line, where the gradient is 15%.

The electric traction system is DC 3000 V in all Czech sections of the RFC 9. The maximum speed of freight train is 100 km h^{-1} in the track section Hranice na Moravě–Praha, and 120 km h^{-1} in the track section Praha–Lovosice.

On the basis of technological track section characteristics, classification of freight train, i.e., type of locomotive, type and number of loaded/empty wagons for transported commodity, can be suggested. In the RFC 9 corridor there are track sections where maximum gross weight of a train is lower than 1200 tonnes. In these tracks we calculated with bank locomotive and a maximum train gross weight of 1600 tonnes on all tracks.

Table 11 Characteristic of track section RFC9 in the Czech Republic

Track section	Length of a track section (km)	Number of track	Max. length of train (m)	Max. gross tonne of a train (t)	Load on track	Max. gradient (‰)
Horní Liděč št. hr.–Hranice na Moravě	76	2	600	1000	D4	↑18 ↓15
Hranice na Moravě–Prosenice	18	2	690	2000	D4	↑0 ↓4
Prosenice–Olomouc	30	2	700	2000	D4	↑7 ↓7
Olomouc–Česká Třebová	86	2	700	1800	D4	↑7 ↓11
Česká Třebová–Pardubice	60	2	666	1800	D4	↑8 ↓2
Pardubice–Kolín	42	2	666	2000	D4	↑4 ↓4
Kolín–Poříčany	22	2	666	2000	D4	↑4 ↓4
Poříčany–Praha	40	2	697	1800	D4	↑7 ↓7
Praha–Lovosice	84	2	595	2000	D4	↑2 ↓2

In our case study the Siemens ES 64 F4/189 locomotive was chosen (Fig. 24) because a train's path can continue on to a different electric traction system. This locomotive is a multi-system locomotive for standard gauge.

Basic characteristics of locomotives are following [42]:

- Track gauge—1435 mm
- Maximum speed—140 km h^{-1}
- Weight—87 tonnes
- Weight for axle—22 tonnes
- Width—3019 mm
- Height—4245
- Length over buffers—19,580 mm
- Voltage system—AC 25 kV/50 Hz, AC 15 kV/16.7 Hz, DC 1.5 kV, DC 3 kV
- Continue rating—6000 kW at DC 3 kV
- Speed at continue rating—approximately 80 km h^{-1} at DC 3 kV

Fig. 24 Eurosprinter ES 64 F4. Four-system high performance locomotive. *Source* [42]

- Wheel arrangement—Bo'Bo'
- Starting tractive effort—300 kN
- Continue tractive effort—270 kN

In the RFC 9 corridor there are track sections where the maximum gross tonne of a train is lower than 1200 tonnes. They are as follows:

- Kuzmice–Slanec—length of track section is 11 km
- Spišská Nová Ves–Liptovský Mikuláš—length of track section is 84 km
- Púchov–Horní Lideč—length of track section is 84 km.

In these tracks we calculated with bank locomotive and maximum gross tonne of a train 1600 tonnes in the all tracks. Using the bank locomotive increases some cost items such as crew costs, locomotive costs, railway infrastructure costs, and traction fuel and energy costs.

6.2 Intermodal Transport

The costs for intermodal transport depend on technology and the gross weight of containers. Our previous study showed a decrease in the weight of intermodal trains. We analysed intermodal trains of different operators of railway freight transport using the RFC 9 corridor or some other railway line in this corridor. The analysis showed that as the length of trains increases, the gross weight of containers decreases [43]. In view of this fact we modelled the costs for intermodal transport into two variants:

- 1. variant—maximum gross weight of TEU (container ISO 1C) is 5 tonnes
- 2. variant—TEU and wagon capacity utilization

For intermodal transport we chose the Sggnss 80 wagon (Fig. 25). The wagon is suitable for transportation of High cube containers, High cube pallet wide containers and ISO containers 20′, 26′, 30′, 40′, 45′ classified in UIC Leaflet 592-2, Class I.

Technical characteristic of wagon Sggnss 80′ [44]:

- Gauge—1435 mm
- Tara—21.5 t
- Mass of loaded wagon—90 t
- Loading mass—68.5 t
- Max. speed at axle load of 20 t—120 km h^{-1}
- Max. speed at axle load of 22.5 t—100 km h^{-1}
- Height of loading level above top of rail—1155 mm
- Length over buffers—25,940 mm
- Loading length—24,700 mm
- Type of bogie—Y25Ls1-K
- Loading and unloading—vertically by crane.

Fig. 25 Intermodal wagon Sggnss 80′. *Source* [44]

On the basis of technical parameters of railway lines and the technical parameters of wagons, the maximum number of wagons in the train can be calculated. The train length is the restrictive factor in the first variant.

In the RFC 9 corridor is the minimum train length in the track section Praha–Lovosice—is 595 m. The number of wagons in the train may be calculated according to the following equations:

$$n_w = \frac{\max l_t - l_l}{l_w} \tag{1}$$

where

n_w number of wagons
$\max l_t$ max. train length
l_l length of locomotive over buffers
l_w length of wagon over buffers

The maximum of number of wagons in the train in the 1. variant is:

$$n_w = \frac{595 - 19.58}{25.94} = 22.18 \cong 22 \, \text{wagons}$$

In the second variant the restrictive factor is the gross weight of the train. The number of wagons in the train is calculated as the ratio of the maximum gross weight of the train and the gross weight of the loading wagons.

Costs for Intermodal Transport—1. Variant

Cost calculations take into account all direct costs related to transportation and the share of indirect costs. Calculation of direct cost was realized by multiplying performed transport output by the rate of the transport unit [45]. The costs of railway infrastructure were calculated separately for the Slovak and Czech Republics according to rules and rates defined in the Network Statement ŽSR and the Network Statement SŽDC, respectively.

In the Czech Republic we used supply prices (55% from the basic prices for usage of transport infrastructure by train movement). Table 12 shows costs of railway infrastructure in the Slovak Republic and Czech Republic for maximum train length.

The aim of calculation was to identify the change of unit cost according to the change of train composition (carrying capacity). The result is depicted in Fig. 26.

The comparison of unit costs for different train carrying capacities illustrates decreasing unit costs. Using 100% carrying capacity of a train decreased the unit cost about 28% compare to using 50% carrying capacity.

Table 12 Cost of railway infrastructure—Intermodal transport 1. Variant

Slovak Republic		Czech Republic	
Charge for	Fee in €	Charge for	Fee in €
Ordering and allocation of capacity	8.4042	Infrastructure capacity allocation	4.6997
Traffic management and organization	388.9480	Railway infrastructure usage for a freight train —S_{1E}	346.3611
Ensuring serviceability of railway infrastructure	532.2660	Railway infrastructure usage for a freight train —S_{2E}	472.3368
Use of electrical supply facility for traction current supply	105.5600		
Access to marshalling yards and train formation facilities and to freight terminals	56.5370		
Total	1091.7152		823.3976

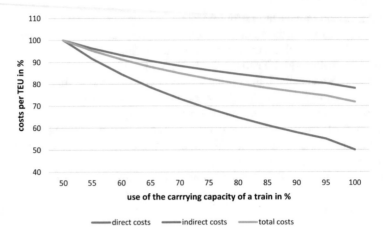

Fig. 26 Comparison of the unit costs—1. variant

Costs for Intermodal Transport—2. Variant

Costs in the second variant were calculated on the same principle as the first variant. In the track section where a bank locomotive was used, we calculated additional costs (crew costs, locomotive costs, railway infrastructure costs, and traction fuel and energy costs).

Table 13 shows costs of railway infrastructure in the Slovak and Czech Republics for a maximum weight train.

Figure 27 presents the comparison of the direct, indirect and total unit costs for different carrying capacities of a train.

Table 13 Cost of railway infrastructure—Intermodal transport 2. Variant

Slovak Republic		Czech Republic	
Charge for	Fee in €	Charge for	Fee in €
Ordering and allocation of capacity	8.4042	Infrastructure capacity allocation	4.6997
Traffic management and organization	388.9480	Railway infrastructure usage for a freight train —S_{1E}	346.3611
Ensuring serviceability of railway infrastructure	822.2697	Railway infrastructure usage for a freight train —S_{2E}	725.5274
Use of electrical supply facility for traction current supply	161.2957		
Access to marshalling yards and train formation facilities and to freight terminals	56.5370		
Total	1437.4546		1076.5882

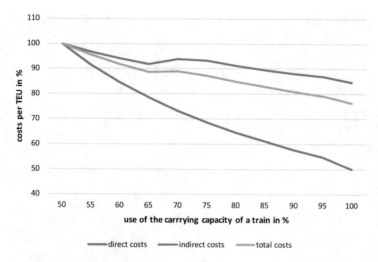

Fig. 27 Comparison of the unit costs—2. variant

Figure 27 shows that unit costs of TEU are affected not only by the carrying capacity of a train, but also by the technical parameters of a track. When the carrying capacity of a track is higher than 70% (for this variant it is 15 wagons in the train), the unit direct cost and the unit total costs change by step. Using 100% carrying capacity of a train decreases the unit cost about 23%. Compared with the first variant, unit costs per TEU are approximately 39 EUR higher.

6.3 Bulk Substrates Transportation

For bulk substrate it is advantageous to use the wagon which has a ratio between the loading weight of the wagon and its deadweight as high as possible. On that basis of we chose the Eanoss wagon. The wagon is designed for transportation of scrap iron, coal, coke, ore, sand, gravel and similar loose materials (Fig. 28). It is solid metal with a straight integral floor, firm fronts and two double-wing doors in each sidewall.

Technical characteristic of wagon [46]:

- Track Gauge—1435 mm
- Tara—24.6 t
- Mass of loaded wagon—90.0 t
- Carrying capacity at 22.5 axle load—65.4 t
- Maximum axle load—22.5 t
- Max. speed at axle load 22.5 t—100 km h^{-1}
- Max. speed at axle load 20 t—120 km h^{-1}
- Loading length—14,492 mm

Fig. 28 Covered and high-sided wagons. *Source* [46]

- Loading height—2100 mm
- Loading volume—82.7 m³
- Length over buffers—15,740 mm
- Type of bogie—Y25 Ls(s)

The train weight is the restrictive factor in the bulk substrate transportation. Maximum number of wagons in a train may be calculated according to the following equations:

$$n_w = \frac{norm_{gt}}{t_w + l_m} \qquad (2)$$

where:

n_w = maximum number of wagons in a train,
$norm_{gt}$ = train weight norm in gross tonnes,
t_w = mass of empty wagon,
l_m = carrying capacity in tonnes

Since the restrictive factor for bulk substrates is the weight of a train, we must take into account using a bank locomotive on the same track sections as in the intermodal transport 2. variant. The maximum number of wagons in the train (by Eq. 2) is 17.

The cost calculation was realized on the same principle as intermodal transport. Table 14 shows costs of railway infrastructure in the Slovak and Czech Republics for a maximum weight train.

The cost of railway infrastructure for bulk substrates in the Czech Republic is higher than for intermodal transport (2. variant). This is caused by a discount for intermodal trains. This discount cannot be applied to bulk substrate transportation.

The minimum number of wagons respective to the minimum of goods in a train is an important criterion for transportation effectiveness. Figure 29 presents the change of unit costs according to the carrying capacity of a train. Costs were

Table 14 Cost of railway infrastructure—bulk substrate transportation

Slovak Republic		Czech Republic	
Charge for	Fee in €	Charge for	Fee in €
Ordering and allocation of capacity	8.4042	Infrastructure capacity allocation	4.6997
Traffic management and organization	388.948	Railway infrastructure usage for a freight train —S_{1E}	629.7475
Ensuring serviceability of railway infrastructure	869.6414	Railway infrastructure usage for a freight train —S_{2E}	1392.4670
Use of electrical supply facility for traction current supply	170.6905		
Access to marshalling yards and train formation facilities and to freight terminals	56.5370		
Total costs	1494.2111		2026.9142

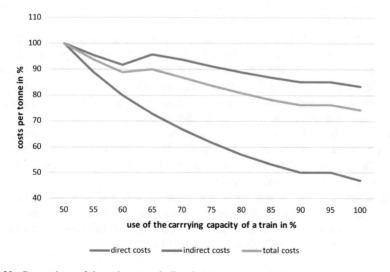

Fig. 29 Comparison of the unit costs—bulk substrate transportation

calculated by the number of wagons in the train and then they were converted by the net tonne.

Even in the case of bulk substrate transportation bank locomotive must be used from the carrying capacity of a train 65%, i.e., 11 wagons in a train. The unit costs per tonne change by step as well in intermodal transport—2. variant.

Knowing the change in unit costs with different carrying capacities of a train is very important to determine the price which should take into account the unused carrying capacity risk.

7   Conclusion

The volume of international rail freight transport has a significant share in the performance of rail transport operators in the Slovak Republic. The geographical location of the Slovak Republic and its transport flows create opportunities for the development of rail transport at the external borders of the European Union and the Schengen area. There are two cross-border stations on the border with Ukraine, which allow shipment between third countries and the European Union.

The transit of freight wagons across the border between Slovak and Ukrainian railway infrastructure has some problematic points. The first limitation is technical. The Slovak railway infrastructure has normal-gauge tracks (1435 mm) while the Ukraine railway has wide-gauge tracks (1520 mm). When crossing the border, shipments must be reloaded from wide-gauge wagons to normal-gauge wagons.

The second limiting factor that influences the shipment of consignments on the border between the Slovak Republic and Ukraine, is the existence of two different international transport laws. The railways operators in the Slovak Republic use the COTIF agreement for international rail transport, mainly appendix B—the CIM common law rules. Ukraine's railway companies in the international transport regime use transport law of the SMGS agreement. They are two different transport rules and they have their own transport conditions.

Despite these differences, there exist possibilities for cooperation in the east-west flow of goods transported by rail. According to the foreign trade analysis (Chapter "Sustainable Development of Transport Systems for Cargo Flows on the East-West Direction") and the infrastructure possibilities (Chapter "Analysis and Development Perspective Scenarios of Transport Corridors"), there is potential for an increase in transport performance in railway transport.

With the optimal development of logistics transport chains and better coordination between countries (e.g., the Russian Federation, Ukraine and China), by using the common CIM/SMGS consignment note, there is more opportunity for Euro-Asian transcontinental transport.

Maritime transport is still more economically efficient than rail transport in Euro-Asian traffic flows, but by optimizing rail transport operations, transport time can be shortened between Asian and European countries. In comparable cases the delivery time of goods by rail was cut in half [47, 48].

Acknowledgements This research was supported by the VEGA Agency, Grant No. 1/0019/17 "Evaluation of regional rail transport in the context of regional economic potential with a view to effective use of public resources and social costs of transport", at Faculty of Operations and Economics of Transport and Communication, University of Žilina, Slovakia.

Bibliography

1. Nozicka J (2016) Development of rail way transportation between China and the European Union countries. In: International scientific conference on knowledge for market use 2016—our interconnected and divided world, Societas scientiarum olonucensis II, Olomouc, pp 324–333
2. Eger KM, Jorgensen-Dahl A (2011) Short way, high risk the Northern Sea route: A Euro-Asian shipping lane? Osteuropa 61:179–193
3. Fan LX, Luo MF, Wilson WW (2014) Pricing joint products in liner shipping. Int J Shipp Transp Log 6:371–386
4. Bykov Y, Fadeeva V (2015) Transport corridors in Euro-Asia Communication. In: New industrialization and urbanization development annual conference—forum on new industrialization development in big data era. Science Press, Beijing, pp 464–467
5. Xu HT (2016) Domestic railroad infrastructure and exports: evidence from the Silk Route. China Econ Rev 41:129–147
6. Zunder TH, Islam DMZ, Mortimer PN (2012) Pan-European Rail freight transport; evidence from a pilot demonstration result. In: Papaioannou P (ed) Transport research arena 2012. Procedia Social and Behavioral Sciences, vol 48. Elsevier, Amsterdam, pp 1346–1355
7. Fagan M (2017) Incoterms rules in international trade. Diploma thesis, Pan-European College in Bratislava FEP UMP
8. Kozlejová D (2012) Medzinárodný obchod na rôzne spôsoby (International trade in different ways). Metodicko-pedagogické centrum v Bratislave, Prešov. Online. Available at: https://mpc-edu.sk/shared/Web/OPSOSO%20III.%20kolo%0vyzvy%20na%20poziciu%20Odborny%20poradca%20vo%20vzdelavani/OPS_Kozlejova%20Danka%20-%20Medzinarodny%20obchod%20na%20rozne%20sposoby.pdf. Accessed 22 January 2017
9. Jančíková E (2014) Medzinárodné obchodné financovanie (International market financial). Sprint dva, Bratislava
10. Eurostat, International trade in goods (2016). Online. Available at: http://ec.europa.eu/eurostat/statistics-explained/index.php/International_trade_in_goods
11. Majercak P, Majercak J, Majercakova E (2016) Logistics performance index for Slovak Republic. In: EBMEI international conference on humanity and social science (EBMEI-HSS 2016). Lecture Notes in Management Science, pp 58–62
12. Tkáčiková M (2016) Comparative analysis of transport-tariff conditions in the rail and road freight transport in the session Puchov—Hannover, Bachelor thesis, University of Žilina
13. Statistical Office of the Slovak Republic. Yearbook of Transport, Posts and Telecommunication (2017). Online. Available at: https://slovak.statistics.sk/wps/portal/ext/themes/sectoral/transport/publications/. Accessed 12 September 2017
14. Bukova B, Brumercikova, E, Kondek P (2016) Determinants of the EU transport market. In: Proceedings of the 2016 international conference on engineering science and management (ESM). AER-Advances in Engineering Research, vol 62, pp 249–252
15. David A, Piala P, Stupalo V (2016) Cargo containerisation and its impact on the development of maritime transport. In: 3rd International conference on traffic and transport engineering (ICTTE). Scientific Research Center LTD Belgrade, pp 306–311
16. Review of Marine Transport (2015) UNCTAD, Layout and printed at United Nations, Geneva, 1522632 (E), Oct 2015. Online. Available at: http://unctad.org/en/PublicationsLibrary/rmt2015_en.pdf. Accessed 21 August 2017

17. David A (2013) Innovation of handling systems in the world container ports and their terminals. In: 17th International conference on transport means. Kaunas University of Technology Press, Kaunas, Lithuania, pp 250–253

18. Chovancova M, Klapita V (2016) Draft model for optimization of the intermodal transport chains by applying the network analysis. In: 20th international scientific conference on transport means. Kaunas Univ. Technology Press, Kaunas, Lithuania, pp 112–116

19. Seidl P (2016) Analýza prepravných reťazcov na reláciách Ďaleký východ – stredná Európa, Bachelor thesis, University of Zilina

20. Cerna L, Zitricky V, Danis J (2017) The methodology of selecting the transport mode for companies on the Slovak transport market. Open Eng 7(1):6–13

21. Zitricky V, Cerna L, Ponicky J (2016) Possibilities of development of railway sidings in the Slovak republic. In: 3rd international conference on traffic and transport engineering (ICTTE). Scientific Research Center LTD Belgrade, pp 454–461

22. Lizbetin J, Ponický J, Zitricky V (2016) The throughput capacity of rail freight corridors on the particular railways network. Naše more 63(3):161–169

23. Regulation (EU) No. 913/2010 of the European parliament and of the council—concerning a European rail network for competitive freight is one possibility to increase of transport volume in railway transport (2017). Online. Available at: http://www.cer.be/topics/freight/external-links/regulation-eu-9132010-european-rail-network-competitive-freight. Accessed 24 September 2017

24. Islam DMZ, Ricci S, Nelldal BL (2016) How to make modal shift from road to rail possible in the European transport market, as aspired to in the EU Transport White Paper 2011. European Transport Research Review. 8, art. no. 18

25. Saeedi H, Wiegmans B, Behdani B, Zuidwijk R (2017) European intermodal freight transport network: Market structure analysis. J Transp Geogr 60:141–154

26. Zitrický V, Černá L, Abramovič B (2017) The proposal for the allocation of capacity for international railway transport. Proc Eng 192:994–999

27. Gasparik J, Abramovic B, Halas M (2015) New graphical approach to railway infrastructure capacity analysis. Promet Traffic Traffico 27(4):283–290

28. Abramović B, Zitricky V, Mesko P (2017) Draft methodology to specify the railway sections capacity. Logi Sci J Transp Log 8(1):1–11

29. Cerna L, Bukova B, Zitricky V (2013) Analýza výkonov a pracovísk sekcie VSP Železničnej spoločnosti Cargo Slovakia a.s. (Analysis performances and workplaces Železničnej spoločnosti Cargo Slovakia a.s.). Železničná doprava a logistika. Vol. 9. No. 2. pp 66–75

30. Holczreiter O (2016) Racionalizácia prekládkových kapacít prekládkovej stanice Čierna nad Tisou (Rationalization of transhipping capacity of the Čierna nad Tisou station). Master thesis, University of Žilina

31. Masek J, Camaj J, Belosevic I (2016) Improving the transport capacity of the intermodal train and track based on different types of wagons. In: 3rd international conference on traffic and transport engineering (ICTTE). Scientific Research Center LTD Belgrade, pp 409–416

32. Sevcikova S (2016) Komparatívna analýza terminálov intermodálnej prepravy v SR a v zahraničí (Comparative analysis of intermodal terminal in the SR and abroad). Bachelor thesis, University of Zilina

33. Zitrický V (2010) Racionalizácia technologických procesov v medzinárodnej železničnej deprave (Rationalization of technological processes in international railway transport). Dissertation thesis, University of Zilina

34. Gasparik J, Gaborova V, Ľuptak V (2016) Process portal for railway cargo operator with CRM support. In: 20th International scientific conference on transport means. Proceedings of the international conference. Kaunas University of Technology Press, Kaunas, Lithuania, pp 245–249

35. Convention concerning International Carriage by Rail as amended by the Vilnius Protocol (COTIF) (1999) Online. Available at: http://www.otif.org/fileadmin/user_upload/otif_verlinkte_files/04_recht/03_CR/03_CR_24_NOT/COTIF_1999_01_12_2010_e.pdf. Accessed 25 September 2017

36. Agreement on International Goods Transport by Rail—SMGS (2015) Online. Available at: http://osjd.org/doco/public/ru?STRUCTURE_ID=5038&layer_id=4581&refererLayerId= 4621&id=885&print=, Retrieved 8 October 2015. Accessed 28 August 2017
37. Abramovič B, Zitricky V, Biškup V (2016) Organisation of railway freight transport: case study CIM/SMGS between Slovakia and Ukraine. Eur Transp Res Rev 8(4):1–13
38. CIM/SMGS Consignment Note Manual (GLV-CIM/SMGS) (2015) Online. Available at: http://www.cit-rail.org/files/Documentation_EN/Freight/GLV-CIMSMGS/GLV_CIM-SMGS_ EN_2017-01-01.pdf?cid=678. Accessed 05 May 2017
39. Corridor RFC9, ŽSR (2013) Online. Available at: http://www.szdc.cz/rfc9. Accessed 15 August 2017
40. Line characteristics table, ŽSR (2015) Online. Available at: http://www.zsr.sk/slovensky/ zeleznicna-dopravna-cesta/marketing/tabulky-tratovych-pomerov.html?page_id=353. Accessed 15 August 2017
41. Line characteristics table, SŽDC (2015) Online. Available at: http://gvd.cz/cz/data/TTP/. Accessed 15 August 2017
42. Siemens. Eurosprinter ES64F4. Technical data (2003) Online. Available at: https://www. mobility.siemens.com/mobility/global/SiteCollectionDocuments/en/logistics/cargo-transport/ reference-list-locomotives-en.pdf. Accessed 12 September 2017
43. Dolinayová A, Buková B, Zitrický V (2012) Comparative analysis of parameters of intermodal trains at selected operators of combined transport in Central Europe. In: Euro–Zel 2012. 20th International Symposium. 5–6 June 2012. Žilina Slovakia, Tribun EU
44. Tatravagónka Poprad, Intermodal wagons Sggnss 80′ (2015) Online. Available at: http:// tatravagonka.sk/wagons/sggnss-80/. Accessed 12 September 2017
45. Dolinayová A, Ľoch M, Kanis J (2015) Modelling the influence of wagon technical parameters on variable costs in rail freight transport. Res Transp Econ 54:33–40
46. Tatravagónka Poprad, Covered and high-sided wagons (2013) Online. Available at: http:// tatravagonka.sk/wagons/sggnss-80/,   http://tatravagonka.sk/inc/uploads/2016/06/eanoss.pdf. Accessed 12 September 2017
47. Podberezkin A, Podberezkina O (2015) The silk road renaissance and new potential of the Russian-Chinese partnership. China Q Int Strateg Stud 1:305–323
48. Burdzik R, Ciesla M, Sladkowski A (2014) Cargo loading and unloading efficiency analysis in multimodal transport. Promet Traffic Transp 26:323–331

Transnational Value of the Republic of Kazakhstan in International Container Transportation

Aleksander Sładkowski, Zhomart Abdirassilov
and Amangeldy Molgazhdarov

Abstract Kazakhstan occupies a key position on routes of transcontinental cargo transportation along the East-West routes. The chapter deals with the containers transportation through the territory of Kazakhstan. Technical and economic aspects related to the possibility of accelerating such transport are being studied. Various types of cargoes and possible routes for this delivery are considered. Particular attention was paid to the development of communication between Chinese port terminals and railway stations in Kazakhstan.

Keywords Containers transportation · Chinese port terminals · Railway stations in Kazakhstan · Technical and economic aspects

1 Development of Transit Capacity Through the Territory of Central Asia

Kazakhstan does not have access to the sea, if do not consider the Caspian as sea. But this problem is compensated by its geographical location on the path of rapidly developing land freight traffic between Europe and Asia. This gives to the country a number of transport and logistics advantages. Kazakhstan's policy is aimed at developing modern infrastructure as the basis for the country's transit potential, as

A. Sładkowski (✉)
Department of Logistics and Aviation Technologies, Faculty of Transport,
Silesian University of Technology, Krasinskiego 8, 40-019 Katowice, Poland
e-mail: aleksander.sladkowski@polsl.pl

Z. Abdirassilov · A. Molgazhdarov
Department "Organization of Transport, Traffic and Transport Operation", Faculty "Logistics and Management", Kazakh Academy of Transport and Communications Named After M. Tynyshpaev, Shevchenko 97, 050012 Almaty, Kazakhstan
e-mail: zhomart23@mail.ru

A. Molgazhdarov
e-mail: opdet_kafedra@mail.ru

© Springer International Publishing AG, part of Springer Nature 2018
A. Sładkowski (ed.), *Transport Systems and Delivery of Cargo on East–West Routes*, Studies in Systems, Decision and Control 155,
https://doi.org/10.1007/978-3-319-78295-9_4

well as for strengthening good-neighborly partnership relation with China, one of the most powerful economic giants in the world. In alliance with other interested countries, Kazakhstan is working on the concept of the revival of the Great Silk Road [1].

Kazakhstan, Kyrgyzstan and Tajikistan are the CIS republics (Commonwealth of Independent States), which border directly with China. However, only the Republic of Kazakhstan (RK) from this list is the most "transit" country. From the source [2] it was noted that parties more often than other states conclude agreements on resource provision.

The success of Kazakhstan in the field of transit traffic was noted by the Chairman of the People's Republic of China (PRC) Xi Jinping, who described the republic as a "champion" in the transit of transcontinental transportations. The close cooperation of the countries also confirms the words of President Nursultan Nazarbayev that both states have already started coordinating the programs of Nurly Zhol and the Economic belt of the Silk Road (ESRP), jointly developing options and analyzing the technical and economic opportunities for their development [2].

For China, the issue of ESRP is strategically important, because the development of a new transport infrastructure will facilitate the promotion of goods from the less economically developed western (Xinjiang Uygur Autonomous Region, Tibet Autonomous Region, Qinghai) and the northeastern provinces (Inner Mongolia, Heilongjiang) to the economically developed east, which will overcome the imbalance in economic development internal regions of China. At the same time, the second key point in the development of ESRP will be the growth of freight traffic to the West [3].

In recent years, the Government of the Kazakhstan has paid special attention to the development of transport infrastructure, it is known that Kazakhstan is a country with a rather complex logistics system, connected with long distances and correspondingly high logistical costs. This is the problem now solved by logistics companies, reduce costs and increase the speed of delivery of goods.

A prerequisite for the realization of the country's transport and transit potential is the development of close coordination cooperation with neighboring countries on the creation and modernization of transport corridors and freedom of movement of goods, incl. elimination of administrative barriers. We should strive to improve the level of service and efficiency of managing the transport infrastructure in general [4].

President N. A. Nazarbayev in his message "Strategy" Kazakhstan-2050 "The new political course of the state" set the goal of Kazakhstan becoming one of the 30 most developed countries in the world by 2050. This implies an increase in the efficiency and scale of the activities of the Samruk-Kazyna Fund and its companies.

In this connection, the Fund has developed and is implementing the Transformation Program. One of the key objectives of the program is to increase the effectiveness of the use of labor, material and intangible resources by the Fund and its companies.

Thus, the Fund sets the following strategic efficiency factors (JSC) "Kazakhstan Temir Zholy–Freight transportation" (JSC "KTZ–GT") to the achievement of the following strategic efficiency factors:

1. Economic profit;
2. Customer satisfaction;
3. Rating of corporate governance;
4. Transport security;
5. Rating of social stability;
6. Environmental impact.

The structure of the Fund's assets includes JSC "KTZ–GT"—the largest transport and logistics holding of state significance. On the basis of JSC "KTZ–GT", a national logistics operator was formed with a full range of assets and competencies for which the seaport of Aktau, ICBS (International Center for Border Cooperation) "Khorgos" and Special Economic Zone "Khorgos–East Gate ", a network of airports. As a national logistics operator, JSC "KTZ–GT" solves the tasks of the Kazakhstan-2050 Strategy on developing the transit potential and increasing transit traffic through Kazakhstan by 2 times by 2020, and by 1050 by 2050 [4, 5].

The external and internal network of Transport and Logistics Centers (TLCs) is developing in the centers of consolidation and distribution of freight traffic. This allows the creation of new multimodal logistic schemes for the transport of goods on the Eurasian continent, along the following routes: (1) from the East Coast and the internal provinces of China to the EU countries (the northern corridor of TAR); (2) the TRACECA corridor—the countries of Central Asia, the Caspian and Black Seas (South Caucasus, Turkey) and Iran; (3) North-South-Iran, countries of the Middle East, India [5].

A third of the modern Silk Road is part of the grand transnational project "Western Europe–Western China", passes through the territory of Kazakhstan—and is an important component of the transport infrastructure. The existing Dostyk-Alashankou and Altynkol-Khorgos crossings, located on the border with China, are currently the links of the Trans-Asian land transport corridor and are used to promote the freight flows Europe-China-Southeast Asia, Central Asia-China-Southeast Asia. At these points, technological operations are carried out to switch from a wide gauge of 1520 mm to a narrow 1435 mm.

This corridor passes through China and through rail networks of Kazakhstan, Russia and other CIS countries goes to Europe. The total length of the route from the port of Lianyungang to the western European borders is about 10 thousand km, of which more than 4 thousand km pass through China, 2.783 thousand km (via the Altynkol border crossing and with the docking with Russia at Iletsk) and 3.025 thousand km (through the border crossing Dostyk and Iletsk)—through the territory of Kazakhstan.

The competitiveness of this route is due to the fact that it reduces the distance of transportation compared to the sea route, i.e. if using the sea corridor, the transit time reaches 45 days, according to Transib, 14 days, then along the corridor

"Western China–Western Europe", from the port of Lianyungan to the borders with European states, the travel time will be about 7 days [6].

Despite the slowdown in the pace of the world economy, and, correspondingly, the decline in freight turnover, container shipments in Kazakhstan show steady growth. Thus, over the past 5 years, the growth of container traffic, incl. in new transit corridors, increased many times and allowed to increase the share of revenues from transit traffic to 25%, while the total container turnover to more than 500 thousand TEU (twenty foot equivalent). Substantial growth of container transit traffic has been achieved in the direction of the PRC-EU-PRC.

According to the source [7] for 2013 only for the group of companies JSC "KTZ–GT" traffic of container trains amounted to 1653 convoys, which is 24% more than a year earlier. The total container traffic in all types of messages reached 593 thousand TEU in 2013, exceeding by 10% the indicator of 2012, the expected traffic volume in this direction in 2018 will more than double the level of 2013 and almost 40 times—the volume of 2012 of the year.

2 Container Transportation

At present, the countries of Southeast Asia have become the main world factory for the production of a wide range of products from various industries. This is due to various economic reasons, the main of which is cheap labor. According to the information in the article [8], most of the world's ships operate on container routes between the ports of Europe and Asia (Singapore, Shanghai, Hong Kong and Shenzhen are leading by a large margin).

In fact, container shipping is the most effective way of transporting goods in Eurasian transit. The container provides safety of cargo, standard sizes, reduced costs for packaging for goods, accelerated rates of loading and unloading operations, unified transport documentation and forwarding operations [9]. Analysis of the flow of goods along the EU-EEA—China axis indicates the transport of goods by land in 20- and 40-foot containers as the most promising [3].

Transportation through Kazakhstan has appreciably fallen in price in the last two years in dollars—the weakening of the tenge helped: the competitiveness increased accordingly. At present, freight transportations on land routes along the China-Europe axis are economically less efficient than offshore ones, but on the short shoulder—to Moscow, the Urals, and Kazakhstan, opportunities for their reduction in cost are seen.

According to experts, container transportation of goods with a high cost per kilogram of weight can be promising, if you put goods in the container worth $50–60 thousand [3].

To reveal the potential of overland routes, systemic efforts are needed to develop container traffic and to eliminate bottlenecks in the infrastructure of Kazakhstan.

The main goal for all interested parties is to solve internal problems of transport and logistics infrastructure, containerization of economies and optimization of industry regulation, customs administration, etc. This will lead to intensive growth of interregional cargo transportation, increase regional cohesion, improve the logistics position of regions that do not have access to the sea, as well as the whole of Central Asia [3, 10].

In order to realize the potential for container transit growth of 1.7 million TEU, the following initiatives are needed:

- Strengthening marketing and sales functions in China to establish direct relations with shippers, explaining the advantages of rail transport in a small difference in price with sea transportation and shorter delivery times;
- Ensuring a reduction in the cost of transportation for the shipper jointly with all countries participating in the transit corridor to a competitive level with maritime transport;
- Increase the coefficient of return load to the average for each level of the level through the intensification of efforts to sell to shippers in Europe;
- Optimization of transportation costs through the further introduction of a cost reduction program with increased goals to save and optimize [11] the flow distribution, taking into account the use of electrified paths and areas with the least load to reduce the requirements for increasing the capacity;
- Strengthening of positions in consolidation and deconsolidation of cargoes to increase control over flows;
- Preservation of competitive delivery terms in 7–10 days with a significant increase in the volume of transit to the target levels;
- Ensuring quality monitoring of the implementation of planned measures for the development of transit traffic in conjunction with the involved structural units and subsidiaries [5].

Shippers are beginning to use land routes more actively, reacting to changes in the price situation.

The development of the ESRP program will promote progress in the Kazakhstan economy, development of its "containerization". Currently, in the system of JSC "KTZ–GT" container transportation accounts for only 2% of the turnover and 6% of the value volumes. The unrealized potential of containerization is largely related to infrastructure constraints. Transport and logistics infrastructure has a small reserve of transit capacity. With the growth of freight traffic, its efficiency will decrease, and this will affect the preferences of shippers.

To solve this problem, it is necessary to create modern container terminals on the territory of Kazakhstan. At the same time, it is necessary to build and reconstruct railways (and to a lesser extent road roads), which will make it possible to increase the total transit capacity of Kazakhstan by 3–5 times, depending on the directions. According to expert estimates, the construction of 3–4 basic infrastructure facilities (modern container hubs) will enable Kazakhstan to increase its transit capacity by more than 2 times and reduce the cost of domestic logistics by 40% [3].

3   Accelerated Container Trains

In the sphere of world freight transportation, container transportation accounts for more than 55% of the total volume of cargo transportation, and, according to experts, this figure will increase to 70% in the near future. Statistics confirm that the most progressive technological form of organization of container transport are container trains [12].

Accelerated container trains follow to the destination without reformation, with a minimum of stops in transit. Transportation by accelerated container trains allows to reduce the time of delivery of goods, exclude the sorting and breaking of trains at marshalling yards, thereby ensuring the speed and safety of the delivery of the goods to the buyer [13].

The organization of plying fast track container trains through the territory of the Kazakhstan is a new type of container transportation for Kazakhstan, effective and promising. Its advantages over transportation by other means of transport are obvious. These are, first of all, more attractive and competitive tariffs, a significant reduction in the terms of transportation, the implementation of prompt delivery to the destination, simplified border crossing procedures, customs clearance, and a high degree of security for the transport of goods. The conclusion is obvious: the organization and development of container transportation in the republic is one of the most urgent problems related to the improvement of the transport complex in the near future [12, 13].

For Kazakhstan, the development of container transportations and their compliance with international standards can cause increased investment in the railway industry, more efficient distribution of financial and material resources of the transport industry, accelerated formation and technical development of transport communications, including those entering international corridors. Container transport can attract large transit freight traffic to the territory of the country and intensify competition between wagon and container transportations, which will maximize the transit transport potential and increase the competitiveness of the transport sector [12].

In this respect, according to experts from the source [6], the future is behind the development of traffic technology on schedule. This will make it possible to compose a real competition for road transport, especially on long routes.

Another characteristic feature of the market for container rail transportation is the formation of strategic alliances and associations. This trend is manifested in the creation of transport and logistics groups that provide a wide range of services. As a result, interstate cooperation is growing. An example of such cooperation is the project on the creation of the United Transport and Logistics Company of Kazakhstan, Russia and Belarus.

It should be noted also the accelerated development of port capacities for the processing of containers. In particular, in 2016 construction of three terminals for transshipment of general types of cargo, containers and grain in the port of Aktau was completed.

The increase in the number of container trains by the China–Europe–China communication is facilitated by the accelerated schedule of their passage through the territory of Kazakhstan [14].

Before the railroad was the task to reach a speed of 1200 km/day. For seven months of 2015 the average route speed of container trains was 1112 km/day. This is one of the best indicators on the route China–Europe.

For the accelerated and safe passage of these trains, their idle time at the stations of technical and commercial inspections, the replacement of locomotives and crews were optimized as much as possible, and time was shortened for inspecting the trains and testing the auto brakes. Thus, at Dostyk station, the duration of the container train station was reduced to 3 h 55 min, at Iletsk station—up to 2 h and 15 min, with the norm for freight trains according to the current technological process of 6 h and 3 h 30 min, respectively.

On the Dostyk–Iletsk site, for the performance of technical and technological operations, there are parking lots only at 12 stations for a period of 10 min for the replacement of locomotive crews and 25 min for locomotive change.

JSC "KTZ–GT" implements measures aimed at increasing the speed of delivery as the most complex indicator that evaluates all technological elements when the goods are shipped from the sender to the recipient [11].

At Altynkol station, the second railway checkpoint (checkpoint) with China (the first at Dostyk station) is equipped. He will ensure the passage of cargoes that are coming to Russia, Belarus, the Baltic States and further to Europe. A line will be used that will connect Kazakhstan with Turkmenistan, Iran, the Persian Gulf countries, i.e. with a direction that joins the south and east. According to experts, it will be possible to transport 15 million ton of cargo per year along the north-west corridor.

The Trans-Caspian International Transport Route (TITR), launched jointly with the administrations from Azerbaijan, Georgia and Turkey, also started. To do this, JSC "KTZ–GT" extends the capacity of the seaport of Aktau, builds capacity through the construction of a ferry terminal in the port of Kuryk in the Caspian Sea.

In June 2016, the pilot container train "Nomad Express" was launched for the first time [15], which for 6 days made its way from China through Kazakhstan to Azerbaijan. Currently, a pilot project is being worked out with Turkey, where the destination of the goods will be Istanbul. Land transportation of cargoes allows to provide reliability, stability and efficiency of their delivery by regular container trains [1, 16].

One of the most promising logistics solutions for partners is multimodal transportation based on the "Rail Air" project. This scheme provides transportation of container trains by rail from China to Kazakhstan, with further air transportation of goods from Kazakhstan to Europe. This multimodal scheme will allow the cargo to be delivered over a distance of 10,000 km in 6–7 days [1].

Since 2015, the following popular express container trains have been operating by rail from China from 3 main stations: Chongqing, Chengdu or Zhengzhou, using the accelerated container trains:

1. Chongqing–Duisburg. This is the most popular container train following the territory of Kazakhstan, which crosses the border with Russia in Iletsk, follows through Moscow, crosses the territory of Belarus and Poland, arriving at the final destination—the German city of Duisburg. The journey takes 7 days. The trains run 18 times a month.
2. Zhengzhou–Hamburg. Accelerated container trains follow the same route, only the final destination in Germany is changing. Travel time is also 7 days. Trains run 4 times a month.
3. Chengdu–Lodz. The train moves along a similar route to the final station Lodz in Poland. Travel time also reaches 7 days. Trains run 3 times a month [17].

In general, there is no doubt today that Kazakhstan every year more confidently occupies its niche in the market of trans-Eurasian container transit, actively using the whole arsenal of tools for increasing efficiency. Economic and geographical factors are the best for this [7].

For 12 months of 2016, 1939 loaded container trains were organized and transported through the territory of Kazakhstan in the export, import and transit traffic (Fig. 1), when compared with the same period in 2015 the number increased by 404 trains or 26%.

Similarly, the volume of transport in the export, import and transit traffic amounted to 191,498 TEU, in the container trains of China–European countries for 12 months of 2016 compared with the same period in 2015 increased by 2 times and amounted to 70,435 TEU.

For 12 months of 2016, 1939 loaded container trains were organized and transported through the territory of Kazakhstan in the export, import and transit traffic (Fig. 1), when compared with the same period in 2015 the number increased by 404 trains or 26%.

Similarly, the volume of transport in the export, import and transit traffic amounted to 191,498 TEU, in the container trains of China–European countries for 12 months of 2016 compared with the same period in 2015 increased by 2 times and amounted to 70,435 TEU.

From June 2016, the European container trains Chongqing–Duisburg/Herne started running through Khorgos–Altynkol. So, for 12 months of 2016 the volume

Fig. 1 Analysis of container trains for the 12 months of 2015 and 2016

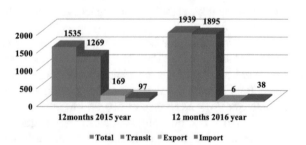

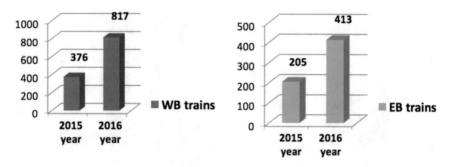

Fig. 2 Diagram of container trains for 12 months of 2015 and 2016

of transportations through this reloading point was 5216 TEUs that passed as a part of 63 container trains. Transported cargo—prefabricated (Fig. 2).

The positive dynamics of the growth of transit traffic through Khorgos is also seen in the opposite direction (Fig. 2) from the countries of Europe to China (266 TEU in the 3 trains of Duisburg–Chongqing) Table 1.

The route of the trains on the territory of Kazakhstan passes through the following stations: Iletsk–Aktobe—1–Saksaul–Kyzyl-Orda–Almaty—1–Altynkol. The length of the route is 2783 km.

Also from the middle of January 2016, the trains following the route of China–European countries–China began to ply along the new branch of Zhezkazgan-Saksaul (the total distance of the new route is 2750 km).

Table 1 Number of trains along the routes

Route	Number of trains		Dynamics, %	
	2015	2016	±	%
Chongqing–Duisburg	146	204	58	+40
Chengdu–Lodz	61	159	98	+161
Chengdu–Nuremberg	10	89	79	+790
Wuhan–Pardubice	23	11	−12	−52
Zhengzhou–Hamburg	49	47	−2	−4
Wuhan–Hamburg	62	99	37	+60
Chongqing–Altynkol–Duisburg/Herne	0	63	63	max
Total	351	672	321	+91
Total China–European countries	376	817	441	+117

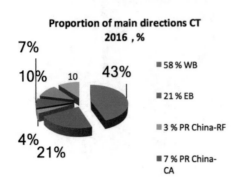

Fig. 3 Main directions of container trains

Thus, for 12 months, 795 trains passed through the new branch along the route China–European countries, along the route of European countries–China—408. On the northern branch of Astana–Iletsk passed 13 trains of China–European countries.

In 2016, cases of the disruptions in the transshipment of container trains on the route of China–European country were noticed because of the lack of fitting platforms for the Dostyk/Altynkol IPBP (international public border point). Thus, during the 12 months of 2016, 41 train delays were recorded (Fig. 3), including:

- through the IPBP Dostyk—37 trains: Yvu–Madrid—3 trains, Chongqing–Duisburg—7 trains, Changdu–Nuremberg—2 trains, Changdu–Lodz—15 trains, Wuhan–Hamburg—4 trains, Hewei–Hamburg—2 trains, Xi'an–Hamburg—1 train, Zhenchdou–Hamburg—2 train, Changdu–Tilburg—1 train;
- through IPBP Altynkol—4 trains (Chongqing–Duisburg).

A significant part of the transit (more than 30% of the total volume of transported goods) falls in the direction of China, exports average 11–12% of the total volume, imports 16–17%, transit—about 30% (Fig. 3). The increase in traffic volumes is currently due to an increase in the volume of export-import exchange between the states of Central Asia and Southeast Asia. Transportation of goods is carried out by sea from the ports of Southeast Asia to the ports of China with further access via the domestic railway network to the TransAsia railway (TAR), starting on the Pacific coast of China, in the port of Lianyungang.

During the period September–December 2016, 367 container trains (Fig. 4) passed through the territory of Kazakhstan in the transit route along the route of China–European country, of which 242 (that is 34%) with violations. The percentage of violations is shown in Fig. 5.

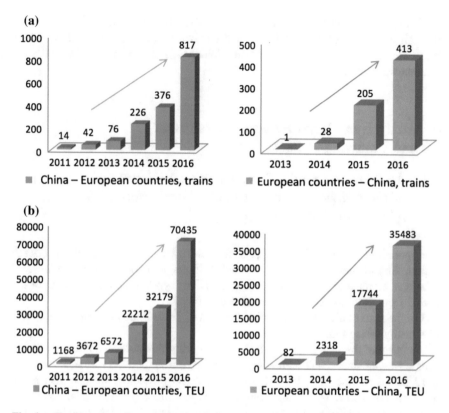

Fig. 4 **a** Total number of container trains. **b** Total number of transported containers

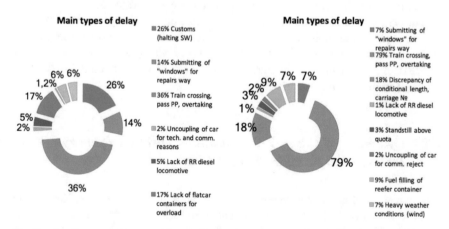

Fig. 5 Diagram of the following of container trains on the route China–European countries and European countries–China

During the period September–December 2016, 186 container trains proceeded through the territory of Kazakhstan in the transit route along the Europe–China route, out of which 87 (that is 53%) with violations. The percentage of violations is also shown in Fig. 5.

4 Analysis of Cargo Between Kazakhstan and Lianyungan Port

4.1 Analysis of Freight Transportation Volumes of the Railway Transport of the Kazakhstan in Terms of Communication Types

Transportation by rail in the Kazakhstan is carried out in interrepublican (local), export, import and transit traffic. Actual volumes of cargo transported by JSC "National Company" Kazakhstan Temir Zholy" by types of communication for the period from 2004 to 2016 are presented in Table 2.

For the sake of clarity of dynamics, the data of Table 2 are presented in the form of a histogram in Fig. 6.

Analysis of the data presented shows that during this period, traffic growth has been observed since 2002 with a slight decrease during 2009, a further increase to 2012 and a decline in 2013–2014. From the data presented, we can also see a relatively constant ratio of traffic volumes by type of message in the total volume of goods transported. Thus, the largest volume falls on local transport, then by volume —exports, imports and transit (approximately in equal shares). The share distribution of traffic for each type of message is presented in Table 3.

For clarity, this distribution is represented also on the histogram (Fig. 7).

The analysis shows that during the period under review, there is a tendency for a relatively constant ratio of volumes of local transportation (more than 50%), exports (more than 30%), with a slight increase in the share of transit from 3.4 to 5.8% of total rail traffic, the rest is imported.

4.2 Dynamics of Cargo Transportation Through Railways Between the Kazakhstan and China

The dynamics of the actual volumes of transportation by rail between Kazakhstan and China (period 2010–2015) are presented in Table 4 and for clarity in Fig. 8.

From 2010 to early 2014, traffic growth is observed at 28.1%. Since 2014—a decrease of 21.2%. In 2015, it is increased the volume of transportation to 16,355 thousand ton. In Fig. 9 the dynamics of the volume of traffic through border crossings between Kazakhstan and China with a division by transitions, thousands

Table 2 Dynamics of cargo transportation by JSC "NC" KTZ " by types of communication (period 2002–2014)

Type of conveyance	Years												
	2002	2003	2004	2005	2006	2007	2008	2009	2010	2011	2012	2013	2014
	Million ton												
Local	99.4	113.6	122.5	129.8	135.0	140.3	140.0	131.3	140.9	148.7	158.8	156.3	154.2
Export	65.2	72.3	72.2	69.0	83.8	84.8	93.4	85.7	96.0	97.9	100.9	101.6	88.6
Import	7.9	10.1	12.7	15.0	17.7	22.3	20.1	15.9	16.9	17.9	18.8	19.7	16.5
Transit	6.1	6.7	8.0	8.9	10.3	13.2	15.5	14.8	14.0	15.1	16.3	16.0	16.0
Total	178.7	202.7	215.5	222.7	246.9	260.5	268.9	247.7	267.7	279.6	294.7	293.6	275.3

Source Analysis of operational activities of JSC "NC" KTZ " (2002–2014)

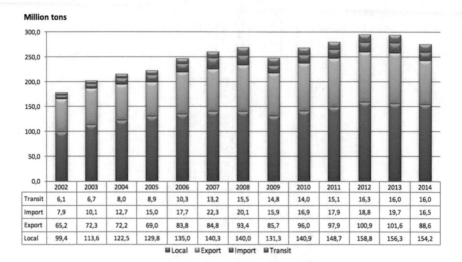

Fig. 6 Dynamics of cargo transportation volumes of JSC "NC" KTZ " by types of communications (period 2002–2014)

Table 3 Shared distribution of volumes of transported cargoes of JSC "NC" KTZ " by types of communications (period 2002–2014)

Type of conveyance	Years												
	2002	2003	2004	2005	2006	2007	2008	2009	2010	2011	2012	2013	2014
	Share, %												
Local	55.6	56.1	56.9	58.3	54.7	53.8	52.1	53.0	52.6	53.2	53.9	53.2	56.0
Export	36.5	35.7	33.5	31.0	33.9	32.5	34.7	34.6	35.9	35.0	34.2	34.6	32.2
Import	4.4	5.0	5.9	6.7	7.2	8.6	7.5	6.4	6.3	6.4	6.4	6.7	6.0
Transit	3.4	3.3	3.7	4.0	4.2	5.1	5.7	6.0	5.2	5.4	5.5	5.4	5.8

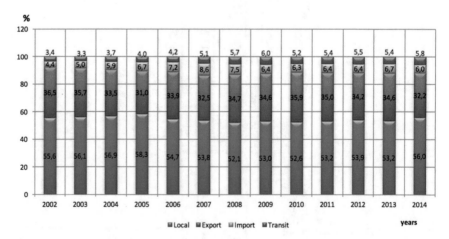

Fig. 7 Shared distribution of transported cargoes volumes of JSC "KTZ" by types of communications (period 2002–2014)

Table 4 Dynamics of traffic volumes through border crossings between Kazakhstan and China, thousands of ton (period 2010–2015)

Name of railway crossing	Direction of transportation	2010	2011	2012	2013	2014	2015
Dostyk–Alashankou	In the direction from RK to PRC	10,151	9651	10,543	12,265	8392	8023
	In the direction from PRC to RK	5004	5593	6044	5486	4870	5730
Total		15,155	15,244	16,587	17,751	13,262	13,753
Altynkol–Khorgos	In the direction from RK to PRC	0	0	0	68	49	332
	In the direction from PRC to RK	0	0	0	1588	1987	2270
Total		0	0	0	1656	2036	2602
Total for transitions	In the direction from RK to PRC	10,151	9651	10,543	12,333	8441	8355
	In the direction from PRC to RK	5004	5593	6044	7074	6857	8000
TOTAL		**15,155**	**15,244**	**16,587**	**19,407**	**15,298**	**16,355**

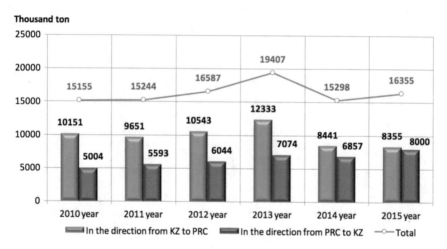

Fig. 8 Dynamics of the volume of transportation through border crossings between Kazakhstan and China, thousands of ton (period of 2010–2015)

of ton (the period of 2010–2015) is presented and the dynamics of the share distribution between the transitions in Fig. 9.

It can be seen that the decrease in traffic through the transits in 2014 was affected by the decrease in traffic through Dostyk–Alashankou, while an increase is observed across the Altynkol–Khorgos crossing (Fig. 9).

As can be seen (Fig. 10), before the beginning of 2013, 100% of the volume of transportation was transported through the Dostyk–Alashankou crossing. With the opening in 2012 and the operation of the Altynkol–Khorgos transition, starting in 2013, the redistribution of traffic volumes begins. A noticeable increase in the share of shipments to Altynkol–Khorgos: in 2015 the share of this transfer is 15.9%.

thousands tons

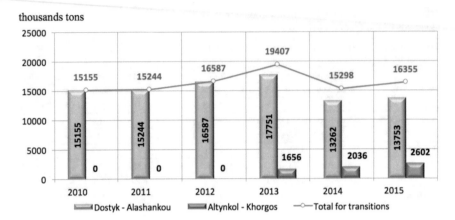

Fig. 9 Dynamics of traffic volumes through border crossings between Kazakhstan and China with a breakdown by transitions, thousands of ton (period 2010–2015)

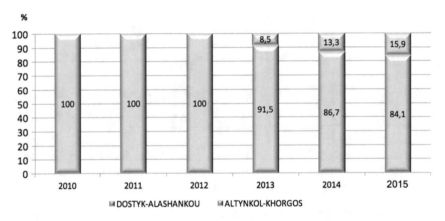

Fig. 10 Shared distribution of traffic between the border crossings of Kazakhstan and China (period 2010–2015)

The distribution of the total volume of traffic through the railway crossings of Kazakhstan and China according to the types of messages for the period 2010–2015 is presented in Table 5, in a graphical form in Fig. 11, with the separation according to the transitions in Tables 6 and 7.

During the period under review, it can be seen that a large volume of traffic with changes in values is accounted for by transportation in the export direction; there is an increase in imports and transit from China; the transit in the direction of China in terms of volumes is relatively uniform.

The distribution of freight traffic through the Dostyk–Alashankou railway border crossing in the context of the Southeast Asian countries, Japan, Korea and China in the example of 2011 and 2012 is presented in Table 8.

Table 5 Dynamics of freight flows through railway crossings of Kazakhstan and China according to the types of messages for the period 2007–2015, thousand ton

Type of conveyance	2007	2008	2009	2010	2011	2012	2013	2014	2015
Exports from the Kazakhstan	6234	5893	9963	9279	8716	9723	11,499	7811	7155
Import to the Kazakhstan	2736	2906	1977	2263	2677	2995	3298	2856	3900
Transit to China	766	542	824	872	935	820	834	630	1200
Transit from China	2309	3272	2747	2741	2916	3049	3776	4000	4100
Transit total	3075	3814	3571	3613	3851	3869	4610	4630	5300
Total	12,045	12,613	15,511	15,155	15,244	16,587	19,407	15,297	16,355

thousands tons

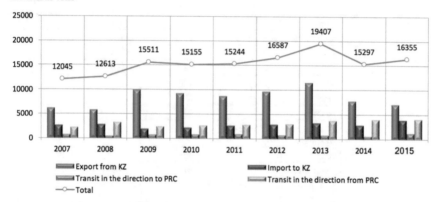

Fig. 11 Dynamics of freight traffic through railway crossings of Kazakhstan and China according to the types of messages for the period 2007–2015, thousand ton

Table 6 Dynamics of freight traffic through the Dostyk–Alashankou railway crossing by types of messages for the period 2010–2015, thousand ton

Type of conveyance	2010	2011	2012	2013	2014	2015
Exports from the Kazakhstan	9279	8716	9723	11465.0	7785.0	7095
Import to the Kazakhstan	2263	2677	2995	2686.0	2252.0	3330
Transit to China	872	935	820	800.0	608.0	928
Transit from China	2741	2916	3049	2800.0	2617.0	2400
Total	15,155	15,244	16,587	17,751	13,262	13,753

Table 7 Dynamics of freight traffic through the Altynkol–Khorgos railway crossing by types of messages for the period 2013–2015, thousand ton

Type of conveyance	2010	2011	2012	2013	2014	2015
Exports from the Kazakhstan	–	–	–	34.0	26.0	60
Import to the Kazakhstan	–	–	–	612.0	604.0	570
Transit to China	–	–	–	34.0	23.0	272
Transit from China	–	–	–	976.0	1382.0	1700
Total	–	–	–	1656.0	2035.0	2602.0

Table 8 Distribution of freight traffic from/to the countries of Southeast Asia, Japan, Korea and China through the railway border crossing Dostyk–Alashankou in 2011 and 2012

Type of conveyance	Country	2011	2012	Share in 2012
		Thousand ton	Thousand ton	%
Export	China	7509	8377	86.2
	Southeast Asia, Japan, Korea	1207	1346	13.8
	Total	8716	9723	100
Import	China	1537	1720	57.4
	Southeast Asia, Japan, Korea	1140	1275	42.6
	Total	2677	2995	100
Transit	To China	930	817	21.1
	To the countries of Southeast Asia, Japan, Korea	5	3	0.1
	Total	935	820	21.2
	from China	2913	3039	78.5
	from the countries of Southeast Asia, Japan, Korea	3	10	0.3
	Total	2916	3049	78.8
	Total transit	3851	3869	100
TOTAL		**15,244**	**16,587**	

Source JSC "KTZ"

Analysis of the data in Table 8 showed that in 2012 the share of freight traffic to/from Southeast Asia in the total volume of cargo traffic through the Dostyk–Alashankou transition was 15.9%. Accordingly, the share of freight traffic from/to the PRC was 84.1%.

China's share in exports is 86.2% of the total exports through Dostyk–Alashankou. Accordingly, the share of Southeast Asia and other countries was 13.8%. The share of China in imports amounted to 57.4% of the total volume of imports, imported through Dostyk–Alashankou. Accordingly, the remaining countries account for 42.6%.

Transit through Kazakhstan towards China was 21.1% of the total volume of transit through Dostyk–Alashankou. The remaining countries in the direction of the West–East account for 0.1%. A similar distribution of transit in the direction of East–West (from the PRC to Kazakhstan): the share of freight traffic from the PRC was 78.5%, from other countries—0.3% of the total volume of transit through the Dostyk–Alashankou transition. The percentage of total cargo traffic between destination/departure countries will be used in further calculations.

4.3 The Structure of Transported Goods Through Railway Crossings Between Kazakhstan and China

The structure of freight flows for the year 2014 through railroad crossings by kind of cargo in the proportion ratio in the context of modes of transportation is shown in Figs. 12, 13, 14 and 15.

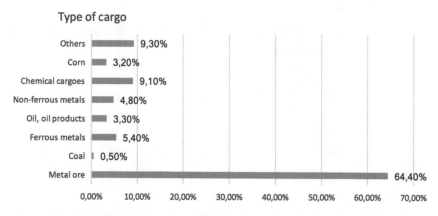

Fig. 12 Structure of cargo in the export direction (from Kazakhstan to China), 2014

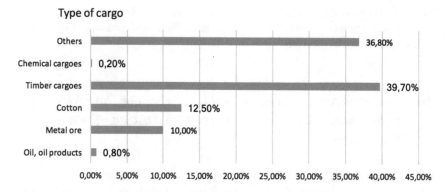

Fig. 13 The structure of goods in the transit direction (through RK towards China), 2014

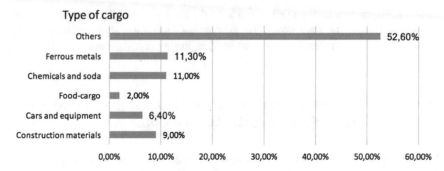

Fig. 14 Structure of cargoes in the import direction (from China in Kazakhstan), 2014

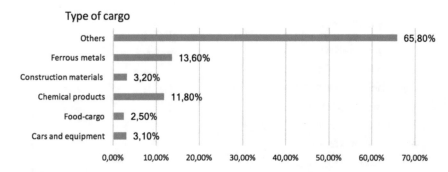

Fig. 15 Structure of cargoes in transit (from China through Kazakhstan), 2014

As can be seen from the histogram, in exports, a large proportion accounted for metal ore—64.4%, about 40% of cargo in transit to China falls on forest products, 12.5—cotton, about 37% for other cargoes, in imports to the Kazakhstan from China, other goods predominate—a total of 52.6%, more than 10%—ferrous metals and chemical products and in the direction of China through the Kazakhstan, other cargoes predominate—about 66% in total, more than 10% accounted for ferrous metals and chemical products.

4.4 The Structure of Freight Traffic to/from the Countries of Southeast Asia, Japan, Korea and Australia from/to Kazakhstan

Below, in the example of 2012, the structure of export and import freight flows of Kazakhstan—the countries of SEA, Japan, Korea and Australia by kinds of cargoes

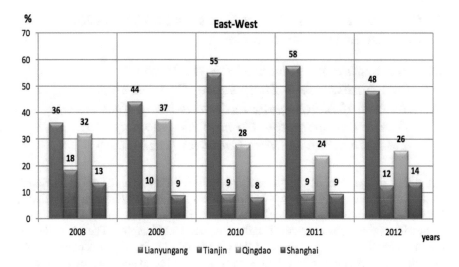

Fig. 16 Dynamics of the distribution of cargo traffic in the East-West direction between the four ports of China (period 2008–2012)

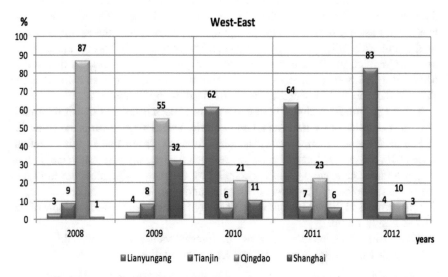

Fig. 17 Dynamics of the distribution of cargo traffic in the direction of the West-East between the four ports of China (period 2008–2012)

(Figs. 16 and 17) with the nomenclature of goods in each of the specified types of messages is presented (source—Customs Control Committee under the Ministry of Finance of the Kazakhstan).

5 Cargo Forecast for Project Years (2020, 2025, 2030) Between Kazakhstan and China

5.1 Forecasting the Volume of Freight Traffic from Kazakhstan to Lianyungang Port, Through the Port of Lianyungang by Destination in Kazakhstan, Transit Transportation Through Lianyungang Port and Kazakhstan to Other Countries (2015–2030)

According to the data of the "The Third Railway Survey and Design Institute" Group Corporation LLC, the whole flow of cargoes to/from/through Kazakhstan from/to the countries of Southeast Asia, Japan, Korea and Australia is distributed among four ports: Lianyungang, Tianjin, Qingdao and Shanghai. The Lianyungang port accounts for the largest share of the total freight traffic, which is explained by the relatively unloaded infrastructure, as well as by the availability of tax preferences from the government of the China, designed to attract cargo traffic to the developing region.

This freight traffic includes both imports to Kazakhstan and transit through it to the countries of Central Asia, the CIS countries and Europe. The analysis shows that in the East–West direction the Lianyungang port occupied the leading position on the admission of freight traffic every year. The second place was occupied by the port of Qingdao, the third and fourth with a slight difference between themselves occupied by the ports of Shanghai and Tianjin. In general, there is an increase in the share of the port of Lianyungang in the traffic flow in this direction.

This freight traffic includes both exports from Kazakhstan and transit through it from the countries of Central Asia, CIS countries and Europe towards the countries of Southeast Asia, Japan, Korea, Australia. Analysis shows that in this direction, since 2010, the port of Lianyungang has retained the leading position on the flow of freight traffic. The second place is occupied by the port of Qingdao, the third and fourth are occupied by the ports of Shanghai and Tianjin. In this direction, there is also an increase in the share of the port of Lianyungang from the total volume of the missed freight traffic.

Thus, using the available raw data, the forecast values are calculated.

The following initial parameters are set for calculation:

1. The share of the port of Lianyungang in the export and import flow in 2012 is 60%, from 2015—100%;
2. The share of the port of Lianyungang in the transit flow to both directions— 100% for the entire period;
3. The share of containerization of export and import flows in 2012 is 72%, 2015–2030—100%;

4. The share of containerization of transit to 2015–2030—100%;
5. The average weight of TEU in the export message is 10 ton (according to the analysis of JSC "KTZ");
6. The average weight of TEU in import and transit traffic is 7 ton (according to the analysis of JSC "KTZ").

Forecasting of transportation through the port of Lianyungang is considered.

Forecasting of freight turnover for the project years was carried out according to the following forecast options:

1. Based on the traditional freight traffic through the railway. Transfers from the PRC through the port of Lianyungang (export, import, transit);
2. On the basis of the traditional freight traffic and the reorientation of freight traffic from the Russian ports to the railway. Transfers from the PRC through the port of Lianyungang for each type of communication (export, import, transit);
3. Based on the traditional freight traffic, the reorientation of freight traffic from the Russian ports, as well as the reorientation of freight traffic from the Chinese ports of Tianjin, Qingdao and Shanghai to the port of Lianyungang.
4. Based on the above-mentioned traffic flows and the attraction of domestic freight flows from the provinces of China, adjacent to the port of Lianyungang.

Below in the tabular (Tables 9, 10, 11 and 12), the forecasted total volumes of freight traffic for each variant of the forecast and with the division by types of messages are presented.

In the graphics (Figs. 18, 19, 20, 21 and 22), the total volumes of cargo transportation in containers for each of the forecast variants are presented.

Graphically, the structure of the export forecast of the total volume of freight flow for the project years from Kazakhstan is presented in Figs. 23, 24, 25 and 26.

Table 9 Forecast volumes of freight traffic through the Lianyungang port (Traditional cargo traffic)

	Units	2012	2013	2014	2015	2020	2025	2030
Import	thous. ton	4.2	4.3	4.2	4.0	107.2	270.9	678.3
	thous. TEU	0.4	0.4	0.6	0.4	15.3	38.7	96.9
Export	thous. ton	399.1	430.9	434.8	710.2	732.2	1740.0	3597.0
	thous. TEU	39.9	43.1	43.5	71.0	73.2	174.0	359.7
Transit from Southeast Asia	thous. ton	0.7	0.1	1.5	255.1	326.9	682.1	1254.2
	thous. TEU	0.1	0.0	0.2	36.44	46.71	97.45	179.17
Transit to Southeast Asia	thous. ton	0.0	21.3	0.7	21.4	13.1	34.1	68.0
	thous. TEU	0.1	3.1	0.3	39.5	48.6	102.3	188.9
Total	thous. ton	404.0	456.7	441.2	990.7	1179.4	2727.2	5597.4
	thous. TEU	40	47	47	147	184	412	825

Table 10 Forecast volumes of freight traffic through the Lianyungang port (Traditional cargo traffic taking into account the reorientation of freight traffic from Russian ports)

	Units	2012	2013	2014	2015	2020	2025	2030
Import	thous. ton	12.9	11.7	7.7	4.0	110.0	273.7	681.1
	thous. TEU	1.3	1.2	0.8	0.4	11.0	27.4	68.1
Export	thous. ton	458.1	438.4	454.2	710.2	1196.4	2204.2	4061.2
	thous. TEU	45.8	43.8	45.4	71.0	119.6	220.4	406.1
Transit from Southeast Asia	thous. ton	150.0	181.4	261.3	255.1	581.8	937.0	1509.0
	thous. TEU	21.4	25.9	37.3	36.4	83.1	133.9	215.6
Transit to Southeast Asia	thous. ton	73.0	68.7	33.0	21.4	34.5	55.5	89.4
	thous. TEU	31.9	35.7	42.0	39.5	88.0	141.8	228.3
Total	thous. ton	693.9	700.1	756.1	991	1922.6	3470.5	6340.7
	thous. TEU	100	107	125	147	302	523	918

Table 11 Projected volumes of freight traffic through the Lianyungang port (Traditional cargo traffic taking into account the reorientation of freight traffic from Russian ports and freight traffic from the Chinese ports of Tianjin. Qingdao and Shanghai)

	Units	2012	2013	2014	2015	2020	2025	2030
Import	thous. ton	12.9	11.7	7.7	4.0	110.0	273.7	681.1
	thous. TEU	1.3	1.2	0.8	0.4	11.0	27.4	68.1
Export	thous. ton	458.1	438.4	454.2	710.2	1196.4	2204.2	4061.2
	thous. TEU	45.8	43.8	45.4	71.0	119.6	220.4	406.1
Transit from Southeast Asia	thous. ton	150.0	181.4	261.3	255.1	581.8	937.0	1509
	thous. TEU	21.4	25.9	37.3	36.4	83.1	133.9	215.6
Transit to Southeast Asia	thous. ton	73.0	68.7	33.0	21.4	34.5	55.5	89.4
	thous. TEU	31.9	35.7	42.0	39.5	88.0	141.8	228.3
Freight from China (Tianjin, Qingdao, Shanghai)	thous. ton	737.4				1736.6	2796.8	4504.3
	thous. TEU	54.6				128.6	207.2	333.7
Total	thous. ton	1431.4	700.1	756.1	991	3659.2	6267.3	10845
	thous. TEU	155	107	125	147	430	731	1252

Table 12 Projected volumes of freight traffic through the Lianyungang port to/from "domestic" China

	Units	2020	2025	2030
Freight traffic in China	Thousand ton	1200	3690	9480
	Thousand TEU	120	369	948
	Thousand TEU	120	369	948

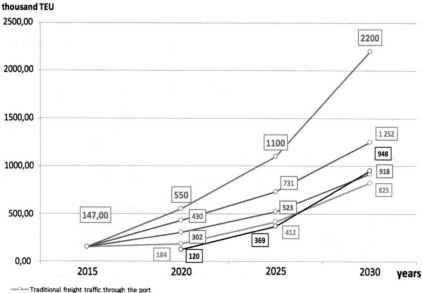

thousand TEU

—○— Traditional freight traffic through the port

—○— Freight traffic (traditional taking into account the reorientation from the ports of the RF)

—○— Freight traffic (traditional taking into account the reorientation from the ports of the RF and from chinese ports (Tianjin, Qingdao, Shanghai))

—○— Total freight traffic (traditional taking into account the reorientation from the ports of the RF and from chinese ports (Tianjin, Qingdao, Shanghai), and also from/to "domestic" China)

—○— Freight traffics via SCO LZ from/to "domestic" China)

Fig. 18 Projected volumes of freight transit through the Lianyungang port

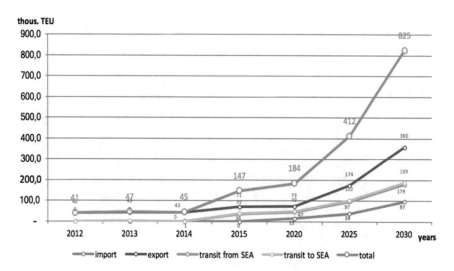

thous. TEU

—○— import   —○— export   —○— transit from SEA   —○— transit to SEA   —○— total

Fig. 19 Projected volumes of freight traffic through the Lianyungang port (Traditional cargo traffic)

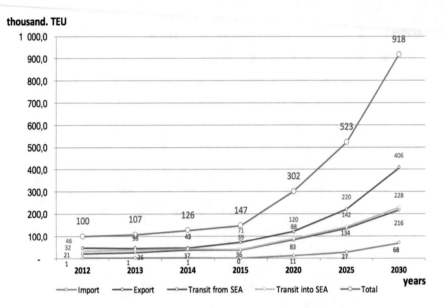

Fig. 20 Projected volumes of freight traffic through the Lianyungang port (Traditional cargo traffic taking into account the reorientation of freight traffic from Russian ports)

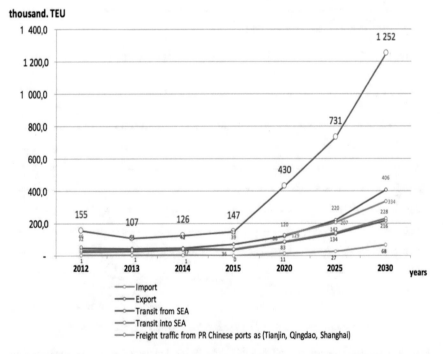

Fig. 21 Projected volumes of freight traffic through the Lianyungang port (Traditional cargo traffic taking into account the reorientation of freight traffic from Russian ports and freight traffic from the Chinese ports of Tianjin, Qingdao and Shanghai)

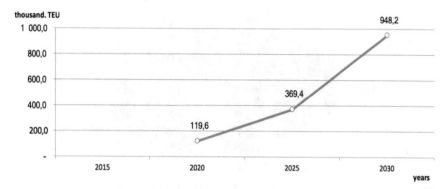

Fig. 22 Projected volumes of transportation through the SCO LZ from/to "domestic" China

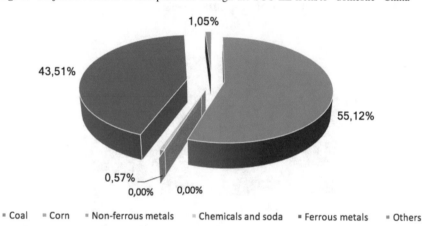

- Coal - Corn - Non-ferrous metals - Chemicals and soda - Ferrous metals - Others

Fig. 23 Structure of export freight traffic for 2015

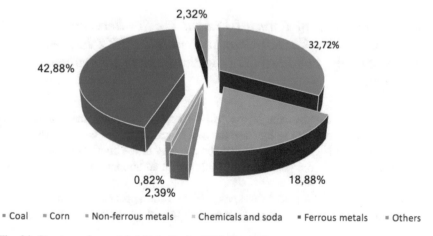

- Coal - Corn - Non-ferrous metals - Chemicals and soda - Ferrous metals - Others

Fig. 24 Structure of export freight traffic for 2020 (forecast)

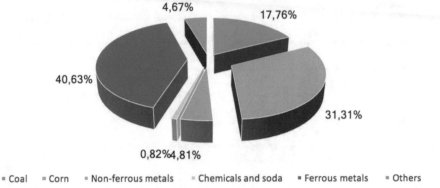

Fig. 25 Structure of export freight traffic in 2025 (forecast)

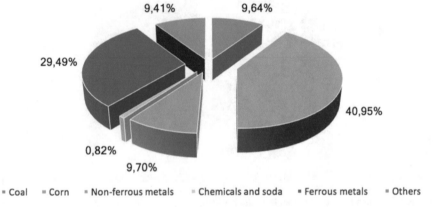

Fig. 26 Structure of export freight traffic for 2030 (forecast)

5.2 Distribution of Forecast Freight Traffics Between the Existing Kazakh-Chinese Terminal ("KTZ Express" Company) and the SCO Logistics Zone in Lianyungang Port

According to the management of the "KTZ Express" company, the container terminal is currently starting to increase the rate of growth in the volume of work. In the near future, it is planned to increase the capacity of the crane farm and the fleet of vehicles. There is the possibility of building an additional railway track to two existing routes.

Attention should be paid to the possibility of the appearance of an additional volume of container processing, in connection with the construction of the second stage of a grain (silo) terminal in the port located next to the terminal of the company.

As you know, when developing a feasibility study for the container terminal "KTZ Express", volumes of non-containerized freight traffic (electronics, chemicals, machinery and equipment, metals and metal structures, etc.) were laid. At the terminal, covered warehouses for storing, transloading and sorting cargoes (for example, ores) were designed. These objects are not currently built.

Thus, the volume of the container terminal "KTZ Express" at the present time is the transfer of containers from the Western direction and the Kazakhstan through the port of Lianyungang towards the countries of Southeast Asia, and back, storage and sorting of containers at the terminal.

But given the lack of opportunities for expanding the terminal's territory in the future due to the proximity of residential quarters and road junctions, the potential volume of the container site of "KTZ Express" is limited.

According to the company's management, the potential volume of container processing at the terminal in the near future may amount to about 300 containers per day, or about 110 thousand TEU per year.

Taking into account that automation and optimization systems at the terminal can be introduced in the future, as well as an additional increase in the capacity of the crane farm and the fleet of vehicles, the potential volume of container processing at the terminal can be about 170–210 thousand TEU for 2020, at 2025 and 2030 years—respectively, 320 and 450 thousand TEU per year.

Thus, the volume of container processing—170–210 thousand TEU (for 2020), 320 thousand TEU (for 2025) and 450 thousand TEU (for 2030) can be considered the volume of work of the container terminal of the company "KTZ Express"!

For these reasons, Table 13 shows the maximum volumes of cargo handling in containers for the terminal of "KTZ Express" and for the future container terminal of the SCO Logistics Zone.

Table 13 Distribution of maximum volumes of cargo processing in containers through the port of Lianyungang for project years (thousands of TEU per year)

Container terminals	Years		
	2020	2025	2030
	Thousand TEU		
"KTZ Express" company	210	320	450
SCO logistics zone	340	780	1750
Overall volume	550	1100	2200

6 Dry Port KTZE-Khorgos Gateway

Academic studies in the field of dry ports are actively developing now, as noted in special studies in the field of dry ports in the maritime economy and logistics. The first mention of dry ports in the academic literature dates back to the 1980s [18–20].

The initial definition of the essence of the dry port concept is "an inland ground terminal to which shipping companies write their own bills of lading for expert cargoes, assuming full responsibility for the price and condition and from which the shipping companies issue their own bills of lading for export cargoes" [21].

According to the definition, a dry port is an inland land transport and logistics terminal that has a direct link to the seaport through a specially organized infrastructure. Communication with the seaport can be realized with the help of road, rail and river transport. Thus, the dry port serves as a transshipment point, expanding the possibilities and simplifying the implementation of sea freight for participants in foreign trade activities located territorially at a considerable distance from the ports. In addition, the organization of dry ports can significantly relieve the terminals of seaports, which effectively affects the functioning of the entire transport and logistics infrastructure. Another important function of dry, as well as sea, port is to fulfill the role of a customs terminal.

Dry ports are widely used all over the world, their services are in great demand —the movement of fast track container trains that take cargo from ports to inland regions is carried out on a regular schedule. The presence of a developed network of rear terminals is a sign of the effective functioning of large seaports [19, 22].

The construction of the dry port KTZE-Khorgos Gateway strategically located on the Kazakh-Chinese border, in the territory of Kazakhstan, located in the center of the Special Economic Zone "Khorgos-East Gate", is connected with the opening of the Kazakhstan terminal in the sea port of Lianyungang. The port of Lianyungang is the largest sea harbor of China in the Yellow Sea, and is included in the group of 30 largest ports of the world with transshipment volumes of more than 200 million ton and 5 million containers a year. The port of Lianyungang is one of the twelve regional hub ports and one of the three main ports in the cluster of ports in the Yangtze River Delta. It is connected with many ports of the countries of Southeast Asia, Japan, Korea, which facilitated the opening of the terminal in the port. This terminal will allow us to develop the sphere of logistics services in Kazakhstan, consolidate freight traffic from/to/through the RK to/from the Pacific region, expand the client base. The construction and operation of its own terminal will give Kazakhstan a chance to profit as a transit country between the Pacific region and the countries of Western Europe. After carrying out all the necessary operations in the seaport, the cargo will follow further by rail in the territory of the PRC to the transshipment ports (Dostyk and Altynkol) [23, 24].

The total area of the Dry Port is 102.8 ha, it is integrated with the logistic (224.9 ha) and industrial (224.6 ha) zones. This project provides the optimal solution for container handling and opens up new opportunities for further industrial processing of imported goods. FEZ participants receive a full range of tax and customs preferences within this perimeter [25].

The operator of the Dry Port is "KTZE-Khorgos Gateway" LLP, a subsidiary of "KTZ Express", in joint management with DP World, the third largest largest port operator in the world. Experience and management of DP World Company allowed KTZE-Khorgos Gateway to be a world-class project providing services in accordance with international standards based on efficiency, safety and environmental protection [24, 25].

KTZE-Khorgos Gateway promotes the development of the country at the regional level, and also positions Kazakhstan as a key country in the formation of the New Eurasian Economy.

Within the framework of the international forum "One Belt One Road", which was launched in Beijing in May 2017, JSC "KTZ–GT" signed an investment agreement with COSCO Shipping Corporation and Lianyungang Port on the joint development of the free economic zone "Khorgos–Eastern Gate", with the purchase 49% of the share in the dry port of Khorgos Gateway. FEZ "Khorgos–Eastern Gate" will become part of the logistics services architecture provided by COSCO Shipping, and the Kazakhstan transit corridors will be integrated into the international logistics network, "the information says. COSCO Shipping is the world's largest maritime operator and provider of logistics services. The corporation has more than 1100 sea vessels with a total capacity of over 1.6 million containers, operates more than 300 international and domestic sea routes connecting 254 seaports in 79 countries of the world [26, 27].

China intends to allocate an additional 100 billion yuan ($14.5 billion) to the Silk Road Fund, as President Xi Jinping said at the opening ceremony of the One Belt One Road Forum in Beijing on May 15, 2017. The PRC chairman added that two Chinese banks will provide another 380 billion yuan ($55 billion) in loans for the development of the One Belt One Road initiative, which was established in 2014 to provide financial support for infrastructure projects in developing countries. This will allow China to create an open platform for cooperation and support the development of an open world economy [28].

In 2017, JSC "KTZ–GT" plans to increase the volume of traffic in the communication between Kazakhstan and China to 21.5 million ton, including through Dostyk—17.5 million ton, through Altynkol—4 million ton. By 2020, the volume of traffic between the two countries is planned to increase to 38 million ton.

7 Conclusion

The chapter analyzes the current state of the issue of cargo transportation along the East-West routes. The unique geographical position of Kazakhstan, which allows to be a "bridge" between the countries of the producers of various kinds of goods, first of all, China, and the countries consumers of this product, is noted. This situation makes it possible to derive significant economic benefits, however, in order to be permanent, it is necessary to develop transport infrastructure, which in turn requires a reasonable investment approach. Currently, one of the most significant brakes on the development of container transport is the price factor. Despite much shorter delivery times for containers by rail, the main traffic flow follows the sea. Thus, investments should contribute to improving the transport infrastructure, reducing the price of delivery of goods, ensuring the regularity of cargo delivery, which in general should reorient the traffic flows to shorter routes that pass through Kazakhstan.

Such an approach will contribute not only to the extraction of financial profit, but also to the development of regions tied to servicing the railway communication. Analysis of the situation shows that the leadership of Kazakhstan and regional railways understands the benefits of the situation and makes significant steps in this direction.

Bibliography

1. Главная тема: Соединяя восток и запад (2017) http://transexpress.kz/ru/magazines.php?id= 494 [In Russian: The main theme: connecting East and West]
2. Кому с КНР « по «Шелковому пути»? (2017) http://sng.today/platon/2904-komu-s-knr-po-shelkovomu-puti.html [In Russian: Who is with the PRC "on the Silk Road"?]
3. Винокуров Е (2017) Шелковый путь: успех в решении логистических проблем https:// kapital.kz/expert/58535/shelkovyj-put-uspeh-v-reshenii-logisticheskih-problem.html [In Russian: Vinokurov E (2017) Silk road: success in solving logistical problems]
4. Послание президента Республики Казахстан Нурсултана Назарбаева народу страны «Стратегия «Казахстан-2050» Новый политический курс состоявшегося государства» (2017) http://www.akorda.kz от 31 января 2017 г. [In Russian: The message of the President of the Republic of Kazakhstan Nursultan Nazarbayev to the people of the country "Strategy "Kazakhstan-2050" The new political course of the held state"]
5. Стратегия развития акционерного общества«Национальная компания«Қазақстан темір жолы»до 2025 года (2015) Утверждена решением Совета директоров АО«НК« ҚТЖ»от 26 ноября 2015 года, № 11 https://KTZ-GT.kz/upload/strategiya_razvitiya_KTZ. pdf [In Russian: Development strategy of the joint stock company "National company "Kazakhstan temir zholy" until 2025. Approved by the decision of the Board of Directors of JSC" NC "KTZh" on 26 Nov 2015]
6. Развитие транзитного потенциала http://railways.kz/ru/node/969 [In Russian: Development of transit potential]
7. Контейнерные перевозки: Ставка на транзит (2014) http://transexpress.kz/index.php? Itemid=739&catid=101&id=132:kontejnernye-perevozki-stavka-na-tranzit&option=com_ content&view=article [In Russian: Container transportation: bet on transit]
8. Сладковски А (2011) Контейнерные перевозки Запад – Восток, Восток – Запад. In: Миндур М (ed) Транспорт в товарообмене между Европой и Азией. Варшава – Радом: ИеЕ – PIB, pp 254–283 [In Russian: Sladkowski A (2011) Container shipments West–East, East–West. In: Mindur M (ed) Transport in the exchange of goods between Europe and Asia. Warsaw–Radom]
9. Паршина РИ (2004) Развитие транзитных и международных контейнерных перевозок. Экспедирование и логистика, pp 14–18 [In Russian: Parshina RI (2004) Development of transit and international container transportation. Forwarding and logistics]

10. Lee D-H, Jin JG, Chen JH (2012) Schedule template design and storage allocation for cyclically visiting feeders in container transshipment hubs. Transp Res Rec 2273:87–95
11. КТЖ добился лучших показателей по скорости контейнерных поездов на маршруте Китай - Европа – Китай (2015) http://www.inform.kz/ru/KTZ-dobilsya-luchshih-pokazateley-po-skorosti-konteynernyh-poezdov-na-marshrute-kitay-evropa-kitay_a2811543 [In Russian: KTZ has achieved the best rates for the speed of container trains on the route China-Europe-China]
12. Эффективную методику организации контейнерных поездов Казахстана предложили ученые КазАТУ им. М.Тынышпаева URL: http://www.inti.kz/ru/news/effektivnuyu-metodikuorganizacii-konteynernyh-poezdov-kazahstana-predlozhili-uchenye [In Russian: The scientists of KazATU named after M.Tynyshpaev proposed effective method of organization of container trains in Kazakhstan]
13. Ускоренные контейнерные поезда. Available at: http://swiftrus.ru/uslugi/uskorennye/ [In Russian: Accelerated container trains]
14. Wang D-Z (2013) Thoughts on development strategies of container railway international inter-model transportation. China Railway Transp Econ 2013(09):9–17. http://www.doc88.com/p-5009881121055.html
15. Транскаспийский контейнерный поезд «Nomad Express ≫ от китайского порта Ляньюнган до Стамбула за 15 дней (2016) http://www.railways.kz/ru/node/9969 [In Russian: Trans-caspian container train "Nomad express" from the Chinese port of Lianyungang to Istanbul in 15 days]
16. Chen JH, Lee D-H, Cao JX (2012) A combinatorial benders' cuts algorithm for the quayside operation problem at container terminals. Transp Res Part E Logistics Transp Rev 48(1): 266–275
17. Ускоренные контейнерные поезда из Китая (2017) http://baltteco.ru/uskorennye-poezda [In Russian: Accelerated container trains from China]
18. Roso V (2005) The dry port concept—applications in Sweden. Logistics Res Network 379–382
19. Roso V, Woxenius J, Olandersson G (2006) Organisation of Swedish dry port terminals. Meddelande 123, Division of logistics and transportation. Chalmers University of technology, Göteborg
20. Галин АВ (2014) Сухие порты как часть транспортной инфраструктуры. Направления развития. Вестник Государственного университета морского и речного флота имени адмирала С.О. Макарова 2(24):87–92 [In Russian: Galin AV (2014) Dry ports as part of the transport infrastructure. Directions of development. Bulletin of the State University of Marine and River Fleet named after Admiral S. O. Makarov]
21. Конвенция ООН по морскому праву (1982) http://www.un.org/ru/documents/decl_conv/conventions/lawsea.shtml [In Russians: United Nations Convention on the Law of the Sea]
22. Li Y, Dong Q, Sun S (2015) Dry port development in China: current status and future strategic directions. J Coast Res. Special Issue 73—Recent developments of port and ocean engineering, pp 641–646 http://www.jcronline.org/doi/abs/10.2112/SI73-111.1?code=cerf-site
23. Жардемов ББ, Куанышев БМ, Карсыбаев ЕЕ et al (2013) Строительство логистического терминала в морском порту Ляньюнган (КНР). Технико-экономическое обоснование. Т.1 Алматы: КазАТК, 230 p [In Russian: Zhardemov BB, Kuanyshev BM, Karsybayev EE et al (2013) Construction of a logistics terminal in the seaport of Lianyungang (PRC). Feasibility study]
24. Справочная информация о проекте Сухой порт «KTZE–Khorgos Gateway» (2015) http://www.railways.kz/sites/default/files/content_nc_ktz_finalrus.pdf [In Russian: Reference information about the project Dry Port «KTZE—Khorgos Gateway»]
25. Сухой порт СЭЗ «Хоргос – Восточные ворота». Available at: http://khorgosgateway.com/sections/%D0%BE_%D0%BD%D0%B0%D1%81 [In Russian: The dry port of the SEZ "Khorgos—Eastern Gate"]

26. Китайские инвесторы планируют купить 49% доли участия в сухом порту «Хоргос» (2017) https://liter.kz/ru/news/show/32488-kitaiskie_investory_planiruyut_kupit_49_doli_ uchastiya_ [In Russian: Chinese investors plan to buy a 49% stake in the dry port of Khorgos]
27. Chinese companies buy stake in dry port in Kazakhstan (2017) http://www.chinadaily.com. cn/bizchina/2017-05/15/content_29355527.htm
28. Китай выделит $14.5 млрд Фонду Шелкового пути (2017) http://www.interfax.ru/ business/562414 [In Russian: China will allocate $14.5 billion to the silk road fund]
29. Кофман А, Крюон Р (1965) Массовое обслуживание. Теория и приложения. Москва: Мир, 302 p [In Russian: Kofman A, Kruon R (1965) Mass service. Theory and applications. The World, Moscow]
30. Акулиничев ВМ, Кудрявцев ВА, Корешков АН (1981) Математические методы в эксплуатации железных дорог. Москва: Транспорт, 223 p [In Russian: Akulinichev VM, Kudryavtsev VA, Koreshkov AN (1981) Mathematical methods in the exploitation of railways. Transport, Moscow]
31. Барышникова НЮ, Зубарев ЮЯ (2009) Моделирование переходных процессов переработки контейнерных грузов. Информационные технологии и системы: Управление, экономика, транспорт, право: вып, 1 (7). Санкт-Петербург: ООО «Андреевский издательский дом » pp 7–10 [In Russian: Baryshnikova NU, Zubarev YuYa (2009) Modeling of transient processes of container cargo processing. Inf Technol Syst: Manage Econ Transp, Law: 1(7). St. Petersburg: ООО "Andreevsky Publishing House"]

Economic Aspects of Freight Transportation Along the East-West Routes Through the Transport and Logistics System of Kazakhstan

Zhanarys Raimbekov, Bakyt Syzdykbayeva
and Kunduz Sharipbekova

Abstract In modern global conditions, transport plays a key and, in some cases, a decisive role in the industrial development of countries. To achieve this goal, Kazakhstan is investing new infrastructure projects in transport infrastructure and systems. Nevertheless, some studies indicate serious problems due to lack of legislation, lack of coordination among government agencies, implementation errors and deficiencies in the sphere of implementation of infrastructure projects. The chapter explores the problems of the development of freight traffic and the state of the modern transport and logistics infrastructure on the territory of Kazakhstan. As active development of transport, together with logistics, will make significant changes in the state of transit, export and domestic transportation of goods between Europe and China through Kazakhstan. The problems are analyzed and the reasons preventing the development of the transport and logistics system (TLS) in the sphere of freight transportation along the international transport corridors passing through Kazakhstan are revealed, ways of increasing their efficiency are suggested. The ways of further development of Kazakhstan's economy have been determined, which will depend to a large extent on the ability to diversify and create modern transport infrastructure facilities, improve the functioning of the transport and logistics system along international transport corridors as a transit country between China and European countries.

Keywords Cargo transportation · Transport system · Transport infrastructure
Transport corridor · Transport and logistics system · Transit potential
Silk road · Logistics efficiency

Z. Raimbekov · B. Syzdykbayeva · K. Sharipbekova (✉)
Eurasian National University named after LN Gumilyov,
Satpayev Str. 2, 010000 Astana, Kazakhstan
e-mail: kunduz.sharipbekova@gmail.com

Z. Raimbekov
e-mail: zh.raimbekov@gmail.com

B. Syzdykbayeva
e-mail: bakyt_syzdykbaeva@mail.ru

© Springer International Publishing AG, part of Springer Nature 2018
A. Sładkowski (ed.), *Transport Systems and Delivery of Cargo
on East–West Routes*, Studies in Systems, Decision and Control 155,
https://doi.org/10.1007/978-3-319-78295-9_5

1 Introduction

The geographical position of Kazakhstan in the center of the Eurasian continent between the largest trading partners—China and Europe—provides an absolute advantage in terms of delivery of goods to become a major transport and logistics hub and develop its own transport and logistics system. The volume of foreign trade between the EU and the PRC in 2014 amounted to about $800 billion dollars. It is expected that by 2020 this figure will increase to a level of 1.2 trillion, and the volume of freight traffic in the communication between China and Europe—up to 170 million ton. At present, Kazakhstan accounts for less than 0.3% of the total volume of trade between these trading partners [1].

The transport and logistics system is the main instrument for the realization of economic ties between the regions of Kazakhstan, the realization of the transit of goods through Kazakhstan, and also the main conductor of the export of Kazakhstani goods to world markets.

The transport and logistics system of the Republic of Kazakhstan (RK) serves the domestic needs of the economy, and is the driver of the development of the TLC of the Single Economic Space that ensures the implementation of transit potential, providing a wide range of transport and logistics services at the level of world standards.

The volume of export operations of Kazakhstan by 2020 can grow by 1.5 times—from 96 to 147 million ton, which will require the TLS to handle additional freight flows to Russia, China, South Korea, Europe, and Central Asia [2]. It is also expected that the volume of trade between neighboring countries will grow by 1.5 times and reach $1 trillion by 2020, which creates the potential for transit through the RK.

At the same time, the lack of access to the sea, huge distances to the main product markets require an accelerated development of the industrial infrastructure, primarily transport.

The share of transportation costs in the cost of final products is relatively high and is at 8 and 11% respectively for domestic rail and road transport, in developed market economies this figure is 4–4.5% [1].

In this regard, one of the key roles in achieving these goals is to be allocated to an efficient transport and logistics system that should ensure not only high and efficient transport connectivity within the country, but also the necessary level of Kazakhstan's integration into the global transport and logistics network.

The urgency of developing a long-term transport and logistics strategy of Kazakhstan until 2030 is due to the need for a comprehensive review of approaches to the management of the transport complex of Kazakhstan, which implies a departure from the traditional "narrowly transport" approach and the application of the new modern paradigm of TLC [3]. Solving the problem of developing the Strategy requires using the best world experience in managing similar systems of leading countries, in particular Germany, America, China, USA, Canada, UAE, Singapore and many others [4–9].

Given the external and internal challenges and associated opportunities, in the framework of the implementation of the Strategy, Kazakhstan sets the following key goals before 2030.

To become the main logistics hub and transit country of the Eurasian region. A key focus will be on attracting transit between China and Europe (the new Silk Road), China and Russia, Europe and Central Asia; to maximize the export potential and domestic needs of the national economy; to increase mobility of the population and realize the tourist potential of the country [1, 10].

To implement these tasks, it is necessary to develop a transport system and infrastructure.

2 Development of the Transport System in Kazakhstan

At present, a modern transport system has been created and is generally stable in Kazakhstan, which is the most important component of the industrial and social infrastructure that meets the needs of the national economy and population in transport services, the territorial integrity, economic and geopolitical security of the country.

The transport system of Kazakhstan currently includes all types of modern transport, providing internal interregional communications, long-distance, local and intracity transportation of passengers and cargo, as well as export-import and international transit traffic. Favorable geographical position of the country allows Kazakhstan to receive significant revenues from the export of transportation services, including the transit transportation of foreign countries through its communications.

The extremely important role of transport in the RK is due to such factors as the vast territory of the country (2,724,900 km^2), stretching from west to east by about 3000 km, from north to south by almost 2000 km; low population density—5.5 people per 1 km^2; the population is 17.6 million; considerable distance of cargo transportation; the nature of products that require long-range transport (coal, iron ore, oil products, metallurgical products and agriculture (grain, wool, meat)); transport and geographical location of the country, through which there are significant flows of transit cargo.

The basis of the transport system is its infrastructure, which in inter-city communication is formed by railways and roads, inland waterways, pipelines, railway junctions, sea and river ports, civil aviation airports (Table 1).

The length of the country's highways is 116.3 thousand km. Of these: 96,353 kilometers (km) of highways; 15,530 km of railways; 4151 km of inland waterways; up to 61,000 km of air routes. The length of the main pipelines is: gas pipelines—12,318 km and oil pipelines—7920 km. Automobile and railway transport performs 93.4% of the total volume of cargo transportation. Nevertheless, transport networks are in poor condition, with outdated infrastructure and outdated

Table 1 Characteristics of the transport infrastructure of Kazakhstan

	2012	2013	2014	2015	2016	Growth 2016/2012, %
Operational length of public railways, km	15,333	15,341	15,341	15,341	15,530	101,3
Length of public roads, km	97,418.0	96,873.0	96,421.0	96,529.0	96,353.0	98.9
Length of public roads with hard cover, km	87,140	86,581	86,419	86,244	87,028.7	99.9
Density of roads with a hard surface of public use, kilometers per 1000 km^2	32.0	31.8	31.7	31.7	31.9	99.8
Length of inland waterways of general use, km	4150.9	4150.9	4150.9	4150.9	4150.9	100.0
Length of main pipelines, km	20,238	20,238	23,196	23,276	23,271	115.0
Mains pipes—all including:	20,238	20,238	23,196	23,276	23,271	115.0
Gas pipelines	12,318	12,318	14,895	15,265	15,256	123.8
Oil pipelines	7920	7920	8301	8011	8015	101.2
Number of airports, units	22	22	23	23	23	

Source Committee on Statistics of Kazakhstan: https://stat.gov.kz

Table 2 Density of communication routes on the territory of Kazakhstan, per 1000 km^2 of territory, km

Indicator	2012	2013	2014	2015	2016
Length of the operational length of railway lines	5.5	5.5	5.5	5.5	5.7
Length of main pipelines	7.4	7.4	8.5	8.5	8.5
The length of all exploited navigable internal public roads	1.5	1.5	1.5	1.5	1.5
Length of public roads with hard surface	32.0	31.8	31.7	31.7	31.9

Source Committee on Statistics of the Kazakhstan: https://stat.gov.kz

technology [11, p. 25]. Transportation costs account for 8–11% of the final cost of goods, while in industrialized countries this figure is 4–4.5%.

By location and structure, the network of transport communications corresponds to transport and economic relations and passenger traffic. However, the state of many infrastructural objects does not meet the requirements.

Automobile transport. According to its configuration and length, the network of public roads in the republic is mainly formed. The density of roads in Kazakhstan is 31.7 km per 1000 km^2 and is relatively low (Table 2). In a number of countries that could be compared with Kazakhstan according to their length, the density of motor roads is higher: in Russia—44 km, Canada—90.5 km, Australia—105.6 km, USA—670 km per 1000 km^2 [12].

In 2016, the volume of passenger transportation by road amounted to 21.2 billion people, passenger turnover—214.9 billion pkm. The volume of cargo transportation in 2016 increased and reached the value of 3.18 billion ton, cargo turnover—163.3 billion ton.

Transit traffic occurs mainly between the republics of Central Asia, Russia, China. The bulk of road transport transit (about 86%) is formed in China. Significantly less is the share of Central Asian republics (about 3.8%) and Russia (4.6%). The share of other countries in transit traffic is about 6.3%.

In road transport, the main problems are:

(1) poor condition of highways of regional and regional significance, as well as bridges; (2) absence of direct automobile communication between the western and northern, western and central regions of the country; (3) low level of security and deterioration of road maintenance equipment in operating organizations on the national and local road network; (4) low throughput of a number of major highways.

Railway transport. The operational length of the railways in Kazakhstan is 15.5 thousand km, including double-track lines—4.8 thousand km (34%), electrified lines—4.1 thousand km (29%).

The park of rolling stock in Kazakhstan has about 2 thousand units of locomotives and more than 90 thousand freight cars, which are not enough to cover all the need for freight. Kazakhstan owns 5.2 thousand grain carriers (half of the need). Missing grain carriers are leased in Russia.

However, in recent years, transportation volumes by other modes of transport have grown faster than the volumes transported by rail. Reducing the share of cargo transportation by rail, starting from 2001, indicates the need to improve the competitiveness of Kazakhstan's rail freight.

Analysis of Kazakhstan's availability of a network of railways in comparison with other countries of the world shows a significant backlog in the density of the network per 1000 km^2 of the territory (5.5 km per 1000 km^2). In the Czech Republic, this figure is 120 km per 1000 km^2, Belgium—117.3 km, Germany—93.8 km, Japan—53.3 km, South Korea—36.4 km, the United States—23.5 km [1].

The regions with the greatest restrictions in the railway transport infrastructure are: West Kazakhstan Region, where the density of railways is 51.5%, Kyzylorda Region—60.7%, East Kazakhstan Region—77.6%, Karaganda Region—82.4%, Aktobe Region—87.3% of the average republican level.

The prospects for the development of rail transport in Kazakhstan presuppose the formation of a ramified transport infrastructure and the construction of new high-speed roads, as well as the improvement of the status of existing roads in order to increase their speed regime.

In the field of railway transport, the main problems are: (1) significant physical and moral wear and tear of the main means of railway transport (63%); (2) shortage of fleet of rolling stock (locomotives, electric locomotives, diesel locomotives, grain trucks, wagons); (3) some sections of the railway lines pass through the territory of Russia; (4) low speed of movement of goods across the country (50 km/h); (5) high costs for the transport of goods; (6) lack of high-speed rail network; (7) weak development of container transportations. Air transport. Nowadays, in Kazakhstan there are 23 airports, 17 of which are permitted to perform flights. Since 2011 there has been an increase in passenger traffic, in 2016 6.5 million people were transported by air. The volume of cargo transportation in 2016 amounted to 18.0 thousand ton per year, which is 4.5% more than in 2015.

Inland water and sea transport. Inland water transport takes a small share in the total volume of transport. The length of the waterways of Kazakhstan, open for navigation, is 4151 km. Waterways suitable for navigation are the Irtysh, Syrdarya, Ural, or Ishim, Bukhtarminsky, Ust-Kamenogorsk, Shulbinsky, Kapchagai reservoirs, Balkhash and Zaisan lakes. Cargo transportation is mainly carried out in Pavlodar, East Kazakhstan, Karaganda regions. In 2016, 1.2 million ton of cargo were transported by sea.

Existing problems of water transport:

(1) lack of own fleet of Kazakhstan, sufficient for conducting trade; (2) lack of sufficient traffic volumes; (3) absence of Kazakhstani ship-repair bases; (4) the absence of a training system for water transport in Kazakhstan that meets international standards.

The experience of developed countries having the best transport network shows that the formation of a competitive transport infrastructure should develop with the introduction of modern innovative materials and technologies.

In railway transport, it is necessary to form a ramified transport infrastructure and build new high-speed roads, as well as improve the condition of existing roads in order to increase their speed limits. At the same time, it is expedient to develop container transportation of cargoes at an accelerating pace, and to update the loading and unloading equipment.

Priority in the automotive industry of Kazakhstan should be given to the development of a network of international transport corridors, construction and reconstruction of roads of the republican, regional and regional importance.

World practice confirms that water transport is traditionally valued as the least energy-intensive, therefore the cheapest, allowing to reduce significantly the transport component, as well as having the least environmental burden on nature. The development of the infrastructure of waterways and river ports serves to provide transportation along international transport corridors. Therefore, in the field of water transport, first of all, the reconstruction of river ports and the reform of port activity are required by creating specialized port facilities for developing new types of cargo flows [11].

The place and significance of transport in the economy of the RK is also shown by its significant share in the country's fixed production assets (8% in 2016), the share of transport services in the gross domestic product (in 2016—7.1%), investments in (in 2008—10.4%), in the number of employed workers (201.1 thousand people employed in the sector—2.4% of the total number of employed in the economy), as well as in energy consumption, metal and a number of other important indicators that characterize the country's economy.

The scale of the transport complex is characterized by such data for 2016 [12]:

- transport complex unites about 619.5 thousand employees (including 91% of permanent workers, 75%—aged 25–54 years) or 7.2% of economically active population;
- the cost of the fixed assets of the complex is 11% of the value of the country's funds;
- the share of transport in GDP is 7.1%.

An important indicator characterizing the work of the transport complex is the transport capacity of the gross domestic product as the ratio of the gross added value of transport services to GDP—12% in 2015.

The volume of cargo transportation by all types of transport amounted to 3.7 billion ton in 2016, passenger transportations—22.3 billion pers. The main share of cargo and passenger transportation falls on road transport. Such important indicators as the share of investments in transport development also have a tendency to increase—15.0% of the total volume of investments in fixed assets in 2016.

The growth of investments in relation to the previous year in the transport sector made from 2012–2016 on the average 36.7%. The volume of investments, directed in general in the transport and warehousing sector, grew in 2016 in comparison with 2012 on 195.7%. Experts estimate that investments in the industry should be at least not lower than the average level for industry, and for sustainable development, investment is required at a faster rate.

The transport infrastructure of Kazakhstan consists of five main types of transport: rail; automotive; pipeline; river and sea; aviation.

On the national scale, the first three modes of transport have the greatest weight in the aggregate volume of cargo transportation. They account for the bulk of freight both in tonnage and in value terms. Naval vessels are used mainly for transport services of international trade, and more recently also for the transport of goods between foreign ports. According to location and structure, the network of transport communications corresponds to transport and economic ties and flows of the passengers. However, the conditions of the majority of infrastructural objects don't meet the required demands. The transport system of Kazakhstan as of the end of 2016 forms 1817 large and small organizations related to various modes of transport (11—rail, 346—bus, 1125—freight automobile, 27—air, etc.) and forms of ownership (state ownership—8, republican—1, municipal—7, private—1106), specializing in the performance of certain types of transport and other transport services.

Ensuring the well-coordinated work of all these organizations and entrepreneurs in carrying out transportation is one of the most important conditions for the effective functioning of the country's transport system. It includes rational distribution of transport between modes of transport in accordance with the spheres of their preferential application; ensuring maximum coherence of actions of organizations of different modes of transport when carrying out transportation in a mixed message.

The main principles of the functioning of the transport system of Kazakhstan in the conditions of market relations are the transfer of activities of all transport organizations to a commercial basis, the development of healthy competition and the formation of a transport services market in which each user can freely choose the most suitable form of transport services, its price and quality. These principles are enshrined in the legal and regulatory framework, the laws of each mode of transport [13]. At the same time, the regulatory role of the state in carrying out transport activities is envisaged.

At the present stage of its development, the transport complex of the republic is characterized by unsatisfactory state of fixed assets, obsolete and underdeveloped infrastructure and technologies [3, 11].

The formation of international transport corridors, the development of a state transport policy, and the development of rail, road and civil aviation are the subject of consideration at the meetings of the Government of Kazakhstan. Solutions have also been adopted to ensure the integration of the management of the transport complex.

All this creates the prerequisites for the formation of a unified transport system in which the transport of goods and passengers will be optimally distributed among different modes of transport, and each of them will be able, mutually interacting with others, to realize its advantages to the maximum.

At present, eight main factors determine the future development of the transport sector of the country [1]:

- institutional transformation;
- development of transport infrastructure;
- creation of a real market of transport services and competition;
- availability of means of legal regulation;
- rationality of investment policy;
- the degree of integration into the world transport system and the use of transit potential;
- the effectiveness of ongoing measures to protect the environment and ensure the safety of transportation of goods and passengers;
- provision with modern technologies and communication and information means.

Formation of the transport and communication infrastructure that provides for the country's domestic needs, involves the following tasks. First of all, the first task is the creation of an effective network of transport communications, updating and modernization of the fleet of vehicles.

The second task is connected with the development of transport logistics and the improvement of infrastructure in key points of direct interaction of various modes of transport.

The third task is to develop and implement modern mechanisms of state regulation of the private transportation market. The process of introducing new developments requires creative and purposeful work to ensure the quality of the products produced, create the necessary conditions for maintaining technical facilities and technologies in an efficient state, monitoring their use and maintenance.

The main priorities are: improving the system of state regulation; formation and development of modern national transport infrastructure; acceleration of integration processes in the world transport system and development of transit potential; ensuring national security; transport safety and environmental protection; technological and technical renewal of the industry.

Nowadays, the industry has the following problems [3, 14].

The implementation of the transit and export potential of Kazakhstan, as well as the maintenance of economic growth, require the country's transport and logistics system of high integration into key international transport corridors, including to influence the distribution of cargo flows; high speed, timeliness, accessibility and reliability of transportation; convenience of transport services. At the same time, the current state of the transport and logistics system does not allow it to meet any of the requirements. The key problems of freight transport can be divided into the following groups: infrastructure restrictions and rolling stock shortage; absence of systemic management of corridors; low level of logistic service; institutional constraints in the management system; lack of competences and modern technologies.

3 The Development Potential of International Transport Corridors Passing Through Kazakhstan

The strategic goals of Kazakhstan in the development of the transport and logistics system are to become a powerful transport and logistics hub in the Eurasian space, to realize the potential of transit and export, to increase the mobility of the population and to realize the country's tourism potential.

Objective prerequisites create conditions for the aggregate growth of export and transit cargo traffic for Kazakhstan by 1.7 times by 2020.

To achieve these goals, a large-scale development of the network of transport and logistics centers, customs infrastructure and processes, the elimination of non-physical restrictions, the creation of a network of transport and transfer nodes and high-speed corridors of passenger transport.

The concept "One belt–one way" opens the possibility of a consistent integration of the Eurasian Economic Union (EAEU) and the project "New Silk Road". This is the reconstruction of the historical transport corridor for political coordination, interconnection of infrastructure, smooth trade, and free movement of capital and strengthening of closeness between peoples, which corresponds to global trends of globalization [15].

The main advantages and potential of the New Silk Road is its universality. This new network of updated transport routes and new trade hubs will be laid between China and Europe.

The basis for the implementation is the creation of a transport and logistics system based on the 5"S" principles: *"speed, service, security, cost and stability"* announced by the President of the Republic of Kazakhstan on the launch of a large-scale project "New Silk Road".

The development of the main transcontinental routes linking Europe and Asia, become super-tasks in the implementation of the transit potential of Kazakhstan.

Kazakhstan plays a key role in this project. For Kazakhstan, the development of the main transcontinental routes linking Europe and Asia, become super-tasks in realizing the transit potential of Kazakhstan. Several main transcontinental transport routes linking Europe and Asia pass through the RK [10]:

(1) the East-West corridor passing through the territory of Kazakhstan connects the west of China with Russia and Europe (the distance is about 3 thousand km). Cargo: machinery and equipment, chemicals and soda, metal products, building materials, fertilizers, rolled metal, non-ferrous metals, with road and rail.

(2) the TRACECA corridor passing through the territory of Kazakhstan connects the west of China with Southern Europe (TRACECA route). The distance is 5–5.5 thousand km. Cargo: oil and oil products, ore, metals, building materials, chemical products, cotton, consumer goods, with auto, rail and sea transport.

(3) the China-Russia corridor connects the west of China with the Russian Black Sea ports through the RK and gives Kazakhstan the opportunity to enter the open sea (route China-Aksaray-Novorossiysk). The distance is 4700 km. Cargo: grain, building materials, fertilizers, rolled metal, non-ferrous metals, consumer goods with auto and rail.

(4) the China-CAR corridor passes through the south of Kazakhstan, connecting the regions of Western China with Central Asia (Route China-CAR). The distances are 1500–1800 km. Cargo: agricultural products, consumer goods, metals, ore, electrical equipment with auto and rail transport.

(5) the North-South corridor passes through the territory of Kazakhstan from north to south, connecting Northern Europe and Russia with Iran and India (North-South Route). The distance is about 1400 km to the port of Aktau and further through the sea.

(6) the Russia-CAR corridor passes through the territory of Kazakhstan from the north to the south, connecting the regions of the Russian Federation and the Baltic countries with Central Asia (the Russia-CAR route). The distance is 2100 km. Cargo: agricultural products, consumer goods, minerals, cotton, metals, ores, electrical engineering.

Neighboring countries develop land and sea routes bypassing Kazakhstan: through the Suez Canal; North-South, through Azerbaijan; on the Kyrgyz-Chinese Railway; The Trans-Siberian Railway.

The most optimal land routes from China to Europe are routes along the Trans-Asian Railway Route (TARR). Air, railway and automobile transit transportations from China to Europe via the RK are inferior in value to sea transport, but they exceed them in speed (Table 3).

The most promising mode of transport from the point of view of the development of transit freight flows from China to Europe for Korea is rail transport, since

Table 3 Assessment of the various routes from China (Shanghai) to Lithuania (Klaipeda) to 5S

No.	Name of the corridor	Speed, days		Cost, USD		Service	Preservation	Stability
		Current value	Target value[a]	Current value	Target value[a]	Current value (%)	Current value	Current value
1	International transit corridor (ITC) "Western China-Western Europe"	19	11	12,000	11,000	20	7	Low
2	Trans-Asian Railway (Northern Corridor)	16	10	10,000	10,000	20	9	Low
3	Transsiberian Corridor	28		7500		20	8	Low
4	TRACECA	37		11,800		0	7	Low
5	The sea route through the Suez Canal	30		3500		100	10	High
6	Air Trans-Asian Route[b]	5		70,000		100	10	Medium

Source Compiled from materials of Kazakhstan and foreign sources, JSC NC "KTZh", US Chamber of Commerce
Note [a]the target value is set only for those routes where the RK can play the role of an integrator/main operator; [b]route Shanghai–airports of Germany

with the reduction of terms of delivery of goods in 2 times in comparison with sea transport, the cost of delivery increases slightly less than 3 times unlike air transport, where the cost increases almost 20 times.

For the development of transit in road transport, it is necessary to shorten the delivery time, where there is considerable potential. Transit in air transport has its own narrow segment of expensive cargoes and therefore does not compete with other types of transport.

The integral stability index for land routes along the Silk Road, passing through the territory of Kazakhstan, has a significant potential for increase (Table 4).

Of the transport corridors connecting China with Europe, the Trans-Asian route has a comparative advantage in transport costs.

Efficiency and attractiveness of transit routes is determined by five key principles of logistics: *5"S" = Speed + Service + Security + Cost + Stability*.

Kazakhstan occupies a geostrategic position in Eurasia, having access to both nearby and remote markets. This dictates a challenge for the development of the transport and logistics system. In the future, Kazakhstan can become a major transport and logistics hub with the appropriate development of transport and logistics centers and hubs in Kazakhstan and abroad.

For the development of transit in road transport, it is necessary to shorten the delivery time, where there is considerable potential.

Transit in air transport has its own narrow segment of expensive cargoes and therefore does not compete with other types of transport.

Table 4 Integral indicator of stability for land routes along the Silk Road, passing through the territory of Kazakhstan

No.	Route	Modality	Stability level	Comments
1	ITC "Western China-Western Europe"	Auto	Low	Terms that depend on many external factors (weather, time of year, vehicle quality, etc.) Variability of the tariff due to the multitude of players Service level inconsistency Low level of cargo safety
2	Trans-Asian Railway (Northern Corridor)	Railway	Low	Fluctuations in tariffs and service level due to the relative novelty of the route and the inconsistency of operators on the route for high and unstable delivery times
3	Transsiberian Corridor	Railway	Low	A low and stable price relative to alternative land routes costs Constantly low level of service due to difficulties in customs clearance Low level of cargo safety Unstable delivery due to the complexities of customs clearance, monopoly of one container player
4	TRACECA	Multi-modal	Low	Constantly high cost and significant delivery time Constantly low level of service Low level of safety due to a lot of border crossings and cargo handling
5	The sea route through the Suez Canal	Sea	High	Constant and low cost of delivery Long, but, as a rule, stable terms of delivery High level of service and safety
6	Air Trans-Asian Route	Air	Medium	Unreliable tariffs (highly dependent on fuel costs) Stable delivery time High level of service and safety

Source Compiled from materials from Kazakhstan and foreign sources
Note Low—with 3 or more unstable indicators of attractiveness and efficiency, average at 1–2 indicators, high in the absence of unstable indicators

4 Development of Transport Development Indicators in Kazakhstan

Economic and geographical features, including a large territory and export orientation, make the economy of the RK one of the most cargo-intensive in the world, causing high dependence on the transport system.

In recent years, the volume of freight transportation has significantly increased, as well as investments in the transport and storage complex.

Table 5 Dynamics of the transport complex performance of the RK for 2012–2016

Type of transport	2012	2013	2014	2015	2016	Change 2016/2012, %
GDP, billions of US dollars	203.5	236.6	221.4	184.4	137.3	67.5
Cargo transportation by mode of transport, million ton						
Total	3231.8	3508	3749.8	3 733.8	3729.2	115.4
Including:						
Railway	294.8	293.7	390.7	341.4	338.9	115.0
Automotive	2718.4	2983.4	3129.1	3174.0	3180.7	117.0
River	1.3	1.1	1.3	1.2	1.2	92.3
Sea	4	4	3.6	2.5	2.6	65.0
Pipeline	213.2	225.9	225	214.6	205.8	96.5
Air, thousand ton	26.9	23.9	19.1	17.2	18	66.9
Cargo turnover, bln. tkm						
All types of transport	478	495.4	554.9	546.3	518.6	108.5
Including:						
Railway	235.9	231.3	280.7	267.4	239	101.3
Automotive	132.3	145.3	155.7	161.9	163.3	123.4
River	0.06	0.03	0.03	0.03	0.02	33.3
Sea	2.7	2.7	2.5	1.6	1.8	66.7
Pipeline	106.9	116	116	115.4	114.5	107.1
Air, mln.tkm	59.5	63.1	49.3	42.7	42.9	72.1
Load capacity, tkm/1 USD	2.35	2.093524	2.506123	2.96279	3.777727	160.8
Average transportation distance, km	147.9	141.2	148.0	146.3	139.1	940.2

Source Committee on Statistics of Kazakhstan: https://stat.gov.kz

More than half of the goods transported by the public transport in the country are accounted for by road transport, which is a priority for the delivery of goods to the regions of the republic.

Automobile and railway transport accounted for 93.2% of the volume of commercial cargo transported, the share of road transport in 2012–2016 increased by 5.5% and reached 84.1%, while by rail it fell by 3.1% % and reached 9.1%, which indicates an increase in the competitiveness of road transport in certain segments of the transport services market (Table 5).

Analysis of Table 5 shows an increase in the growth rate of freight turnover. If we analyze by type of transport, both in terms of freight turnover and the volume of transport, we observe: growth in the volume of transportation and freight turnover by about 6–9% per year; the growth of freight turnover is observed for all types of transport, except for river and air; there is a decrease in the growth rates of freight traffic and cargo turnover; there is a redistribution of transport. The average distance of transportation is reduced by 21 km or 12.4%, which shows a decrease and an average range of transportation of 1 ton of cargo.

As can be seen from Table 5, the share of rail transport in the total freight turnover exceeds 50%; the share of motor transport in the total freight turnover is more than 27%; the underdevelopment of cargo transportation by air transport is observed, the share in the total freight turnover is less than 1%; and a low share of less than 0.2% of cargo transportation by water transport.

The constant increase in cargo traffic with the constant length of roads leads to an increase in congestion of existing transport routes and further congestion of the logistics infrastructure in Kazakhstan.

The volume of cargo transportation by region is uneven. Comparison of the shares of participation of different regions in the overall transportation of goods and in Kazakhstan's freight turnover shows that high growth rates of transportation are observed in Almaty (+10.0%), Zhambyl (+14.1%), South Kazakhstan (+16.3%), Karaganda (+12.4%), Mangistau (+11.2%) regions and in the cities of Almaty and Astana. The specific weight of cargo transportation is above the average republican level in Karaganda (19.2%) and East Kazakhstan (14.8%) regions.

Railway transport is the main mode of transportation for transportation of such types of bulk cargo as hard coal and coke (over 90% of the volume of their transportation by all types of transport), ore, mineral fertilizers, cement (about 90% for each type), ferrous metals 75%), timber cargo (over 60%), grain cargo (over 50%).

On the third place there are oil cargoes (first of all fuel oil and light oil products).

Of these segments of transportation can be divided into three groups of goods by profitability level [1].

1 **group—low-profit transportations**: hard coal (export, inter-regional communication), ore (export, inter-regional communication), construction cargo (import), grain (export, inter-region communication), other cargo (inter-region
   communication);
2 **group—high-yield transportation**: oil cargo (export), ferrous metals (exports,
   imports, inter-regional communication), chemical and mineral fertilizers
   (exports, imports, inter-region communication), other goods (imports);
3 **group—self-supporting transportation**: hard coal (import), oil cargo (import,
   inter-regional communication), ore (import), construction cargo (export,
   inter-regional communication), grain cargo (import), other cargo (export).

In the total volume of freight traffic in 2016, 50.2% are freight traffic in inter-region traffic, 22.3% in export, 3.4% in import and 17% in transit.

At the same time, revenues from cargo transportation were distributed as fol-lows: in the inter-region communication—36.6%, in the export message—23%, in import—7.9%, in transit—31.2%, i.e. there is a tendency to increase the income from transit freight shipments.

The largest cargo is shipped within the republic, by intercity communication (in order of 50%), by export route in the order of 22.2%, by transit in the order of 19.0%, by import—3.4%, other intraurban and suburban.

Road transport is mainly used to transport small cargo flows over short distances. This is due to the relatively high cost of this type of transport and its low carrying capacity. The advantages of road transport include high speed and the possibility of delivering goods "from door to door" without additional costs for reloading.

In 2016, the volume of passenger transportation by road amounted to 21.2 billion people, passenger turnover—214.9 billion pkm. The volume of cargo transportation in 2014 increased and reached the value of 3.1 billion ton, cargo turnover—155.1 billion ton.

Transit traffic occurs mainly between the republics of Central Asia, Russia, and China. The bulk of road transport transit (about 86%) is formed in China. The share of the Central Asian republics (approximately 3.8%) and Russia (4.6%) is much less. The share of other countries in transit traffic is about 6.3%.

In road transport, the main problems are:

(1) poor condition of highways of regional and regional significance, as well as bridges;
(2) absence of direct automobile communication between the western and northern, western and central regions of the country;
(3) low level of security and deterioration of road maintenance equipment in operating organizations on the national and local road network;
(4) low throughput of a number of major highways.

Sea transport is an important part of the transport system in Kazakhstan. In terms of cargo turnover, it takes the 4th place after the pipeline, railway, and automobile. The total freight turnover in 2016 amounted to 2.6 billion ton of km. It has a leading role in transport services in the Caspian Sea. By 2016, the volume of traffic in the Caspian increased to 65 million ton, and the share of Kazakhstan's exports and transit from the total volume of transported goods increased to 46% and amounted to about 30 million ton.

The main types of cargo transported in the Caspian Sea are oil, steel, timber and paper, the share of which is 72% of the total volume of freight. River transport plays an inconspicuous role in intra-district and inter-district transportation of the country (Pavlodar, East-Kazakhstan Region, West-Kazakhstan Region and North-Kazakhstan Region).

For the period from 2012 to 2016, the value of fixed assets at the beginning of the year in the field of transport and warehousing increased 2.56 times, the degree of their deterioration for the same time decreased by 10.4% (from 26.4 to 23.6%). The significant excess of the growth rates of the value of funds over depreciation is due, first of all, to the faster growth rates of investments in fixed assets "transportation and warehousing," which increased by 17.1% [12].

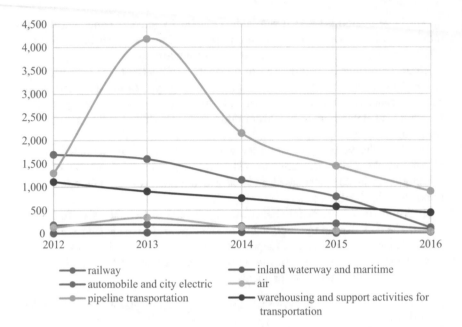

Fig. 1 Investments in fixed assets due to all sources of financing by mode of transport for 2012–2016 (in millions of US dollars)

Investments in the transport sector consist of several receipts; three sources of financing have been identified from the transport development program until 2020. At the expense of the republican budget, external loans of international financial institutions and private investments.

In 2016, investments in fixed assets from all sources of financing by mode of transport of Kazakhstan amounted to 1628.57 million US\$, which is 2 times less than in 2013, which amounted to 7 237,42 million US\$. Of these, in 2016, the first place for investment in transportation through the pipeline was 905 million US\$; storage facilities and auxiliary transport activities ranked second—444 million US\$; on the third, railway transport—124 million US\$ (Fig. 1).

Also, depreciation of fixed assets by transport enterprises for the year of 2016 amounted to 23.6%, which is less by −2.8% than in 2012 (Table 6).

One of the main factors that have a significant impact on the effectiveness of this area is labor resources. The share of the employed population in the sphere of transport and warehousing is in the range of 6.7–7.2% and in 2016 it was 7.2% of the average annual number of employees in the country's economy. The average annual number of employees in the industry increased by 8.5% from 2012 to 2016 and amounted to 619.5 thousand people, the transport industry made up—201.2 thousand people, the share of transport workers from all employed in 2016 was −2, 4%.

Table 6 Degree of depreciation of fixed assets of transport organizations and storage in the RK for 2012–2016

	Unit of measure	2012	2013	2014	2015	2016
Fixed assets by transport enterprises	million US $	32,842.95	36,867.87	35,541.43	35,329.73	25,274.74
Degree of depreciation of fixed assets by transport enterprises	in percentages	26.4	27.2	24.2	24.4	23.6

Source Committee on Statistics of Kazakhstan: https://stat.gov.kz

For the last 5 years (2012–2016), revenues from transportation in Kazakhstan have been growing steadily. For the year 2016, total income exceeded $7.5 billion. At the same time, revenues from cargo transportation were distributed as follows: in the inter-regional communication—36.6%, in the export message—23%, in import —7.9%, in transit—31.2%. There is a tendency of increase in income from transit cargo transportation.

64% of all revenue from transportations falls on the international direction, which indicates the paramount modernization and improvement of the quality of the transport and logistics infrastructure serving international cargo transportation, especially transit.

Considering incomes in the direction of transportation it is seen (Fig. 2) that the international direction is more profitable. 63% of all revenue from transportations falls on the international direction, which indicates the paramount modernization and improvement of the quality of the transport and logistics infrastructure serving international cargo transportation, especially transit.

In this regard, to maximize the use of the country's transit potential, "break-through" projects are needed.

These requirements are met by the new transport corridor "Western Europe-Western China" [16]. The length of the sites to be reconstructed is 2309 km. The cost is 286 billion tenge ($2.3 billion). The project implementation will allow reorienting part of Chinese goods from sea transport to road transport (45 days by sea against 11 days by road).

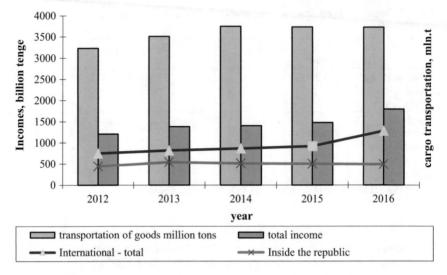

Fig. 2 Dependence of income on cargo transportation volume in Kazakhstan

5 Development and Integration of the Transport and Logistics System of Kazakhstan into the International Transport System

At the moment, the countries of Western Europe and developed Asian regions are confidently leading the development of transport logistics [17].

So, according to the World Bank report on the developed indicator of logistics efficiency (LPI), Kazakhstan in 2016 took 77th place out of 160 surveyed countries, rising 11 points from 88th place. Over the past 10 years, Kazakhstan's performance index has improved by 56 positions (from 133 to 77 places). The level of logistics development in Kazakhstan (77th place in the LPI rating) suggests that the logistics potential of Kazakhstan as a transit country is not used enough.

All this should positively affect the factors affecting the development of transport.

In order to occupy its niche in the world logistics market, it is necessary to improve the competitiveness of the main routes passing through the countries of the Silk Road, to increase the main international logistics ratings, which will improve the investment attractiveness of the industry and increase investments in the development of the logistics infrastructure. Table 6 presents the average value of the level of logistics development (LPI index) in groups of countries where the main routes of the economic belt of the Silk Road pass [18].

Table 7 shows that in terms of logistics development indicators (by index and subindexes LPI) for 2016, the leader among these groups of countries are European countries (3.5 points) and large regional countries (3.06), the worst indicators of

Table 7 Level of development of logistics in groups of countries potentially participating in the route "Economic belt of the Silk Road", 2016

Countries	LPI score, point	LPI rank
European countrie	3.50	34
Large countries of regional level (China, Russia, Turkey, Iran)	3.06	64
European CIS countries (Ukraine, Belarus, Moldova)	2.58	97.6
Countries of Transcaucasia (Azerbaijan, Georgia, Armenia)	2.34	132
The countries of Central Asia (Kazakhstan, Uzbekistan, Kyrgyzstan, Tajikistan, Turkmenistan)	2.32	126.8
Average in the world	2.88	–

Source Compiled by the authors according to the World Bank, 2016

logistics development—in the countries of Central Asia (2.32) and the countries of Transcaucasia (2.34). Above the average world LPI indicators are the EU countries and large regional countries, the remaining groups of countries have values below the world average.

Figure 3 shows the ratings and values of the Logistics Performance Index (LPI) in countries passing through the Silk Road or potentially related to the Silk Road (Fig. 3).

The best average indicators are in European countries, from China, China, India, Saudi Arabia are higher than the world ones.

Kazakhstan and other CIS countries are lagging behind the average global LPI. Kazakhstan is the leader among the CIS countries in terms of the main index of the LPI index both in 2012, 2014, and in 2016, ahead of all CIS countries. According to experts, the similarity of our systems holds the CIS countries at the bottom of the LPI index. Trade relations both within the EAEC and within the CIS are much easier than with other countries of the world, therefore the main difficulties in our countries are customs clearance issues, as well as weak integration of national logistics systems into the common Eurasian and European systems.

Among the CIS countries, the growth potential of LPI is, as transit countries, Kazakhstan, Russia and Belarus, between Europe and China in international trade. To do this, it is necessary to reduce transportation tariffs, fees and charges in ports, train personnel in the best European schools or improve the skills of specialists, introduce modern innovations in technology in logistics, information technology to track and shorten delivery times.

A characteristic trend of the world economy is the disappearance of customs borders in connection with the integration of countries into a single economic space, improving the efficiency of customs operations, as evidenced by the improvement of trade and logistics indicators in Kazakhstan, Russia, Belarus, Kyrgyzstan and Tajikistan after integration into the EAEC. Reducing barriers in customs is one of the key factors in the development of the economy (improving

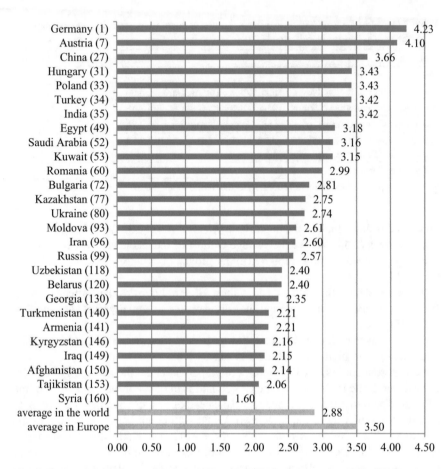

Fig. 3 Rating and evaluation of logistics efficiency (LPI) in the countries of the Silk Road

trade and logistics). For example, the creation of an effective customs in Singapore, as a global trading hub, gave impetus to the development of the economy [5].

It should be separately noted the tendency to improve the quality of logistics service against the backdrop of growing competition in virtually all groups of countries.

Among the CIS countries, a significant increase in the rating of Kazakhstan in terms of the level of logistics performance occurred for all components of the LPI. Nevertheless, over the past 3 years there has been a deterioration in quality and competence, tracking the passage of goods and the timeliness of deliveries.

Despite a certain subjectivity of the study of logistics efficiency, it is possible to single out a number of basic problems of logistics development in the CIS member countries. Among them—a lack of investment in infrastructure development, improving the quality and competence of specialists, increasing the level of international transportation, meeting deadlines and tracking cargo.

6 Increasing the Efficiency of Freight Transportation Through the Transport and Logistics System of Kazakhstan

A real struggle for cargo flows unfolds on the global freight market. Kazakhstan, due to its favorable geographical location, has a number of advantages in this growing competition.

Operational figures for the first quarter of 2017 are positive. The cargo turnover was exceeded to the plan by 9.3% and to the fact of the previous year by 13.9%, loading in tons exceeded the plan by 7.9% and the last year's figure by 15.6%.

In order to develop the transport infrastructure and transport system in 2013, the State Program for the Development and Integration of the Infrastructure of the Transport System of Kazakhstan up to 2020 has been adopted. Its purpose is to form the modern transport infrastructure of Kazakhstan, as well as to ensure its integration into the international transport system and the implementation of transit potential.

To implement this program until 2020, funds will be spent in the amount of about 5.2 trillion tenge (34.2 billion US dollars), including by types of financing: the republican budget—52.2%, the National Fund—4.48%, the local budget—3.4%, the own and borrowed funds of the group of JSC " NC "Kazakhstan Temir Zholy"—29.4%, private investment—5.28%, public-private partnership—2.8%, fees from fees 1.7%.

As can be seen, mainly funding is provided from the state budget and the national fund of the country (specific weight 60.8%), then such sources of financing, as means of off-budget funds, state loans under guarantees, funds of non-state enterprises are absolutely not used. The share of public-private partnership is only 2.8%.

World experience almost unequivocally confirms the appropriateness of using concessionary mechanisms.

In Kazakhstan, concessions are particularly relevant given the high wear and tear of a significant number of infrastructure facilities and chronic underfunding. At the same time, while the level of readiness of the state system for the practical implementation of this progressive tool is not very high.

As the main directions for attracting extra-budgetary funds to the transport and logistics infrastructure, one can consider [14]:

– provision of budgetary financing for non-budgetary funds in the construction and maintenance of transport and logistics infrastructure and the attraction of additional financial resources through public-private partnership mechanisms and the introduction of payment for the use of publicly owned facilities;
– wide involvement of non-state enterprises related to servicing users of transport and logistics infrastructure, design, construction, repair, maintenance of transport infrastructure, production of building materials, road service;

– creation of additional sources of financial resources for the public sector of the industry through loans, loans and government guarantees.

The indicator of the successful implementation of the program will be the improvement of Kazakhstan's position in the World Bank's ranking on the Logistics Performance Index (LPI), which provides for the development of all sectors of the transport sector—the quality of infrastructure, the efficiency of customs, the simplicity of organizing international deliveries of goods, the ability to track cargo, meet deadlines and competencies in logistics.

So, following the results of 2020, Kazakhstan is expected to achieve the 40th place in the above rating, increasing revenues from transit traffic 2 times from 157 million tenge (2012) to 364 million tenge by 2020, and an increase in the volume of transit cargo through the territory of Kazakhstan from 17.8 million ton in 2012 to 35.5 million ton by 2020 [10].

It is expected that by 2030 the volume of transit freight through the RK will almost triple and exceed 46 million ton.

By 2020, the development of the TSC of the RK should enable Kazakhstan to reduce the transport component in the economy to 7.5%; to increase the share of logistics services to 25%; to reach the position of the European countries in the rating of the effectiveness of the TLC.

In turn, to accomplish this task, measures must be taken to:

- simplification of customs procedures in foreign economic activity;
- reduction of permits for export-import operations in the transportation process;
- attraction of world companies or creation of a competitive company capable of providing high-level transport-forwarding services;
- provision of cargo tracking capabilities.

The launch of regular container trains in the main directions of freight traffic with the aim of achieving delivery deadlines.

In 2015, the state program for infrastructural development "Nurly Jol" for 2015–2019 was also adopted [19].

The goal of the program is the formation of a single economic market through the integration of the country's macro regions based on the building of an effective infrastructure on the hub principle to ensure Kazakhstan's long-term economic growth, as well as the implementation of anti-crisis measures to support certain sectors of the economy in the face of worsening of the conjuncture in foreign markets.

One of the tasks is to create an efficient transport and logistics infrastructure on a "beam" basis.

The total amount of financing at the expense of the National Fund of Kazakhstan will amount to $3 billion in 2017, and the indicative amount of co-financing from international financial institutions will be about $ 8.97 billion.

Of these, the amount of financing for the development of transport infrastructure will be 54%.

Thus, the strategic objectives of the development of the transport and logistics system of Kazakhstan for the future are as follows.

1. To become the main logistics hub and transit region.

The main focus will be on attracting transit between China and Europe (the New Silk Road); China and the Russian Federation; Europe and Central Asia.

2. To fully realize the export potential and domestic needs of the national economy. The main task is to ensure:

   - free and timely access to target markets for the main products of the basic sectors of the economy of the RK: grain, coal, metallurgical products;
   - unhindered and timely attraction of necessary resources and raw materials to the market of the RK (especially for high-tech products);
   - to ensure free transport of goods in the domestic market (delivery of goods to consumers).

The strategic positioning of Kazakhstan in 2020 is a powerful transport and logistics hub in the Eurasian space [20].

This will be achieved through: the development of a multi-level system of transport and logistics centers; providing a regular container message; integrated customs infrastructure - Customs clearance centers within the TLC; unified technical, technological, tariff and investment policies; a single information field;

The transport and logistics system of Kazakhstan in the target model will set high standards of logistics service on the basis of the "5S" principles:

(1) Cost—competitive tariffs: reduction of the transport component in the cost of the final product; free formation of tariffs depending on the configuration for all modes of transport; transparent tariffication of competitive domestic and transit routes.

(2) Service—high level of service: high level of containerization; the widespread use of regular container trains; through multimodal freight transport; support from the intellectual transport system; a wide service level for cargo airlines using Kazakhstan as an international air hub.

(3) Speed—high speed: the route through the RK is the fastest and the way of delivery of goods from Asia to Europe; the work of customs points on the border with China and the CAR was maximally coordinated, the customs standard of CIM-SMGS, preliminary declaration of cargo was introduced.

(4) Security—high standards of reliability and safety: reducing the number of accidents in transport; ensuring the safety of goods through the introduction of technological and management innovations (tracking cargo E-Freight, E-Train, TRITON K).

(5) Stability—guaranteed conditions for the transport of goods: fixed time and cost of delivery of goods, customs clearance, provision of wagons, sorting, etc.

7 Conclusion

The features of the geopolitical position of Kazakhstan, located between two dynamically developing world centers of business activity—Europe and Asia are shown. The results of the formation and development of international transport corridors in the territory of Kazakhstan are analyzed, the variants of their positive impact on the performance of the transport and logistics system of the country and macroeconomic indicators are shown;

The level of development of the transport and logistics complex in the regions as a whole is assessed as insufficient for the following reasons;

- incomplete use of transit potential due to the lack of supply of transportation and logistics services that meet the international standards sufficient to cover the growing demand;
- lack of modern transport and logistics technologies for solving within regional problems;
- shortage of qualified specialists.
- For efficient operation of the transport and logistics system, it is necessary;
- to increase capacity and improving the quality of transport links;
- to from an infrastructural basis for a modern intermodal transport and logistics system in Kazakhstan;
- to optimize cargo flows, as well as to decrease the operational and infrastructure component of the cost of transportation;
- to provide rates of growth of investments into a fixed capital ≪ transport and warehousing ≫ not below 15% a year;
- to provide infrastructural rates of growth of a total regional product in regions at a level of not less than 8–10% per year.

The results of the analysis showed that one geographical situation is far from sufficient for the rapid integration of the country into the global logistics system and for the automatic change in favor of the Kazakhstani transit of the main transport corridors of international trade. In order to ensure that transit cargo flows are reoriented to the transport system of Kazakhstan, first of all, it is necessary to significantly raise the overall level of development of logistics efficiency, its reliability and security, modernize and significantly expand the capacity of the Kazakhstan seaport and transport highways, ensure transparency of transportation tariffs and machinery their control and regulation.

The logistics system of the countries lying along the routes of the Silk Road countries, especially Central Asia, the Caucasus, needs restructuring, further integration with the systems of larger countries further and developed countries of Europe. It is necessary to raise the level of the regulatory and regulatory framework governing the industry, to resolve the issues of training highly qualified personnel, introducing new technologies, and improving the quality of the services provided.

Important is also the use of public-private partnerships, as evidenced by the international experience of the advanced countries of the world, leading today in the

LPI rating. All of them actively support public-private partnership. In this respect, their complex approach in the development of transport services, infrastructure and efficient logistics are also important.

The most intensive development of logistics in European countries and major regional countries. The countries of Central Asia and Transcaucasia received less development. Although all conditions for the development of logistics have been created in these countries: favorable business conditions have been created, the investment attractiveness of the logistics industry is increasing, the logistics market is open to foreign companies, a high degree of integration into the world economy, and the competitiveness of the national economy is growing.

Especially the countries of Central Asia should attract large investments to improve the quality of transport and logistics infrastructure, reduce customs barriers when passing cargo, improve the quality of services provided, and reduce logistics costs.

The growth of trade volumes will lead to a significant increase in the volume of cargo transportation across borders. Therefore, the countries of Central Asia, the Caucasus and Russia should reduce the cost of customs and border control in order to create some effective customs. It is necessary to apply various innovative technologies and preferences in this area, as it was done in its time in Singapore.

Bibliography

1. Указ Президента РК (2013) Государственная Программа Развития и Интеграции Инфраструктуры Транспортной Системы Республики Казахстан до 2020 года. No. 725, 76 p [In Russian: Decree of the President of the Republic of Kazakhstan (2013) The state program for the development and integration of the infrastructure of the transport system of the Republic of Kazakhstan until 2020]
2. Джадралиев МА (2009) Транзитный потенциал ЕврАзЭС и региональная транспортная интеграция. Евразийская Экономическая Интеграция. 2(3):130–139 [In Russian: Jadraliev MA (2009) EurAsEC transit potential and regional transport integration. Eurasian Economic Integration]
3. Раимбеков ЖС, Сыздыкбаева БУ (2012) Транспортно-логистическая система Казахстана: механизмы формирования и развития. BG-print, 328 p [In Russian: Raimbekov ZS, Syzdykbaev BU (2012) Transport-logistical system of Kazakhstan: mechanisms of formation and development]
4. Rodrigue J-P, Debrie J, Fremont A, Gouvernal E (2010) Functions and actors of inland ports: European and North American dynamics. Transp Geogr 18:519–529
5. Kwan Tan AW, Hilmola O-P (2012) Future of transshipment in Singapore. Ind Manage Data Syst 112(7):1085–1100
6. Wanga C, Ducruet C (2014) Transport corridors and regional balance in China: the case of coal trade and logistics. Transp Geogr 40:3–16
7. Wang JJ, Cheng MC (2010) From a hub port city to a global supply chain management center: a case study of Hong Kong. Transp Geogr 18:104–115
8. Notteboom T, Rodrigue J-P (2009) Inland terminals within North American and European supply chains. Transp Commun Bull Asia Pac 78:1–39
9. Jevtić M, Radmanovac M (2007) Logistics in European traffic policy. Interdisc Manage Res IV:516–525

10. Указ Президента РК (1994) Закон РК«О транспорте в Республике Казахстан». Ведомость Верховного Совета РК. No. 156 [In Russian: Decree of the President of the RK (1994) Law of the Republic of Kazakhstan "On transport in the Republic of Kazakhstan". Statement of the Supreme Council of the Republic of Kazakhstan]

11. Mariotti I (2015) Transport and logistics in a globalizing world: a focus on Italy. Springer Briefs in Applied Sciences and Technology, 100 p

12. Kabashkin I (2012) Freight transport logistics in the Baltic Sea region. Regional aspects. Transp Telecommun 13(1):33–50

13. Можарова ВВ (2011) Транспорт в Казахстане: современная ситуация, проблемы и перспективы развития. КИСИ при Президенте РК. 216 p [In Russian: Mozharova VV (2011) Transport in Kazakhstan: current situation, problems and development prospects. KISI under the President of the RK]

14. Arvis J-F, Saslavsky D, Ojala L et.al. (2016) Connecting to compete. Trade logistics in the global economy. The logistics performance index and its indicators. The World Bank. 76 p

15. Транспорт в Республике Казахстан (2017) Статистический сборник. 63 p. URL: www. stat.gov.kz [In Russian: Transport in the Republic of Kazakhstan (2017) Statistical collection]

16. Выступление Председателя КНР Си Цзиньпина в Назарбаев университете (2013) http:// kz.china-embassy.org/rus/zhgx/t1077192.htm [In Russian: Speech by President Xi Jinping to the Nazarbayev University (2013)]

17. Постановление Правительства Республики Казахстан (2005) О Транспортной стратегии Республики Казахстан до 2020 года. No. 75. http://adilet.zan.kz/rus/docs/P050000075 [In Russian: Resolution of the Government of the Republic of Kazakhstan (2005) About the transport strategy of the Republic of Kazakhstan until 2020]

18. Экономические и социальные эффекты от реализации проекта. Сайт проекта«Западная Европа-Западный Китай». http://www.europe-china.kz/print/86 [In Russian: Economic and social effects of the project. The site of the project "Western Europe-Western China"]

19. Раимбеков ЖС, Сыздыкбаева БУ (2017) Исследование состояния и путей развития логистики, международной торговли и туризма в странах Шелкового пути. Материалы Евро-Азиатского форума экономики и политики и научной конференции«Китайско-Казахстанское сотрудничество в рамках проекта«Один пояс – Один путь». pp 13–19 [In Russian: Raimbekov ZS, Syzdykbaev BU (2017) Research of the state and ways of development of logistics, international trade and tourism in the countries of the Silk Road. Materials of the Euro-Asian Forum of Economics and Politics and the scientific conference "China-Kazakhstan cooperation within the framework of the One-Belt-One Way" project]

20. Раимбеков ЖС, Сыздыкбаева БУ (2014) Механизмы развития транспортно-логистических кластеров в Казахстане. LAP Lambert Academic Publishing, Saarbrucken. 164 p [In Russian: Raimbekov ZS, Syzdykbaev BU (2014) Mechanisms of development of transport-logistical clusters in Kazakhstan]

21. Указ Президента Республики Казахстан (2015) Государственная программа инфраструктурного развития«Нұрлы жол»на 2015–2019 годы. No. 1030 [In Russian: Decree of the President of the Republic of Kazakhstan (2015) The state program for infrastructure development "Nurly Jol" for 2015–2019]

22. Raimbekov Z, Syzdykbayeva B, Zhenskhan D, Bayneeva P, Amirbekuly Y (2016) Study of the state of logistics in Kazakhstan: prospects for development and deployment of transport and logistics centres. Transp Probl 11(4):57–71

Part II
Technical and Informatics Issues of Transport Systems

Present and Future Operation of Rail Freight Terminals

Marco Antognoli, Luigi Capodilupo, Cristiano Marinacci, Stefano Ricci, Luca Rizzetto and Eros Tombesi

Abstract Rail freight has not progressed coherently to economy: during the last century, the wagonload was the core business of railways, later declining in favor of combined transport, which include the notion of transshipment in an intermediate terminal. Terminals are a key element of transport services and, in this study, the main goal are methods suitable to evaluate the performances of different types of rail freight terminals: Rail to road for long distance and shorter range units transfer, Rail to rail for shunting and/or gauge interchange, Rail to waterways (sea and inland). The evaluation of the performances of terminals and the influence on them of innovative operational measures and technologies is based on a selected combination of tested analytical methods based on sequential application of algorithms and discrete events simulation models, capable to quantify different Key Performance Indicators.

Keywords Freight rail transport · Terminals performances · KPI
Operational costs · Economic analysis · Financial analysis

M. Antognoli · L. Capodilupo · C. Marinacci · S. Ricci (✉) · L. Rizzetto · E. Tombesi
DICEA, Sapienza Università di Roma, Via Eudossiana 18, 00184 Rome, Italy
e-mail: stefano.ricci@uniroma1.it

M. Antognoli
e-mail: marco.antognoli@uniroma1.it

L. Capodilupo
e-mail: luigi.capodilupo@uniroma1.it

C. Marinacci
e-mail: cristiano.marinacci@uniroma1.it

L. Rizzetto
e-mail: luca.rizzetto@uniroma1.it

E. Tombesi
e-mail: eros.tombesi@uniroma1.it

© Springer International Publishing AG, part of Springer Nature 2018
A. Sładkowski (ed.), *Transport Systems and Delivery of Cargo on East–West Routes*, Studies in Systems, Decision and Control 155,
https://doi.org/10.1007/978-3-319-78295-9_6

1 Terminals as a Key Element of Transport Services

The rail freight transport has not progressed in parallel with the World economy.

The single wagon used to be the core business of railways during the last century; today, in contrast to the decline of the conventional rail freight services, the combined transport has shown relevant signs of growth.

On this basis, the rail freight transport spread out in two main typologies of services: conventional rail freight (wagonload) and combined transport, which includes the notion of transshipment and the flow of goods from an origin to an intermediate destination, and from there to another destination.

The terminals are a key element of these transport services and a main research goal is to setup suitable methods to evaluate the performance of different rail freight terminals, flexible and potentially applicable to various families of terminals:

- Rail to road for long distance and shorter range units transfer;
- Rail to rail for shunting and/or gauge interchange;
- Rail to waterways (sea and inland).

The evaluation of their performances is necessary in the present operational situation and under the influence of improvements basing on innovative operational measures and new technologies. Moreover, methods and models to evaluate rail freight terminals are required to calculate relevant Key Performances Indicators (KPI) with acceptable levels of accuracy.

2 State of the Art

2.1 Development of the Rail Market

The transport of freight by rail did not progress in parallel with the economy: Fig. 1 expresses the variation of respective trends of GDP as well as rail and road freight traffic in the period 2004–2013 in the European market.

After the economic crisis in 2008, the freight traffic entered in a depression not yet recovered, both in rail and road fields.

In the same period, the modal share rail vs. road has remained almost unchanged. In the past, the single wagon traffic served both big and smaller markets with various frequency of orders; today small and medium volumes are mostly in the hands of road transport.

On the other hand, full trainloads are almost maintaining their mostly captive markets represented by heavy industry and related business as well as maritime generated traffic from/to ports, which are able to order volumes and frequencies matching the full train offer.

Some recent strategies of the railway undertakings are moving towards an integration of wagonload and trainload systems: e.g. the D. B. Schenker

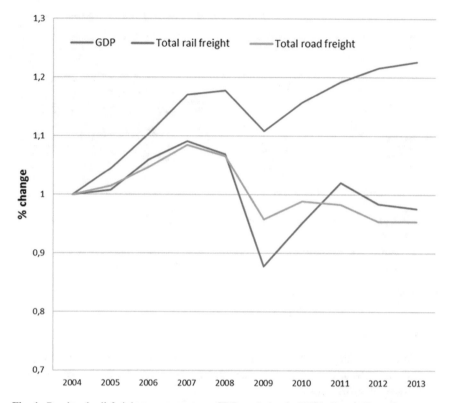

Fig. 1 Road and rail freight transport versus GDP evolution in EU28. *Source* Eurostat

Netzwerkbahn (Fig. 2), treating the conventional traffic as dynamic wagon blocks suitable for coupling and decoupling according to IT based booking systems.

The aim is a better coordination of the timetable to increase the capacity of trains and the frequency of offered services for customers without enough volume to order a full train.

Auxiliary freight stations and marshalling yards are anyway necessary for the production of wagonload services and for the combination of wagons sharing destinations. They are potential time and resources consuming sources, which definitively need relevant efficiency improvements by better operational coordination and larger automation.

The combined transport is the only mode that has really accompanied the economic development, reaching nowadays almost ¼ of the total rail freight volume.

On this basis, the advances in interoperability of systems, in combination with appropriate legislative measures, will further increase the attractiveness of such combined mode.

An 83-companies survey commissioned by UIC clearly stated that almost 50% of the intermodal providers, responsible for almost 80% of the total intermodal volume, are active both in National and International markets.

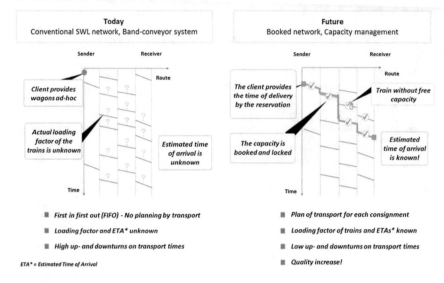

Fig. 2 Netzwerkbahn booking system. *Source* D. B. Schenker

General trends in the field are:

- Worldwide diffusion of Interoperable components and systems, both in vehicle and infrastructure, generally facilitating the cross-border services (e.g. Genoa-Rotterdam corridor);
- Infrastructure managers and railway undertakings are progressively homogenizing criteria for infrastructure use (booking, charges, timetables, etc.), mainly driven by the development of more and more standardized and worldwide diffused IT systems;
- Intermodal services are mainly operated on a corridor basis and on distance long enough to make them economically feasible, regardless of the number of crossed national borders;
- In terms of use of transport units, it can be distinguished traffic of ISO containers on one side and various others, like swap bodies, semitrailers, full Lorries and other domestic units.

ISO container traffic has normally its origin overseas and trains are typically the terrestrial links between seaports and inland terminals.

The most frequent dimensions are 20 and 40 ft (with increasing use of 45 ft units for continental traffic) and in general large units (e.g. TEU/Containers ratio in Rotterdam was growing from 1.45 in 1970 to more than 1.65 nowadays.

The other most common units are:

- Semitrailers (ST), reaching almost 15% of intermodal traffic;
- Swap-Bodies (SB);
- Tank and silo containers (including 26 ft units);

- Other less frequent standard units, including 30, 45 ft and pallet wide containers;
- Full Lorries (accompanied transport).

2.2 Terminals for Multimodal Transport

According to [1] (Fig. 3) Multimodal transport is the most general term when referring to the shipment of goods by at least two modes.

The multimodal transport becomes Intermodal, as soon as the goods are stored in transferred in loading units, without handling the goods as such.

The multimodal becomes Combined, as soon as the majority of the journey is by rail, inland waterways or sea, with only minor initial and/or final legs, if any, by road, according the UN/ECE definition issued in 2001, which also defines a terminal is "a place equipped for the transshipment and storage of loading units".

Based on these definitions, by excluding air traffic terminals, normally playing a limited role in the transfer of freight volumes, the typical intermodal terminals are:

- Sea–Rail and Sea–Road port terminals;
- Rail–Road inland terminals.

These terminals act as land bridges (last hundreds miles) for continental and intercontinental flows in main freight traffic corridors [2].

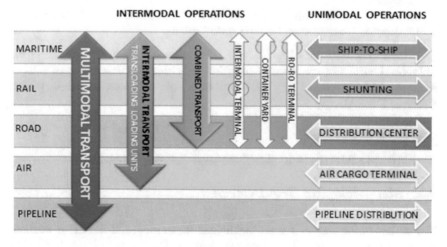

Fig. 3 Terminals for multimodal transport. *Source* [1]

Moreover, as nodes of the global logistic chain, these terminals are a part of a socio-technical system, integrating organizational and market related aspects, as well as infrastructural and technological issues influenced by the following aspects:

- Interaction of private (operators, forwards, etc.) and public (national, regional and local authorities, etc.) stakeholders playing relevant roles in decisions.
- Localization requirements combining the proximity to major flows generating areas (freight corridors, ports, rail and road networks) and the availability of space for operation and storage.
- Handling organization taking into account amount and typology of goods, transshipment technologies and other productivity factors measured by key performance indicators, to carefully identifying.

3 Measurement of Terminals Performances: Key Performance Indicators

In order to quantify the performances of the operated terminals, as well as to assess the implementation of innovative operational measures and/or technologies, it is necessary to identify Key Performances Indicators (KPI) customized to the selected terminal typologies.

The main requirements for these indicators are:

- Measurable by routine data collected during operation;
- Capability to synthetize the terminal efficiency;
- Sensible to potential changes introduced by new technologies and/or operational measures;
- Related to different aspects sketched in Sect. 2.2 (interaction of stakeholders, localization requirements, handing organization).

In the following Tables 1 and 2, with reference respectively to ports and inland rail related terminals, are reported analytical definition and description of a set of selected KPI developed in the framework of European research project CAPACITY4RAIL [3].

4 Suitable Methods to Analyze and Assess Terminals

4.1 Methodological Requirements

The review of the state of the art highlighted the existence of numerous methodologies for the study of railway terminals (e.g. [4, 5]).

Table 1 Sea–Rail port terminals key performance indicators

Definition		Description
Total transit time (ITU and vehicle)	$TTR = \sum_{i=1}^{n} TWi + \sum_{i=i}^{n} TOi$	Time period from the arrival of the freight unit (or vehicle) to the terminal gate from rail or sea to the exit of the unit (or vehicle) from the terminal towards sea or rail • TTR_v = vehicle total transit time (train and truck) • TTR_{ITU} = Unit total transit time • TW = waiting time • TO = operational time **Depending on:** • External infrastructures and services • Technologies • Operational rules • Terminal dimensions
Utilization rate of handling equipment	$Er = \left(\frac{nETr}{nE}\right) Th$	Average number of handling equipment, engaged on a train during the handling time (Equipment rate utilization in handling area) • Er = utilization rate of handling equipment • $nETr$ = number of handling equipment employed per train • nE = total number of handling equipment available in handling area • Th = handling (loading/unloading) time **Depending on:** • Handling technologies • Operational rules • Terminal dimensions
Utilization rate of ITU storage	$s_{ITUi} = \left(\left(\frac{nITU_{in} + nITU_{s(i-1)} - nITU_{out}}{Cs_{max}}\right) Ti\right)$	Influence of the number intermodal units, which transit within terminal, on the storage area capacity • $S_{ITU,i}$ = utilization rate of ITU storage area • $nITU_{in}$ = number of incoming ITUs in terminal • $nITU_{s(i-1)}$ = number of stored ITUs • $nITU_{out}$ = number of departing ITUs from the terminal • T = time gap (day, week, month or year) • Cs_{max} = maximum storage capacity • T_i = time gap **Depending on:** • External infrastructures and services • Technologies • Operational rules

(continued)

Table 1 (continued)

Definition		Description
		• Flow of ITU handled in the terminal
Energy consumption rate	$Ec(ITU) = \frac{Ec(v)}{nITU(v)}$	Energy consumption of handling equipment per ITU • $Ec(v)$ = energy consumption of handling equipment per vehicle • $nITU(v)$ = number of intermodal transport units per vehicle **Depending on:** • *ITU* Throughput • Technologies • Number of handling equipment • Operational rules
	$Ec(ta) = \frac{C}{S}$	Energy consumption of Terminal area compared to its surface: e.g., terminal lighting, office consumption • C = energy consumption of terminal • S = terminal area **Depending on:** • *ITU* Throughput • Technologies • Number handling equipment • Operational rules
Equipment performance	$Ep = \frac{nITU}{h}$	Capacity of handling equipment • $nITU$ = number of handled intermodal transport unit • h = hour **Depending on:** • Handling technologies • Skills of the equipment operator(s)
Equipment haul	$Eh = \frac{Er}{Lr}$	Influence of train length on the path covered by handling equipment • Eh = equipment haul • Lr = train length • Er = length of route for handling equipment in handling area **Depending on:** • Handling technologies • Operational rules • Terminal dimensions

(continued)

Table 1 (continued)

Definition		Description
Truck waiting rate	$TW_{rate} = \frac{Twt}{tTrain}$	Influence of handling time of train on the waiting time of ship • TW_{rate} = Ship waiting rate • $tTrain$ = handling time of train • Twt = truck waiting time **Depending on:** • Handling technologies • Operational rules • Terminal dimensions
Terminal occupancy	$T_{occ} = \frac{nVq}{nV}$	Rate of the number of vehicles in the queue related to the number of vehicles within the terminal • nVq = number of vehicles in the queue • nV = number of vehicles in the terminal **Depending on:** • Technologies • Operational rules • Terminal dimensions
Maintainability indicator	$RAMS_M = \frac{nITU}{nMc}$	Maintainability of the terminal equipment • nMc = maintenance cycles of terminal equipment per year • $nITU$ = number of handled ITU per year **Depending on:** • ITU Throughput • Technologies • Number of handling equipment • Operational rules
Reliability indicator	$RAMS_R = \frac{nITU}{(nIEE + nIB)}$	Reliability of the terminal taking into account interruptions caused by equipment failures or external events (e.g. bad weather conditions) • $nIEE$ = number of interruptions due to external events per year • nIB = number of interruptions due to terminal equipment failures per year • $nITU$ = number of handled ITU per year **Depending on:** • ITU Throughput • Technologies • Number of handling equipment • Operational rules

(continued)

Table 1 (continued)

Definition		Description
System utilization rate	$\varrho = \frac{\lambda}{\mu}$	Queuing theory basic formula, useful to measure the correct sizing of different sidings
		$\bullet$ ϱ = system utilization
		$\bullet$ λ = average rate of arrivals
		$\bullet$ μ = average rate of served
		Depending on:
		$\bullet$ External infrastructures and transport services
		$\bullet$ Technologies
		$\bullet$ Operational rules
		$\bullet$ Terminal dimensions
Personnel distribution rate	$P_r = \frac{n_{um}}{n_{ut}}$	Personnel distribution, useful to measure the number of employees required in the terminal, split by various operation and the possible personnel reduction
		$\bullet$ P_r = personnel distribution
		$\bullet$ n_{um} = number of terminal employees
		$\bullet$ n_{ut} = total number of the employees of the yard
		Depending on:
		$\bullet$ Technologies
		$\bullet$ Operational rules
		$\bullet$ Terminal dimensions
		$\bullet$ Training frequency and level

Table 2 Rail–Road inland terminal key performances indicators

Definition		Description
Total transit time (ITU and vehicle)	$TTR = \sum_{i=1}^{n} TWi + \sum_{i=i}^{n} TOi$	Time period from the arrival of the freight unit (or vehicle) to the terminal gate from rail or road to the exit of the unit (or vehicle) from the terminal towards rail or road • TTR_v = vehicle total transit time (train and truck) • TTR_{ITU} = unit total transit time • TW = waiting time • TO = operational time **Depending on:** • Road and railway network infrastructures and transport services • Technologies • Operational rules • Terminal dimensions
Utilization rate of handling equipment	$Er = \left(\frac{nETr}{nE}\right) Th$	Average number of handling equipment, engaged on a train during the handling time (Equipment rate utilization in handling area) • Er = utilization rate of handling equipment • $nETr$ = number of handling equipment employed per train • nE = total number of handling equipment available in handling area • Th = handling (loading/unloading) time **Depending on:** • Handling technologies • Operational rules • Terminal dimensions
Utilization rate of storage ITU	$S_{ITUi} = \left(\frac{(nITU_{in} + nITU_{s(t-1)} - nITU_{out})}{Cs_{max}}\right) T_i$	Influence of the number of intermodal units, the which transit through terminal, on the storage area capacity • S_{ITUi} = utilization rate of ITU storage area • $nITU_{in}$ = number of incoming ITUs in terminal • $nITU_{s(t-1)}$ = number of stored ITUs • $nITU_{out}$ = number of departing ITUs from the terminal • T = time gap (day, week, month or year) • Cs_{max} = maximum storage capacity • T_i = time gap

(continued)

Table 2 (continued)

Definition		Description
		Depending on: • External infrastructures and services • Technologies • Operational rules • Flow of ITU handled in the terminal
Energy consumption rate	$Ec(ITU) = \frac{Ec(v)}{nITU(v)}$	Energy consumption of handling equipment per ITU • $Ec(v)$ = energy consumption of handling equipment per vehicle • $nITU(v)$ = number of intermodal transport units per vehicle **Depending on:** • ITU Throughput • Technologies • Number of handling equipment • Operational rules
	$Ec(ta) = \frac{C}{S}$	Energy consumption of terminal area compared to its surface: e.g., terminal lighting, office consumption • C = energy consumption of terminal • S = terminal area **Depending on:** • ITU Throughput • Technologies • Number of handling equipment • Operational rules
Equipment performance	$Ep = \frac{nITU}{h}$	Capacity of handling equipment • $nITU$ = number of handled intermodal transport unit • h = hour **Depending on:** • Handling technologies • Skills of the equipment operator(s)
Equipment haul	$Eh = \frac{Er}{Lr}$	Influence of train length on the length of path covered by handling equipment • Eh = equipment haul

(continued)

Table 2 (continued)

Definition		Description
		• Ltr = train length
		• Er = length route for handling equipment in handling area
		Depending on:
		• Handling technologies
		• Operational rules
		• Terminal dimensions
Truck waiting rate	$TW_{rate} = \frac{Twt}{tTrain}$	Influence of handling time of train on the waiting time of trucks
		• TW_{rate} = Truck waiting rate
		• $tTrain$ = handling time of train
		• Twt = truck waiting time
		Depending on:
		• Handling technologies
		• Operational rules
		• Terminal dimensions
Terminal occupancy	$T_{occ} = \frac{nVq}{nV}$	Rate of number of vehicles in the queue and number of vehicles within the terminal
		• nVq = number of vehicles in the queue
		• nV = number of vehicles within terminal
		Depending on:
		• Technologies
		• Operational rules
		• Terminal dimensions
Maintainability indicator	$RAMS_M = \frac{nITU}{nMc}$	Maintainability indicator of the terminal equipment
		• nMc = maintenance cycles of terminal equipment per year
		• $nITU$ = number of handled ITU per year
		Depending on:
		• ITU Throughput
		• Technologies
		• Number of handling equipment
		• Operational rules

(continued)

Table 2 (continued)

Definition		Description
Reliability indicator	$RAMS_R = \frac{nITU}{(nIEE + nIB)}$	Reliability of the terminal, taking into account of interruptions caused by equipment failures or external events (e.g. bad weather conditions) • $nIEE$ = number of interruptions for external events per year • nIB = number of interruptions for terminal equipment failures per year • $nITU$ = number of handling ITU per year **Depending on:** • ITU Throughput • Technologies • Number of handling equipment • Operational rules
System utilization rate	$\varrho = \frac{\lambda}{\mu}$	Queuing theory basic formula, useful to measure the correct sizing of different sidings • ϱ = system utilization • λ = average rate of arrivals • μ = average rate of served **Depending on:** • External infrastructures and transport services • Technologies • Operational rules • Terminal dimensions
Personnel distribution rate	$P_r = \frac{n_{am}}{n_{at}}$	Personnel distribution, useful to measure the number of employees required in an intermodal rail–road terminal split into various operations • P_r = personnel distribution • n_{am} = number of terminal employees • n_{at} = total number of the employees of the yard **Depending on:** • Technologies • Operational rules • Terminal dimensions • Training frequency and level

Many methods are not suited to evaluate the performances of the terminal as a whole, because they are appropriate to evaluate a single aspect or not sufficiently flexibles and generalizable to different terminal typologies.

Finally, the requirements for suitable assessment methodologies are:

- To be generalizable to different rail freight terminal typologies;
- To allow the assessment of an as large as possible set of terminal performances;
- To be sensible to the introduction of new technologies and innovative operational measures.

The next sections describe some options to fulfil these requirements: a generalized approach, based on an analytical method, as well as a simulation procedure.

4.2 Analytical Methods Based on Sequential Algorithms

The constraints of the problem will derive from minimum requirements, particularly to be able to manage traffic perturbations due to congestion and/or technical/human failures.

Moreover, the methodological approach must be able to evaluate and compare conceptual innovations as well as technological improvements (e.g.: automatic units transfer for co-modal transshipment, wagons coupling with or without multi-function connections and human interventions, automatic marshalling for single wagon management optimization, electric self-powered freight vehicles, advanced information management systems to be interfaced with tracking and tracing systems).

Normally, the assessment of terminal operation is necessary from various viewpoints (operators, final customers and Community); therefore, the provided indicators should be flexible enough and effective for such varieties of perspectives.

The operational times inside the terminal represent the primary indicators for the multi-criteria assessment of their performances and key components to quantify many other KPIs, as well as the costs by the concerned stakeholders (terminal and vectors operators as well as Community).

Therefore, the quantitative analysis is a strategic activity, both in the terminal planning and operation and in the entire logistic chain organisation.

The global operational time include both deterministic and stochastic components, which increase significantly the problem complexity.

IT is a period from the arrival of the single freight unit to the terminal, by an external transport service, to its exit from the terminal itself, towards a different transport service.

A model finalized to the determination of the global time is basing on the following general formalisation [6, 7] (Fig. 4):

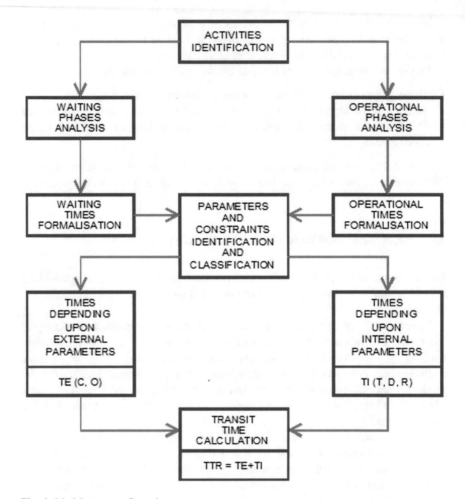

Fig. 4 Model structure flow-chart

$$T_{OG} = T_{EXT}(I, S) + T_{INT}(E, D, R) \qquad (1)$$

where

- T_{EXT} depends upon external constraints formalized in two sets of variables:

I   Infrastructures carrying capacity (e.g. railway lines and nodes bottlenecks),
S   Services operation planning (e.g. traffic density and timetable structures);

- T_{INT} depends upon internal constraints formalized in three sets of variables:

E Equipment performances parameters (e.g. check-in/out and units transfer technology);
D Dimensions of operational areas (e.g. distances between transfer and stocking areas, number of tracks);
R Rules to ensure safe operation (e.g. speed limits, maximum loading weights).

On this basis, for m generic activities it is possible to calculate a waiting time (TW) and for n $\geq$ m generic activities a corresponding operational time (TO) [8].

Therefore, the formalization of the global time spent in the terminal is the following:

$$T_{OG} = \sum_{i=1...m} T_{Wi} + \sum_{j=1...n} T_{Oj} \tag{2}$$

For a generic terminal, the following single or multiple activities may be further split into more elementary actions according to the required level of detail:

- Vehicle entering;
- Unit or vehicle check-in;
- Unit or vehicle transfers;
- Unit or vehicle check-out;
- Vehicle exiting.

Moreover, in each intermodal terminal it is possible to identify two classes (V′ and V″) of vehicles.

In general, the vehicles can transport various amounts of freight units; nevertheless, the following macroscopic rules exist [9]:

- Rail–road terminals (e.g. inland terminal): NU′ (truck) < NU″ (train);
- Rail–waterway (e.g. maritime terminal): NU′ (train) < NU″ (ship).

Figure 5 shows the single activities performed by a freight unit (e.g. a container) from Rail to Road in an inland terminal.

Three typologies of activities are there:

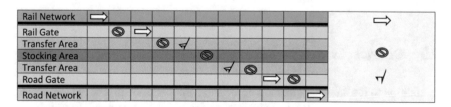

Fig. 5 Schematic representation of the Train–Truck flow in an inland terminal

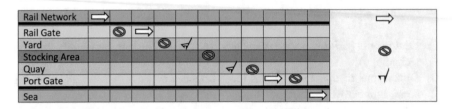

Fig. 6 Schematic representation of the Train–Ship flow in a port terminal

   (i)   Ro-Ro movements on-board a vehicle (train and truck);
  (ii)   Lo-Lo transfer from/to vehicles and stocking area;
 (iii)   Waiting for the following activity on-board or in the stocking area itself.

In case of direct train–truck transfer, the second transfer and the corresponding waiting phase are missing.

A similar representation is obviously feasible for the opposite Road to Rail flow.

Figure 6 shows the single activities performed by a freight unit (e.g. a container) from Rail to Sea in a port terminal.

The typologies of activities are there:

   (i)   Movements on-board a vehicle (train or ship);
  (ii)   Transfer from/to vehicles and stocking area;
 (iii)   Waiting for the following activity on-board or in the stocking area itself.

In case of direct train-ship (tracks located on the quay), the second transfer and the corresponding waiting phases do not apply.

A similar representation is obviously feasible for the opposite Sea to Rail flow.

The main performances of the model relate with the possibility to calculate key parameters concerning the operation of freight terminals.

It allows evaluating the development of internal activities, to quantify the duration of waiting and operational phases, to estimate the utilization rate in comparison with the capacity of single sub-stations and the whole terminal.

In a wider context, it is possible to assess alternative operational framework, including innovations capable to modify the state-of-the-art conditions, based on innovations in technologies and/or operational measures, by the quantification of the operational changes induced by them.

4.3 Discrete Events Simulation Models

In addition to the analytical method, to evaluate the performances of the terminals it is possible to setup a suitable simulation model and the corresponding tools.

In the literature, the most appropriate simulation processes are basing on the discrete events, with elements corresponding to the individual operative phases in the terminal [10].

The main modelled elements are normally:

- ITUs, trucks and trains in the rail–road freight terminals;
- ITUs, trailers, ships and trains in the rail–water freight terminals.

An example of model, developed by the authors and basing on the Planimate® software, allows the building of discrete event micro simulative models.

Thanks to its flexibility, it is particularly suitable for simulating complex systems, which use large amounts of data and sub-processes, with parallel and synchronized loops, ensuring an easy monitoring of the evolution of the system, with the capability to model the time flow.

The simulation model includes four main phases related to the design of the following elements:

- Objects;
- Flows;
- Interactions;
- Graphics.

The result of these phases is a multiple graph, which represents the static properties of the system, while the dynamic properties are depending upon the operational rules of the network, in particular:

- An event occurs as soon as all the pre-conditions are enabled;
- The occurrence of an event disable the pre-conditions and enables the post-conditions.

The basic elements of the simulation tool are the following (Fig. 7):

- Objects: fixed entities within the system, able to change the properties of entities that run through them during the simulation or to retain these properties for a certain period of time;
- Items: dynamic entities (such as, for example, orders, customers, operations, etc.), moving within the system and coming out of it, moving from one Object to another one.

The state of the system corresponds to the set of active conditions, while the Items, which can move from one Object of the network to another one through paths that represent a logical sequence of events between two or more Objects, determine the evolution of the system.

Fig. 7 Graphic interface for objects and items

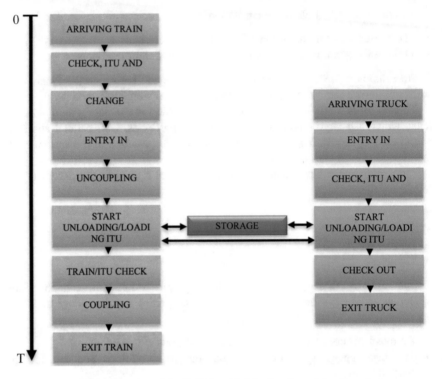

Fig. 8 Schematic representation of the Rail–Road freight flow in an inland terminal

Therefore, once identified the Objects necessary for designing the model, it is possible to build the Paths that enable the Items to move from one Object to another one, by creating the succession of steps that are necessary for simulating the system evolution.

For each class of Items it is possible to define a sequence of steps, animated during the simulation, which allows Items to move between Objects.

The set of the Paths represent the Flow, where more Items may move simultaneously during the simulation.

Specialized model for inland intermodal terminals

Generally, the structure of a discrete event simulation of a Rail–Road intermodal freight terminal is similar to the scheme in Fig. 8.

The model can reproduce both the rail side and the roadside of the terminal.

The model includes subsystems in order to characterize all the phases described above [11], such as:

- Trucks arrivals area;
- Trucks check in area;
- Train arrivals area;
- Train holding track;

- Operative module, including;

  - Operative rail track under gantry;
  - Operative road lane;
  - Storage lane;
  - Departure truck area.

The simulation provides with extensive information about terminal operation (Fig. 9) organized in table elaborated achieving results for different aspects, generally: timing, vehicles, procedures, ITUs, equipment transfer, terminal layout, etc.

(a)

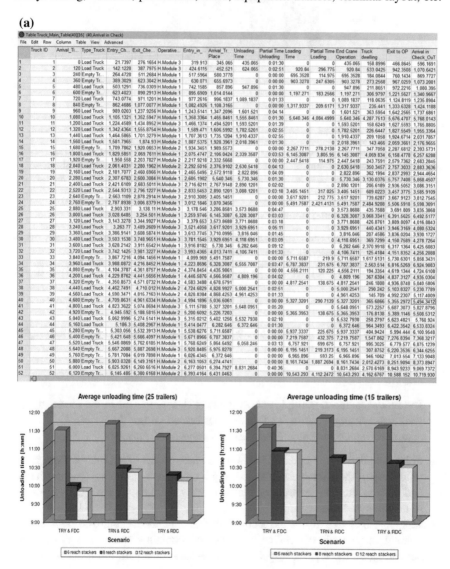

Fig. 9 Elaboration of outputs produced by the simulation model

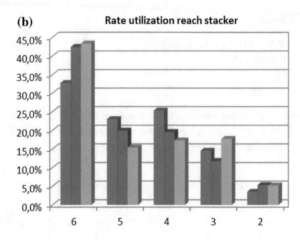

Fig. 9 (continued)

Specialization of model to port terminals

The model simulates a generic container terminal in a port and includes various subnetworks, which reproduce all the functions needed to operate the plant.

After the data collection carried out in the plant itself in order to design the specific simulation model, it is necessary to define the following Items, moving among the various subsystems:

- Truck: it picks up the container from stocks to bring them to their final destination or brings them in stock, if they are to be shipped;
- Trailer: it is the vehicle that transports the containers from the quay to the storage and back;
- Reach stacker: it is the vehicle used for handling containers in the export and in the customs areas;
- Transtainer: it is the hoisting device used for handling containers in the import area;
- Container;
- Ship.

The model includes multiple sub systems [12], reproduced in the model by the Portals, representing a particular function performed within the terminal.

Figure 10 represents some examples of subsystems' portals:

- Ship: includes all the operations that take place on the quay; one of them is the unloading of the containers from the ship by means of portainers (gantry cranes) and their positioning on trailers, that will bring them to storage or, if requested, to customs and vice versa;
- Customs: includes all the inspections of the containers' contents: scanner and manual inspection; the trailer bringing the container enters the Portal and, depending on the type of inspection, addresses it on a path or on the other one;

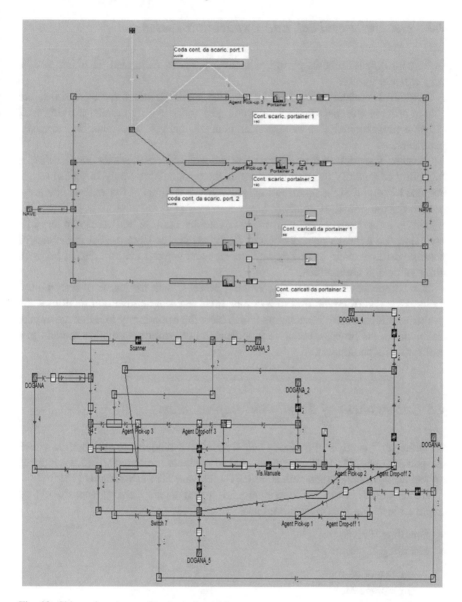

Fig. 10 Ship and customs subsystems (portals)

these operations considerably affect the time needed to unload containers from the ship.

The model allows obtaining multiple output data to process and obtain examples of impact of customs inspections on processing time of ships.

4.4   General Feedback and Application Fields

Quite consolidated models permit, whenever properly calibrated, to obtain results very close to reality.

Analytical models, basing on a generalized approach structured in modules, are able to provide with data on various typologies and size of terminals, working with different transshipment technologies, number of operators and other characteristic parameters.

Another very interesting requirement of such discrete events models is the capability to quantify the performances of the terminals, not only in global terms, but also highlighting the contributions and relative weights of the various operation needed for the transit of goods.

This is a fundamental aspect for an accurate detection of bottlenecks in terminal operation and a proper reproduction of the future processes to predict the effects of possible infrastructural or operational changes and suggest the best choice between different design alternatives.

The complexity of the decision-making process, with its degree of uncertainty, the number and different kind of relations involved, the number and quality of the goals to achieve, the different actors which have the opportunity to influence or take decisions on the process, make very widespread the use of simulation models as a tool for decision support.

4.5   Assessment of Terminals' Improvements

As anticipated, analytical methods and simulation's models are able both to check the present operation in the terminals and to evaluate the impact of improvements on technological and management sides by relevant KPI calculations [13].

Hereafter are some the operative and management elements considered as improvable in the intermodal terminals.

- Handling Typology;
- Handling Equipment:

  - In operative track,
  - In storage area,
  - Positioning and grab,
  - Devices for vertical handling;

- Handling Layout:

  - Track operative length;

- Terminal Access—ICT technologies:
  - ITU/Vehicles Identification and transport documentation exchange;
- Internal Moving Vehicles:
  - Locomotives;
- Technological Systems:
  - Control and security;
- Working periods;
- Conceptual Train Side layout;
- Conceptual Horizontal Handling.

Any innovation may have an impact on operational phases and related input parameters influenced by each improved terminal element [14, 15]: e.g., the track operative length related to the handling layout have an influence on:

- Mean distance between holding track (rail) and handling area;
- Number of transfer equipment;
- Length of train;
- Mean distance between rail track and handling;
- Number of operative track;
- Mean number of loading units per train.

Moreover, a cross check is necessary to check the reciprocal compatibility of innovations and to combine them into effective scenarios.

In Fig. 11 an example of this crosscheck compatibility process managed by a typical matrix approach.

The resulting compatible innovations are suitable to the progressive combination into effective scenarios (e.g. in Table 3).

4.6 Validation of Assessment Methods and Models

Figure 12 represents schematically a typical process for the validation of assessment methods and models by their pilot applications.

In the practical cases, the selection of validation parameters was normally basing on the amount and the reliability of data available for the present operational situation.

As a practical example of the validation process, it is here presented (Fig. 4) the key performance indicators results obtained using both the analytical method described in Sect. 4.2 and the simulation model described in Sect. 4.3 to the Munich Riem inland terminal in the framework of the Capacity4Rail Project [16].

Figure 13 shows the results of an accuracy assessment for validation purposes carried out for the simulation model (Sect. 4.3) basing on 6 years real world data.

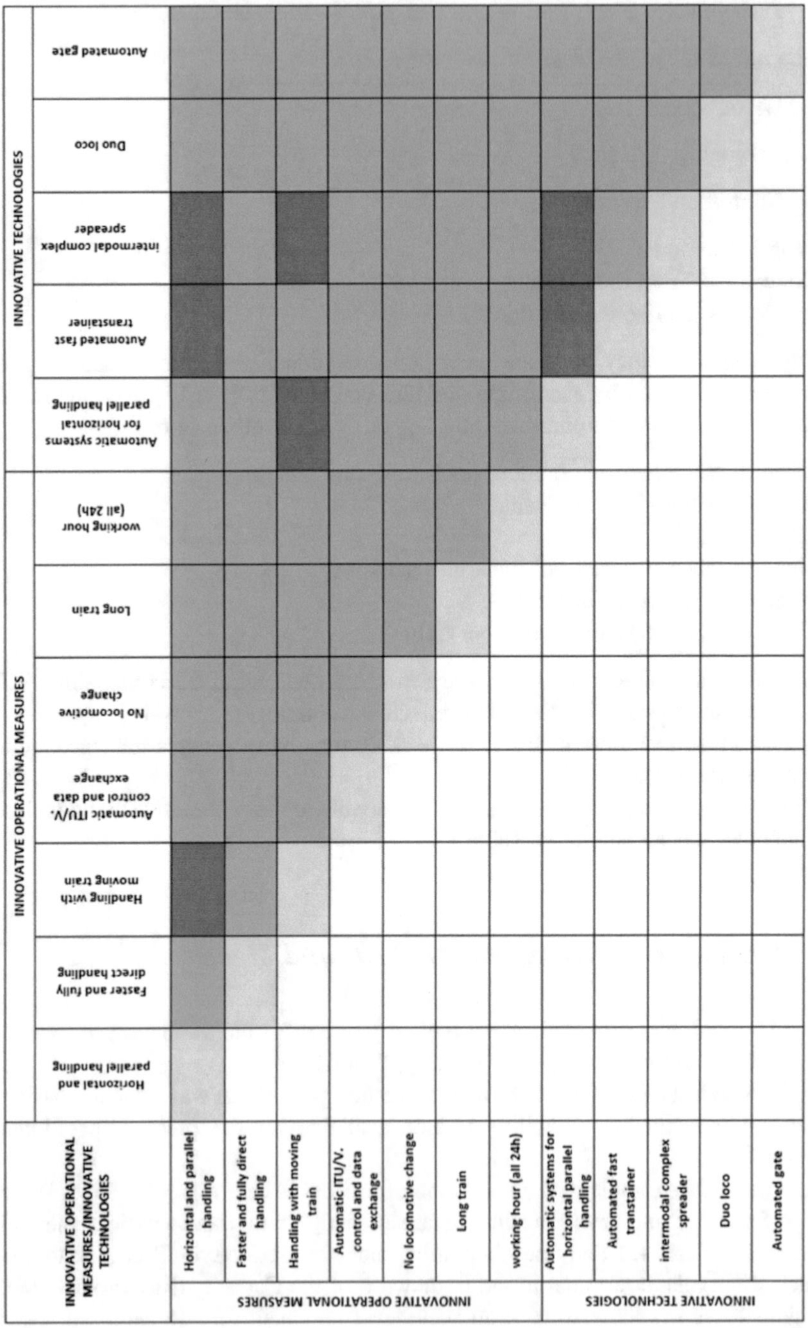

Fig. 11 Compatibility matrix among terminals improvements

Table 3 Example of future scenario for intermodal terminals

Handling Typology			
	Present standard	**Innovations (Step 1)**	**Innovations (Step 2)**
	- Indirect and direct	- Mainly direct	- Faster and fully direct

Handling Equipment 1			
	Present standard	**Innovations (Step 1)**	**Innovations (Step 2)**
	- Transtainer and reach stacker or forklift - Few systems for horizontal transfer	- Fast transtainer - More systems for horizontal transfer	- Automated fast transtainer with moving train - Automated systems for horizontal and parallel handling

Conceptual Layout – Train Side

Present standard	Innovations

A – Line   C1 - Check in   D - Holding   → Exit
B, B1 - Arrival   C2 - Check out   E - Operative   → Entry
B2 - Departure   Terminal area

Conceptual Horizontal Handling

Common standard	System change

The correspondence is 87% with reference to the yearly number of handled ITU and 99% with reference to the truck dwelling time: in this case, the results highlight that the model provides with a good reproduction of terminal operation, which is a solid basis for future scenarios simulation (Table 4).

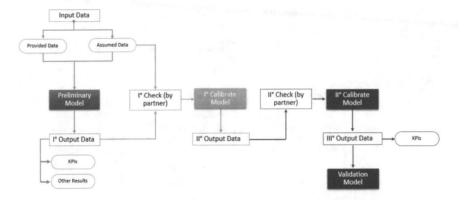

Fig. 12 Schematic process for validation and pilot application of methods and models

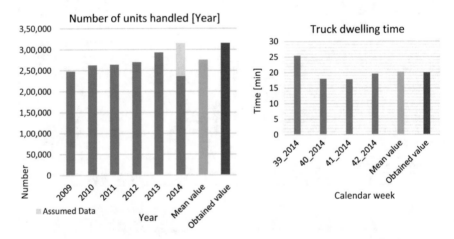

Fig. 13 Estimation of accuracy of simulation's results for Munich Riem inland terminal

Table 4 KPI calculation by analytical method and simulation model for Munich Riem inland terminal

KPI	Analytical method	Simulation model
Total transit time of ITU [h]	Truck–Train = 3.45 Train–Truck = 2.86	Truck–Train = 2.73 Train–Truck = 2.43
Total transit time of vehicles [h]	Train = 3.37 Truck = 0.80	Train = 2.30 Truck = 0.63
Energy consumption [kWh]	9.96	7.73
Equipment performance [ITU/h]	15.15	13.00
Equipment haul [m/m]	1	1
Truck waiting rate	7%	9%
System utilization rate	Train = 78% Truck = 33%	Train = 61% Truck = 38%

5 Carriage Application to Design Terminals' Improvements

5.1 Selection of Scenarios to Be Analyzed

Basing on the compatibility analysis of potential improvements to terminals, combinations of elements allow obtaining the scenarios to be analysed by means of the selected methods and models, taking into account a progressive temporal implementation of some operational measures and technologies.

Therefore, each scenario may be also qualified to represent temporal steps for the implementation of the innovations.

5.2 Key Performance Indicators Calculation by Methods and Models

The application of the selected analytical method and simulation model is able to provide results such as those in achieved in Capacity4Rail project [17] for an inland terminal (Fig. 14) and a maritime terminal (Fig. 15), by comparing the present situation (state of art) with possible improvement scenarios.

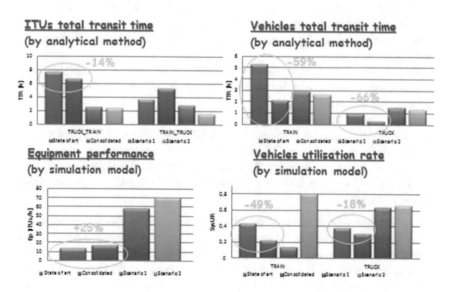

Fig. 14 Examples of results for an inland Rail–Road terminal (Munich Riem)

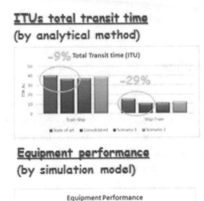

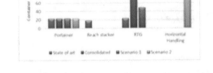

Fig. 15 Examples of results for a maritime Rail–Sea terminal (Valencia Principe Felipe)

Indeed, the selection of the most appropriate methodology is a key activity in such typology of operational assessment.

The validation phase is focusing on this choice, by comparing the reliability of the results achievable by available methods and models.

In principle, both analytical methods and simulation models are able to provide useful KPI calculations.

Nevertheless, the main purposes of the study, as well as the peculiarities of the terminals, can often suggest, by a careful validation activity, the most appropriate way to calculate each KPI.

As an example of this concept, the results showed in Figs. 14 and 15 highlight that the analytical model was selected for a reliable calculation of ITUs and vehicles total transit times, as well as the simulation models demonstrate its higher effectiveness to estimate equipment performances and vehicles utilisation rates.

The potential of such widespread results is very high and can provide useful feedback towards:

- Identification of new terminal typologies (e.g. smaller road–rail terminals, liner traffic terminals, multimodal marshalling stations, more integrated rail–maritime terminals) capable to cover a larger variety of operational contexts;
- Identification of additional scenarios more personalized on typical features of terminals as a result of consistency assessment among scenarios and achieved results;

- Integration of terminals analysis with logistic chains operational requirements;
- Introduction to economic analysis to quantify effects of improvements on operational costs and to depict the most promising scenarios from the point of view of key stakeholders (Society, operators, final customers, etc.).

6 Operational Costs of Terminals: Economical and Financial Analyses

6.1 Background

This section deals with the operational cost and the benefits linked to terminal management.

In many cases, the infrastructures in general and specifically the intermodal terminals can support the local economic development, the employment through productive activities, as well as the satisfaction of the transport needs of the local population.

The first step is to specify whether the concerned terminal is a consolidated infrastructure, a new construction, an extension or an upgrading of an existing one.

Moreover, the appropriate connection of the terminal to the correlated networks (rail, road, waterways) is a key success factor to foster the modal shift from road to the other environmental friendly modes.

Therefore, the economic and financial analyses should include all the relevant investments and operational costs to guarantee the correct functioning of the entire terminal handling process.

Society, operators' and final customers' viewpoints will be basing on investigations whether the terminal will let gain benefits from the investment and the operation (e.g. higher level of automation), so that additional traffic is shifting to rail and waterways based transports.

The users/customers, such as rail freight operators, trucking companies, maritime shipping will benefit from the investment in the terminal by a shorter turn-around time and higher asset utilization, among other aspects.

Moreover, the analysis must include the socio-economic aspects to explore and determine whether the terminal is able to improve the attractiveness of the freight intermodality, basing on the quantification of the present volume of goods traffic and the forecasts for the future pattern of flows.

6.2 Methodological Framework

This section aims to define the methods adopted for economic Cost Benefit Analysis (CBA) and Financial Analysis (FA) of terminals operations.

The calculation of costs and benefits is on an incremental basis, by considering the difference among alternative operational scenarios.

- CBA is basing on the public point of view, by comparing only differential costs and benefits paid or taken by the community;
- FA is basing on the private or business point of view of the subjects who runs the operations (and/or make it commercially feasible).

The main methodological is the Guide to cost-benefit analysis of investment projects of the European Commission issued in 2015 [3].

Cost Benefit Analysis (CBA)

The objective of CBA is to identify and monetize (i.e. to assign a monetary value to) all possible impacts in order to determine the terminal costs and benefits.

The results' are aggregation allows calculating the Net Benefits (Total Costs–Total Benefits).

The analyses are varying at different geographical levels, therefore a border has to be fixed on which costs and benefits should be considered as relevant, which typically depends size and operating range of the terminal, from urban to continental.

The Social Discount Rate (SDR) is towards present future values.

It reflects the social view on how net future benefits should be valued against present ones.

The European Guidelines suggest discounting costs and benefits by the following rates:

- Real SDR;
- SDR benchmark values:

  - 5.5% for Cohesion and IPA countries and for convergence regions elsewhere with high growth outlook;
  - 3.5% for Competitive regions.

Transfer, subsidies, VAT or other indirect taxes are not included in the calculation of future revenues.

Financial Analysis (FA)

The main purpose of the FA is to compute the project's financial performance. This is from the viewpoint of owners' of the infrastructure or operator of the service.

The traditional methodology for it is the Discounted Cash Flow (DCF) analysis, which implies the following calculation pillars:

- Consideration of cash flows only, i.e. the actual amount of cash paid out or received by the terminal operation, by excluding non-cash accounting items, like depreciation and contingency reserves. Cash flows are in the year when they occur and over a given reference period, with an additional residual value when the actual economically useful life of the project exceeds the reference period considered.

- Aggregation of cash flows occurring in different years: the time value of money justifies the discount back to the present of future cash flows using a time-declining discount factor.
- Check of financial sustainability as an important deciding factor. The financial sustainability of operation and investments should be assessed by checking that the cumulated (undiscounted) Net Cash Flows are positive over the entire reference period considered. It should take into account operational and investment costs, all financial resources and net revenues, without residual value, unless the liquidation of the asset is in the last year of analysis considered.

In mathematical terms, the Financial Net Present Value (*FNPV*) is:

$$FNPV = \sum_t B_t(1+i_t)^{-t} - \sum_t C_t(1+i_t)^{-t} - K$$

where is

- B_t = Benefits (inflows) in year t;
- C_t = Costs (outflows) in year t;
- i = Discount rate;
- K = Initial investment.

The Financial Rate of Return (*FRR*) on investment is the discount rate that zeros out the Financial Net Present Value (*FNPV*). A comparative benchmark allows evaluating the terminal performances.

In other words, if the FRR of the investment is the discount rate, the FNPV equals to zero:

$$FNPV = \sum_t B_t(1+FRR)^{-t} - \sum_t C_t(1+FRR)^{-t} - K = 0$$

As discussed in previous paragraphs, the first elements for the analysis are:

- Time horizon;
- Discount rate;
- Geographical scope;
- Benefits/Revenues.

The Financial Rate of Return (*FRR*) is the rate to discount at present future values in the financial analysis to reflect the opportunity cost of capital.

Considering the nature of intermodal terminals, the most common reference values is 5% to assume a higher minimum remuneration rate for the private investor.

The project will generate their own revenues from the sale of terminal services. This revenue depends upon the forecasts of the quantities (number of containers, wagons and trains handled or loaded/unloaded) of services provided and by their competitive prices for different types of cargo as well as services.

Choice of time horizon

The maximum number of years for which operational and infrastructure forecasts regarding the future of the project refers to a period appropriate to its economically useful life of the main assets and long enough to encompass its likely mid-to-long term impact, are reliable.

The forecasts typical time horizon is 15–25 years for short-life equipment/asset. For the analysis of terminal handling assets, the current analysis frequently reaches 30 years.

6.3 Traffic Estimations

The general performance of the freight market is strictly linked to the economic contingency: e.g. in Europe it sustained a growth for many years (2.8% per year on average in the period 1995–2007), a decrease in the years 2008–2011, due to economic crisis, and again a recovery in recent years, with a present share for rail of about 18%.

In this context, it is realistic to estimate different scenarios for freight traffic development, including, at least, a Business As Usual (BAU) scenario and some more developed scenarios, strictly linked to the economic dynamics.

As an example, Table 5 shows the increase factor and the yearly percentage increase for the railway freight scenarios corresponding to EU White Paper intermodal road to rail shift targets.

The consequences on the intermodal terminals is immediate: as examples, Figs. 16 and 17 show the corresponding increase in terms of yearly number of transshipped ITUs in the Munich Riem (Rail–Road) and Valencia Principe Felipe (Rail–Sea) terminals according to [17].

According to the traffic forecast, for all terminals the benefits are in transit timesaving, decrease of external costs and extra traffic revenues.

Table 5 Railway freight increase factors and percentage of increase per year

	Increase factor			Yearly % increase		
	2015–2030	2030–2050	2015–2050	2015–2030 (%)	2030–2050 (%)	2015–2050 (%)
Business as usual	1.16	1.17	1.37	1.0	0.8	0.9
Modal shift low scenario	1.34	1.38	1.87	2.0	1.6	1.8
Modal shift high scenario	1.65	1.84	3.06	3.4	3.1	3.2

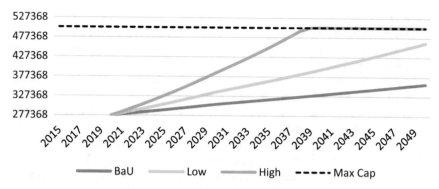

Fig. 16 Forecasted increase of transshipped ITUs in Munich Riem terminal

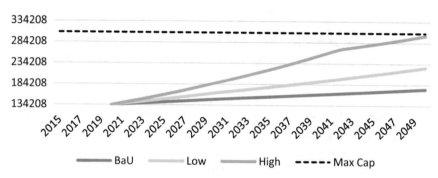

Fig. 17 Forecasted increase of transshipped ITUs in Valencia Principe Felipe terminal

6.4 Costs Analysis

The cost of a transport service, in economic terms, is the value of the resources consumed to operate it and maintain the system in operational conditions, including both pure operational and investment costs.

Operational costs

The operational costs normally include disbursements for the purchase of goods and services consumed within each accounting period.

For the terminals, they will include skilled crane operators, loco drivers, other operational personnel, administrative and management personnel, energy for power and lighting, buildings management, operational management systems, etc.

Costs for shunting engines are not included in the terminal costs as soon as the rail operators provide them.

Investment costs

Investment costs include the procurement of terminal equipment (gates, shunting locomotives, cranes, etc.) to recover as yearly capital costs by depreciation along the operation period.

For basic investment in terminal infrastructures, the depreciation period is normally 30 years, as well as for Gantry cranes.

For reach stackers and other lifts, the depreciation period is normally 5 years, with a residual value of 30%.

As examples, operational and investment costs for Munich Riem terminal calculated in Capacity4Rail EU Research Project are in Tables 6 and 7.

From the same source, the cost distribution by operational and capital costs for various typologies of terminals is in Figs. 18 (total) and 19 (by unit).

The operational costs, which the terminal operator themselves should be able to control, is 40–60% of the total cost, which grows to 60–85% of the total costs by adding the capital costs for technical equipment.

The capital cost for basic terminal investment, not always directly paid by the terminal operator, stands for 15–40% of the total cost.

The cost per unit is more interesting than the total cost because it relates to the actual utilization.

Table 6 Example of operational costs for Munich Riem terminal

DUSS Munich-Riem terminal	Share	Cost €	Source
Annual terminal operational cost components/items	%	Thousands	
Annual transhipment equipment running/hire (excluding procurement) cost	5.8	487	DB
Annual transhipment equipment maintenance cost including procurement of spare parts but excluding major procurement/investment	12.6	1053	DB
Annual Personnel cost (split into salaries + social/health/ pension insurance)	43.1	3585	DB
Annual insurance cost (equipment + operation)	1.7	142	DB
Annual energy cost	4.1	338	DB
Annual terminal hire/rent/mortgage/bank interest cost	3.9	323	DB
Annual infrastructure maintenance cost	9.8	813	DB
Other terminal costs (fuel tanks, truck depots security and others)	9.6	802	DB
Rent	4.2	350	DB
Annual cost for shunting engine	5.2	433	KTH model
Cost in thousand Euros—Total (average for the period 2011–2014, excluding VAT)	100	8326	

Table 7 Example of investment costs for Munich Riem terminal

Investment costs					
	Unit	Cost		Cost	Share
		Number	€/Unit	Thousands €	%
Terminal investment					
Land acquisition (m²)	m²	280,338	25	7108	6.9
Connection track 200 m (5 tracks)—track foundation	m	1000	317	317	0.3
Connection track 200 m (5 tracks)—track structure	m	1000	634	634	0.6
Points (switches) (excluding heaters)	m	45	169,035	7607	7.3
Handling tracks—track foundation	m	9800	317	3106	3.0
Handling track—track structure	m	9800	634	6212	6.0
Shunting tracks—track foundation	m	8000	317	2536	2.4
Shunting tracks—track structure	m	8000	634	5071	4.9
Buffer stop	No.	5	4226	21	0.0
Catenary to the handling tracks (200 m)	m	600	1056	634	0.6
Catenary to other tracks	m	8000	1056	8452	8.2
Road link to the main network	m	2800	53	148	0.1
Fences, gates, barriers	m	2880	37	106	0.1
Security equipment (cameras/alarms)	m	2880	53	152	0.1
Handling and space requirements—dim. 110-tonne axle load	m²	138,171	116	16,057	15.5
Administrative building and maintenance depot (m²)	m²	800	528	423	0.4
Fuel tanks	No.	2	4226	8	0.0
Lighting	m/ track-m	301	1056	318	0.3
Drainage	m	9800	106	1035	1.0
Noise barrier	No.	3	2,112,939	6339	6.1
Crane runway	No.	3	4,014,584	12,044	11.6
Rainwater retention	No.	1	1,584,704	1585	1.5
Forch water	No.	1	316,941	317	0.3
Spill through	No.	1	105,647	106	0.1
Land examination	m²	0	–	–	0.0
IT system	No.	3	306,376	919	0.9
Sum		700	–	81,254	78.5
Technical equipment					
New reach stacker	No.	1	475,411	475	0.5
Used reach stackers	No.	1	158,470	158	0.2
RMG cranes	No.	6	3,486,350	20,918	20.2

(continued)

Table 7 (continued)

Investment costs					
	Unit	Cost		Cost	Share
		Number	€/Unit	Thousands €	%
Locomotive	No.	1	739,529	740	0.7
Sum				22,292	21.5
Total investment costs				103,545	100.0

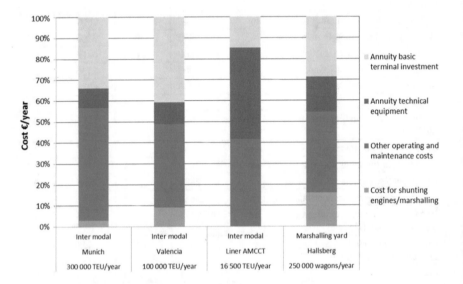

Fig. 18 Distribution of total costs per year

The operational cost per TEU results to be 28 € for Munich, 15 € for Valencia and 6 € for a generic automatic liner terminal (no shunting engines and no personnel).

The operational costs summed to the capital costs for the technical equipment is normally a medium term cost accounting to 33 €/TEU for Munich, 18 €/TEU for Valencia and 12 €/TEU for the generic automatic liner terminal.

This is nearby to the market price for handling units and wagons (e.g. about 30 €/TEU in Munich and Valencia).

The total cost is also including basic investments for building a new terminal: the results becomes 49 €/TEU for Munich, 30 €/TEU for Valencia and 14 €/TEU for the automatic liner terminal. This cost can be a reference whenever the renewal of the terminal is necessary.

As a comparative term, the socio-economic marginal costs for handling wagons on a not intermodal marshalling yard (located in Sweden) are also there:

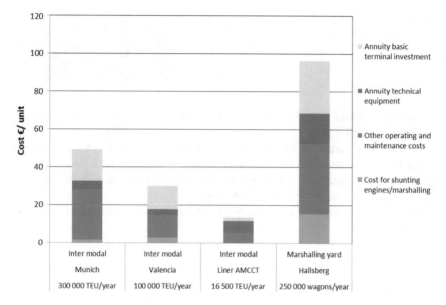

Fig. 19 Cost per unit with distribution of cost types

they account to 15 € per wagon and in Sweden the operator has to pay a fee of 7 € per train or 0.3 € per wagon, which makes an approximate total of 16 € per wagon.

By adding the yearly maintenance and operational cost for the infrastructure manager (52 € per wagon), the total costs will account to 96 € per wagon.

6.5 Economic Assessment

A first class of benefits takes into account the difference in transit time between among scenarios.

The amount of saved transit time is in combination with the correspondent transshipped tonnes and the Value of Time (VOT). Literature reports VOT for a general goods type between 1.0 and 1.7 €/t.

Moreover, for all the terminals, the total transit time of an ITU is normally depending upon the flow direction (truck to train or train to truck).

A second class of benefits takes into account the part of traffic exceeding the reference by assuming the transport where this is coming from.

The external cost variation between concerned transport modes: e.g. in the literature the cost reduction from Road to Rail is estimated in the range between 30.8 and 42.2 €/1000 t km.

The last class of benefits is due to the potential extra traffic revenues from transshipping activities due to the increased handling of ITUs: the most

Table 8 Net present values for different scenarios

Rate of return (%)	BAU	Low	High
2	NPV_{B2}	NPV_{L2}	NPV_{H2}
3	NPV_{B3}	NPV_{L3}	NPV_{H3}
5.5 (EU Guide)	$NPV_{B5.5}$	$NPV_{L5.5}$	$NPV_{H5.5}$

consolidated values in the literature are around 30 €/ITU for a direct transshipment (without storage).

In the Cost-Benefit Analysis (CBA), the Net Present Value (NPV) is calculated using various values of the rate of return: 5.5% is the value fixed by the EU Guide to Cost Benefit Analysis for Investment Project; lower values (e.g. 2–3%) are used in favor of long term upgrading investments.

Table 8 shows a matrix to calculate the NPV across scenarios and values of the rate of return.

6.6 Financial Assessment

The financial feasibility analysis compares costs with potential revenues.

The profit (or loss) is the difference between revenues and costs.

The results provide with feedback about the profitability from the start or from higher volumes only.

This approximates the profitability because there is no perfect market for terminal handling, which is an activity in the transport chain not always priced separately and not even for the full cost considering all capital costs.

In terms of traffic development, the analysed situations are normally including present handled traffic (BAU), normal (Low) and maximum (High) increases.

In conventional terminals, the handling cost will increase because they normally require more human resources, reach-stackers, gantry cranes and finally investments in larger handling area with higher volumes.

Nevertheless, the cost per wagon are normally decreasing by traffic volume.

Bibliography

1. Kordnejad B (2014) Intermodal transport cost model and intermodal distribution in urban freight. Proc Soc Behav Sci 125:358–372
2. Mangone A, Ricci S (2014) Modeling of port - freight village systems and loading units' tracking functions. IF - Ingegneria Ferroviaria 1:7–37
3. European Commission (2015) Guide to cost-benefit analysis of investment projects. Economic appraisal tool for Cohesion Policy 2014–2020, Bruxelles
4. Baldassarra A, Impastato S, Ricci S (2010) Intermodal terminal simulation for operations management. Eur Transp 46:86–99

5. Baldassarra A, Margiotta A, Marinacci C, Ricci S (2012) Containers management simulation in short sea shipping. In: International Research Conference on Short Sea Shipping 2012, Estoril
6. Ballis A, Abacoumkin C (2001) An expert system approach to intermodal terminal design. In: World Congress on Railway Research (WCRR), Cologne, November 25–29
7. Bektas T, Crainic TG (2007) An overview of intermodal transport. Université de Montréal, Publication CRT 07-03, Centre de Recherche sur les Transport, Montréal, Canada
8. Marinacci C, Quattrini A, Ricci S (2008) Integrated design process of maritime terminals assisted by simulation models. In: International Conference on Harbour, Maritime and Multimodal Logistics Modelling and Simulation, vol 1, pp 190–201
9. Ricci S (2014) Systematic approach to functional requirements for future freight terminals. Transport Research Arena, Paris
10. Ricci S, Capodilupo L, Tombesi E (2016) Discrete events simulation of intermodal terminals operation: modelling techniques and achievable results. In: Civil Comp Proceedings, vol 110
11. Davidsson P, Persson JA, Woxenius J (2007) Measures for increasing the loading space utilization of intermodal line train systems. In: Proceedings of the 11th WCTR, Berkeley
12. CEFIC-ECTA (2011) Guidelines for measuring and managing CO_2 emissions from transport. Coop Logistik AB operations. Brussels, Belgium
13. Ricci S, Capodilupo L, Mueller B, Karl J, Schneberger J (2016) Assessment methods for innovative operational measures and technologies for intermodal freight terminals. Transp Res Proc 14:2840–2849
14. Woxenius J (2007) Alternative transport network designs and their implications for intermodal transhipment technologies. Eur Transp 35:27–45
15. Woxenius J (1998) Evaluation of small-scale intermodal transhipment technologies. In: Bask AH, Vepsalainen PJ (eds) Opening markets for logistics. Proceedings of the 10th NOFOMA Conference, Helsingfors, pp 404–417
16. Antognoli M, Capodilupo L, Furiò Prunonosa S, Karl J, Marinacci C, Nelldal BL, Ricci S, Thunborg M, Tombesi E, Woroniuk C (2015) Co-modal transshipments and terminals (Intermediate). CAPACITY4RAIL. Increased Capacity 4 Rail networks through enhanced infrastructure and optimised operations. Deliverable D23.1
17. Antognoli M, Capodilupo L, Islam DMZ, Kordnejad B, Karl J, Marinacci C, Nelldal BL, Ricci S, Tombesi E, Woroniuk C (2016) Co-modal transshipments and terminals (Final). CAPACITY4RAIL. Increased Capacity 4 Rail networks through enhanced infrastructure and optimised operations. Deliverable D23.2

Development of the Silesian Logistic Centres in Terms of Handling Improvement in Intermodal Transport on the East-West Routes

Damian Gąska and Jerzy Margielewicz

Abstract The chapter presents the characteristics of Silesian logistics centres as the most important elements of logistics point infrastructure in the south of Poland, serving connections and integrating various branches of transport in East-West relations. The analysis was carried out on the example of the centre in Gliwice—located in the inland port in the west of the Silesian Province and the centre in Sławków—the point in Europe with the easternmost broad-gauge (1520 mm) railway line. These centres support the second largest in terms of population and the first in terms of industrialization region in Poland—the province of Silesia. The authors focused on presenting the development that has taken place over the last few years in both centres. The expansion and new equipment made it possible to significantly increase the throughput and capacity of handling and storage. The parameters of handling machines for container reloading have been characterized and compared, which in recent years have been installed in two container terminals. In addition, the warehouse space and its equipment, handling capacity, advantages and disadvantages of the applied solutions and development plans were presented. A separate theme is the characterization of the location of both logistic centres and accompanying facilities in the Polish transport system and on the east-west line. As well as the characteristics of the location giving opportunities for the development and increase of intermodality of transhipments and implementation of international and intercontinental transports. As part of this, the comparison of reloading machines and trends in their application and construction are presented. The chapter is concluded with a summary and conclusions regarding the planned increase in cargo transport about the new Silk Road.

Keywords Intermodal transport · Railway transport · Logistics centres
Container terminals · Cranes · East-West routes · Silesia

D. Gąska (✉) · J. Margielewicz
Department of Logistics and Aviation Technologies, Faculty of Transport,
Silesian University of Technology, Krasinskiego 8, 40-019 Katowice, Poland
e-mail: damian.gaska@polsl.pl

J. Margielewicz
e-mail: jerzy.margielewicz@polsl.pl

1   Poland as a Key Country for Transcontinental Logistics

Optimal configuration of logistics infrastructure is the biggest challenge for both businessman's and the authorities of a country or region. It is not possible to develop without the appropriate amount and length and technical condition of road, railway and inland waterway infrastructure in each area. These conditions both the development of a given area, companies and individual territorial units. The most important element, however, is the integration of various transport systems into one reliable network that will enable efficient, economic and ecological transport of various goods necessary in the global economy. Such integration is ensured by the appropriate point infrastructure in the form of logistic centres. The developed and efficient transport system, as a unified transport potential covering all branches of transport, is an important condition conducive to the creation of logistics centres. The logistics centre is located at the crossroads of important (international) transportation arteries and is a point element in transport infrastructure with a high degree of technical and organizational complexity. It is equipped with elements such as: intermodal transport node, modern warehouse space, transhipment platforms, modern office facilities, customs post, gas station for various means of transport, technical service and repair point of transport equipment, IT infrastructure, bank, post office, insurance offices, hotel and catering facilities and others [1].

Modern transport logistics is largely based on the transport, storage and transhipment of ISO containers both in short-distance, international and intercontinental connections. A significant concentration of fixed and investment assets on a limited number of point facilities of transport infrastructure requires adequate provision of large transhipment and storage needs. This can be achieved by closing dispensed check-in desks on the rail or road network and organizing transport between individual logistics centres. Within this system, carriers from various transport sectors (rail, car, water and air) cooperate, to deliver cargo in containers in the simplest relation from the sender to the recipient (so-called "door to door"), by means of one or several transport modes.

The contemporary approach to transport requires a comprehensive look at the entire transport chain. It is necessary to depart from the branch assessment of transport, and look at it as one whole structure. Modern intermodal transport is such a contemporary expression of trends in modern transport. The advantages of intermodal transport are primarily a reduction in transport costs and ensuring a fast and timely delivery of cargo, especially in international transport and reducing the risk of damage to the goods. In addition, increasing the possibility of disposable transport of a larger amount of cargo and an increase in the number of possible modes of transport and ecological aspects resulting from the use of railways and water transport. Basic disadvantages, and at the same time development opportunities for logistic centres is the necessity of using specialized handling machines (e.g. cranes, reachstackers), which provide the possibility of transporting multi-ton cargo units outside the reach of stationary equipment and the need to equip railway terminals with appropriate handling devices.

The key element of intermodal transport is transhipment between various means of transport belonging to separate transport modes. The logistics centre as an integrator of these branches is the key to the efficient implementation of transport and logistics processes in international relations. The scope of services provided by the centres is determined by efficient and very fast transshipment between individual means or modes of transport. The use of individual means of transport in the transport of goods is determined by the distance. On the shortest sections, transport is carried out by road, on further routes by rail and as part of the global economy by means of sea transport. There are, of course, deviations from this rule, but one remains unchanged—the need for transhipment. In this regard, it becomes necessary to equip logistics centres with appropriate technologies and handling machines.

Typical reloading machines in the integration of means and modes of transport are overhead cranes, reachstackers, forklifts and bulk cargo handling machines—tipplers and conveyors. However, transshipment itself is not everything. It is unnecessary to equip the transhipment points with spaces for the storage of individual goods and passive means of transport (e.g. containers) and warehouses for protection against weather conditions. In addition, it is necessary to increase the security of the entire transhipment and storage process. Such points become complicated, both in technical, organizational and IT terms, places that meet the increasing requirements of clients and enable the implementation of most logistic services.

Poland is one of the countries whose transit location determines the flow of cargo in the global economy, especially in the era of globalization. The immediate neighbours of Poland are the countries of the former Eastern bloc like Belarus and Ukraine and in the west Germany. The latter is Poland's main trading partner, with which goods exchange accounts for 40% of the total import and export of goods from Poland and to Poland. On the eastern border there is mainly one-sided exchange—import of goods to Poland with a small share of exports. The reason for this state of affairs is a certain collapse of the Ukrainian market and a flood of cheap products from Asia and energy raw materials from Russia. The majority of rail transport is carried out using containers, whose share in the total number of loading units is estimated at 98%, with 40-foot containers dominating in rail transport with a share of 50–60%. In the last few years, an extraordinary boom in container transport by rail in Poland has been noted.

The second aspect of transport on the East-West line relates to the transit location of Poland as a link between East and West not only in Europe but also in Asia. Currently, most of the goods reach Europe from Asia by sea, but more and more often (both in political and business circles) talk about the development of the land route as a new silk route. The amount of freight carried by rail in this relation is only 2% of the total exchange between Asia and Europe. However, this is to change in the coming years in favour of land transport, especially rail transport. However, this will require appropriate line and point infrastructure in the form of onshore container terminals and logistics centres. The new Silk Road may contribute to changing the 500-year global transport structure. In the past, all the most important countries in the world have based their power on maritime trade. China intends to add a land road to this, which will allow the development of both the Middle Kingdom, former Eastern Bloc countries and Europe, and allow the development and use of logistics centres, including in Poland.

According to information provided by railway carriers, despite the gradual development of the point infrastructure in Poland, intermodal transport is still much less competitive than in other European countries. Poland has a relatively large number of intermodal terminals, located quite symmetrically throughout the country [2]. However, these are small terminals, which do not constitute the strength of intermodal transport in Poland. Logistics centres are located only in some places, including Sławków and Gliwice in Silesia (Fig. 1). These are centres that, immediately after Polish seaports, have the highest capacity and handling possibilities in Poland.

Fig. 1 Container terminals and logistics centres in Poland. 1—Logistics centre in Gliwice (SCL Gliwice), 2—Logistics centre in Sławków (Euroterminal Sławków). *Source* Own elaboration based on [14]

Intermodal transport is indicated by the European Union as the most ecological form of transport in building supply chains. The intermodal transport market in Poland is still a relatively underdeveloped market, which over the last few years has been characterized by a constant development trend. The implementation of the European Union policy in the field of sustainable development requires a properly prepared transport network, because transport is crucial for the proper functioning of logistics processes. Properly developed transport infrastructure is the basis for organizing the flow of resources, such as raw materials, semi-finished products, finished goods or information, and the most important resources, such as employees [3].

For a long time, there have been systematically growing flows of intermodal transports across Poland, including directions:

east-west: China, Russia, Belarus:

- area of Poland (especially harbours in Gdynia and Gdansk container terminals located in Silesia and the western part of Poland),
- Germany (especially the harbours: Hamburg, Bremerhaven/Duisburg)
- The Netherlands (especially the harbour of Rotterdam), Belgium (including the harbour of Antwerp)

Logistics centres in Silesia mainly serve north-south directions in the field of transport and transhipment to Polish seaports and the east-west direction in the scope of:

- The logistics centre in Gliwice (SCL Gliwice) supports transport and reloading mainly towards the west,
- The logistics centre in Sławków (Euroterminal Sławków) supports transport and reloading mainly to the east.

2 Silesian Region and Its Location on the East-West Route

The Silesia region is in the southern part of the country near the border with the Czech Republic and Slovakia. It borders on the following provinces: Łódzkie, Świętokrzyskie, Małopolskie and Opolskie. In the radius of 600 km from Katowice (capitol of Silesia) there are six European capitals: Berlin, Bratislava, Budapest, Prague, Warsaw and Vienna. The Silesia region is located in the node area of two major European corridors that run from the West to the East and from the North to the South of Europe, these are:

- Corridor III—relation: (Madrid–Paris–Brussels) Berlin–Wrocław–Katowice–Kraków–Kiev–(Asia),
- Corridor VI—relation: (Helsinki) Stockholm–Gdańsk–Katowice–Žilina–(Budapest–Athens), with a branch VI B for the relation Częstochowa–Ostrava (Vienna–Venice).

The density of public roads in the Silesia region is by far the largest compared to other voivodships and regions in Poland and amounts to 176.6 km/100 km^2. The value of investment outlays incurred on public roads in the region is the highest among all voivodships in Poland. Through the Silesia voivodship runs: A1 (Gdańsk–Gorzyczki) and A4 (Jędrzychowice–Korczowa) highways and S1 (Pyrzowice–Cieszyn) and S69 expressways (Bielsko-Biała–Myto–Skalité). The point of view of transport is the very favourable location of road infrastructure in the region, including important international roads such as A1 and A4 highways in the north-south and east-west directions (Fig. 2). Despite the constant development

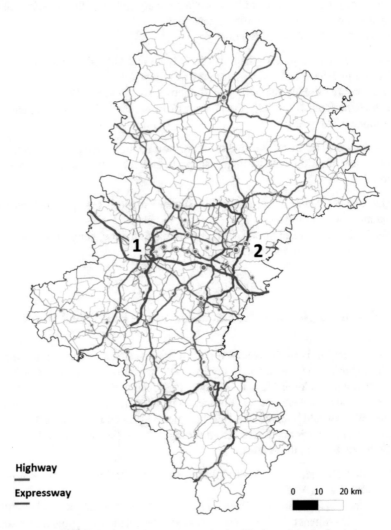

Highway
—

Expressway
—

0   10   20 km

Fig. 2 A network of highways and expressways in the Silesia region. 1—Logistics centre in Gliwice (SCL Gliwice), 2—Logistics centre in Sławków (Euroterminal Sławków). *Source* Own elaboration based on [7]

of road infrastructure on the Voivodship's roads, there are still so-called Bottlenecks, above all north-south in the northern part of the region. The east-west direction is very well connected due to the A4 motorway.

The Śląskie Voivodeship has the densest network of railway lines in Poland (16 km/100 km^2) [4]. It is twice as large as the average value in Poland. Many railways in the region are included in international and national rail transport systems. The Main Central Railway, which is a part of the international transport corridor C-E 65, deserves special attention. In Tarnowskie Góry, one of the largest marshalling yards in Europe is located, which is also the largest freight railway junction in the country. The final point of the non-electrified LHS broad-gauge railway line (1520 mm) is located in logistics centre in Sławków in Silesian region. This line through the Ukrainian railway system has direct access to the Trans-Siberian Railway. It gives the opportunity to connect with the rail system of Ukraine and Russia and create a pan-European Europe-Asia land transport corridor. Due to poor technical condition of the tracks and many years of lack of investment in the linear infrastructure, intensified works aimed at increasing the average speed of trains have been carried out for several years.

There are currently only a few short sections of waterways in the Śląskie Voivodeship. They are part of the Oder Waterway. The following parts belong to this waterway:

- Canal Gliwice (IIIrd class waterway, which enables connection of the Śląskie Voivodeship with Western Europe via Wrocław, Szczecin and inland canals of Germany). This canal has a beginning in the Gliwice inland port which is at the same time the headquarters of the logistics centre,

  - Oder River (km 51.2–98.6, fragment of the river located in the Silesian Voivodeship is a class Ia waterway),
  - Gliwice inland port (it is the beginning of the Oder Waterway and the Gliwice Canal). It is also the place where the Logistics Centre in Gliwice (SCL Gliwice) is located. The port in Gliwice is characterized by high communication accessibility. The A1 and A4 highways are located near the port.

The existence of various modes of transport in the Silesia voivodeship and the even distribution of transport networks have a positive impact on the development possibilities of logistics centres. However, it should be borne in mind that in the light of standards adopted in better developed countries, Poland is still at the stage of creating and developing a network of large and modern logistics centres. In the area of the Silesia region, apart from logistics centres in Sławków (Euroterminal Sławków) and Gliwice (SCL Gliwice), there are about several dozen logistic parks and storage centres, the largest of which have a storage area of up to 250,000 m^2. However, these centres are equipped only with storage infrastructure and they do not have the possibility of integrating transport modes due to the integration with only car roads.

3 Characteristics and Parameters of Silesian Logistics Centres

3.1 Transhipment Possibilities and Development of the Logistics Centre in Sławków (Euroterminal Sławków)

In the seventies of the twentieth century, when Huta Katowice (Katowice Steelworks) was founded in Upper Silesia, the only 394 km section of the broad-gauge railway was built on the territory of Poland, directly connecting the railway network of the then USSR with the Polish steelworks. This solution served an easier (without the need to reload goods or replace trolleys) exchange of materials between the Polish metallurgical plant and the Soviet Union. The broad-gauge route operated for many years under the name Linia Hutnicza-Siarkowa, because one of the transported components was sulphur. In addition, there is a broad-gauge railway siding to transport iron ore, which extends from Sławków to Huta Katowice located in Dąbrowa Górnicza. The total length of the line is approximately 394.65 km, of which about 2.2 km in the Śląskie Voivodeship. As part of the construction there was also a bulk material handling terminal which, after expansion, became a logistics centre under the name Euroterminal Sławków. Due to the location and convenient transport location with the international rail and road network—it is a very attractive place for organizing intermodal reloading.

Euroterminal Sławków benefited from several large EU subsidies for the construction of the International Logistics Centre (modern warehouses and a container slab were build). Thanks to the investments, Euroterminal Sławków diversified the handling capacity and was transformed from a typical bulk cargo terminal, mainly ore, coal and coke, into an intermodal terminal with the simultaneous option of handling and storing standardized goods—on pallets, in big bags and steel constructions and above all reloading and container storage.

Euroterminal Sławków has an excellent location and thanks to investments also a connection in domestic and international traffic. The most important railway routes that allow to connect the logistics centre in Sławków are:

1. Direct connection with the LHS broad-gauge railway line about 400 km long, via the border crossing Izow/Hrubieszów through Ukraine with the Far East (Fig. 3),
2. Access to the lines specified in the AGTC Agreement (agreement on main lines of combined transport), as a result of connections with lines:

   – CE30: Zgorzelec-Wrocław-Katowice-Kraków-Przemyśl-Medyka,
   – CE65: Gdynia-Gdańsk-Warszawa-Katowice-Zebrzydowice,

Fig. 3 Container terminals and logistics centres in Silesia as well as transport corridors and broad-gauge railway line to the border with Ukraine. 1 —Logistics centre in Gliwice (SCL Gliwice), 2—Logistics centre in Sławków (Euroterminal Sławków)

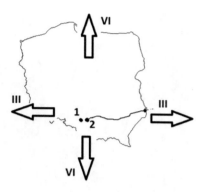

3. Access to the Europe–Asia transport corridors. The location of Euroterminal allows convenient connection both to the Pan-European Corridor VI and III (Fig. 3).

The most important road connections, thanks to the proximity of the main transport routes of Poland are:

1. A1 Highway (Warszawa–Bielsko Biała), distance 8 km from Euroterminal,
2. A4 Highway (Katowice–Kraków), distance 10 km from Euroterminal,
3. E40 Route (Katowice–Kraków), distance 5 km from Euroterminal.

Near Euroterminal there are also two international airports:

1. Kraków Balice, distance 57 km from Euroterminal,
2. Katowice Pyrzowice, distance 44 km from Euroterminal.

The Euroterminal Sławków logistic centre is the furthest west to which trains with broad-gauge railway can reach (1520 mm). Such railway lines are widely used in former USSR countries in opposite to Poland and other European countries with standard-gauge (1435 mm) railway lines. Sławków is therefore designed to handle rail transport to and from Ukraine, Russia and Kazakhstan as well as China and South Korea (Fig. 4). The terminal has permanent intermodal connections with the Polish Baltic ports and the Italian terminal at Maddaloni and the Schwarzheide terminal (Germany) 2 × per week. In diffused mode, containers are sent daily eastward, among others to Ukraine, to Russia or Kazakhstan [5].

Euroterminal estimates its transhipment capacity at almost 285,000. containers per year; it can also reload about 2 million ton of coal, 380 thousand tonne of steel products and approx. 365 thousand ton of other mass goods, such as salt, biomass or grain, and 200,000 tonne of goods on pallets. Part of the terminal area was included in the Katowice Special Economic Zone. Table 1 presents the handling capacity and basic information about the logistics centre Euroterminal Sławków.

In addition to terminal services in intermodal transport, reloading, security and shipment of palletized, bulk, metallurgical products as well as non-standard and oversized goods are also carried out in direct and indirect relations (via storage

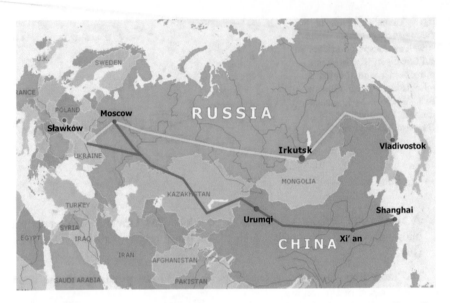

Fig. 4 Railway connections from Sławków are possible thanks to the broad-gauge railway line. *Source* Own elaboration based on [5]

Table 1 Basic characteristics of the logistics centre Euroterminal Sławków

Characteristics of the container plate	Unit	
Area of logistic centre	ha	91
Area of container terminal	ha	4
Loading capacity	TEU/year	284,810
Number of storage layers	–	5
Steel products	t/year	380,000
Goods on pallets	t/year	200,000
Commodities	t/year	2,000,000
Bridge cranes (load capacity 40/50 t)	–	2
Reachstackers	–	3
Storage capacity	TEU	3500
Connections to isotherms	–	90
Warehouse area	m^2	10,000
Distance to the national road	km	4 to DK1/S1 10 to DK4/A4 4 to S96
Distance to the railway line	km	5.7 to RL 665 2.2 to RL 674
Number and length of railway tracks for loading and unloading on terminal	-/m	7/700
Total length of railway tracks (1435 mm)	m	24,256
Total length of railway tracks (1520 mm)	m	17,521
Number of parking spaces for trucks	–	90

yards, warehouses) from broad-gauge railway wagons/standard-gauge railway wagons/road trucks. Transhipments are carried out in three main directions:

1. from broad-gauge railway lines (1520 mm) to standard-gauge railway lines (1435 mm) and in reverse direction,
2. from broad-gauge railway lines (1520 mm) to trucks and in reverse direction,
3. from standard-gauge railway lines (1435 mm) to trucks and in reverse direction.

Trans-shipments are carried out using cranes, jib-cranes, excavators, loaders. The logistics centre is also equipped with two reloading points for unloading loose goods (pellets, salt, grain) from Hopper type lower wagons in relation to trucks.

In recent years, many investments have been carried out at Euroterminal Sławków, the effect of which is the launch of the Universal Warehouse No. 2 (2013) and the expansion of the container deck with a container bridge crane (2013). The landfill of steel products as well as loose commodities were also put into use, as well as technical infrastructure accompanying railway tracks (broad and standard gauge), roads and parking lots as well as water supply, gas network and storm water drainage system. Thanks to these investments, Euroterminal currently has a technical infrastructure that provides comprehensive forwarding and transport services in Poland and abroad using rail and road transport. The container terminal after expansion in the logistics centre in Sławków has a total storage area of 40,000 m^2.

Figures 5 and 6, according to [6], show the quantities of container units reloaded in 2012–2016 and the forecasted value in 2017. Despite the infrastructure conducive to the disruption of transport in corridor III—on the east-west axis—in 2014 there was a breakdown and a definite decrease in the transported containers (Fig. 5). The reason for this is the outbreak of armed conflict in Ukraine. In the years 2012–2013, there was a dynamic increase in the number of transhipments and if this trend was maintained, we could be dealing with amounts comparable to the north-south direction. Currently, the terminal reloads mainly containers within the VI corridor in relation to Polish seaports and terminals in southern Europe (Fig. 6). The total planned quantity of reloaded containers in 2017 is 150,000 TEU, of which only 20% is in the east-west direction. However, this does not change the fact that the

Fig. 5 Total transferred container units in Euroterminal Sławków as part of the III transport corridor in East-West direction

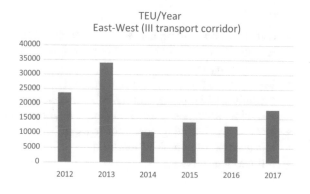

Fig. 6 Total transferred container units in Euroterminal Sławków as part of the VI transport corridor in South-West direction

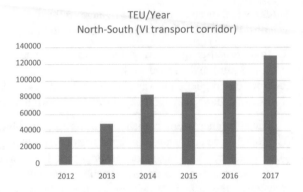

handling capacity is high, and the use of broad-gauge railway is only a matter of time.

In addition to the infrastructure related to container reloading, Euroterminal Sławków also has extensive warehouse infrastructure, also in the field of bimodal warehouses combining the road and railway infrastructure at the same time. There are 4 warehouses in the logistics centre:

1. Warehouse with an area of 2660 m², standard and broad-gauge railway ramps and truck docks. Capacity of 1780 europallets—the possibility of accepting 12 wagons with pallets and a minimum of 576 pallets for trucks daily
2. Warehouse with an area of 4860 m², standard and broad-gauge railway ramps and truck docks
3. 380 m² warehouse with a capacity of 200 europallets
4. 600 m² warehouse with a capacity of 350 europallets

Permanent railway connections carried out by the Euroterminal Sławków logistics centre are connections three times a week to the Deepwater Container Terminal in Gdańsk and in the reverse direction. In addition, twice a week to the logistics centre in Maddaloni and in the return direction, as well as connections in the dispersed traffic with the countries of the former Eastern bloc are served (Fig. 7).

Other services provided at the Euroterminal Sławków logistics centre are as follows:

- container hoist (filling and emptying containers),
- technical assessment and suitability for food products,
- container repairs,
- basic washing of containers and vehicles,
- possibility of crushing, sorting, mixing bulk goods,
- weighing wide-gauge, standard-gauge wagons, trucks,
- receiving and sending goods for transport,
- organization of combined transport,
- full terminal service,

Fig. 7 Bimodal warehouse
—rail-road. *Source* [5]

- customs service of goods in import and export,
- freight forwarding services in Poland and abroad,
- installation of flexi-tanks.

3.2 Reloading Possibilities and Development of the Silesian Logistics Centre in Gliwice (SCL Gliwice)

Silesian Logistics Centre S.A. commenced operations in 1989, as the Silesian Free Customs Area, a company established to establish and organize the Free Customs Area, as well as manage and administer inland ports in Gliwice and Kędzierzyn-Koźle. In 2002, the company's name was changed to the Śląskie Centrum Logistyki Spółka Akcyjna (SCL Gliwice). The city of Gliwice, by contribution of the land and property of the Gliwice port, has become the majority shareholder [7]. The logistics centre in Gliwice is located on the territory of the inland port—at the end of the Gliwice Canal (Fig. 8). For this reason, it is a place that connects and integrates three modes of transport—road, rail and water-inland.

Fig. 8 A view of the Port and logistics centre SCL in Gliwice. *Source* [8]

Port in Gliwice due to the layout and shape of port basins, constructions and length of wharfs (over 2.5 km), reloading facilities, storage yards and handling capacity (about 1.6 million ton per year) is the most modern and universal inland port in Poland. The Gliwice port has a direct connection with the Szczecin–Świnoujście seaports complex and with the entire waterway network in Western Europe (Fig. 9) [8].

Port in Gliwice is the largest and most modern facility of this type in Poland. The reloading of goods takes place here using jib cranes with a lifting capacity of up to 20 t (Fig. 10) or by using the unloading system of bucket wagons type FAS. All cargo can be weighed on site using electronic railway and car weights.

The Gliwice Canal and the port in Gliwice provide national and international transport of goods, especially mass and large-size goods. Due to obsolete devices and hydraulic engineering structures, transport capacity (6 million ton per year) is used only in 50%. Only barges with a maximum tonnage of 500 tonne can flow through the canal. The Gliwice Canal with a total length of 41 km (in the Silesia Voivodship is 22 km at the canal width of 37 m), connects the river port in Gliwice with the Oder River and further with the Szczecin-Świnoujście seaport providing access to the Baltic Sea, and through linking via inland canals waterway transport in Germany—the Elbe and Rhine—gives access to many ports of Western Europe [9].

For several years, the port was not used for water transport and reloading in this area. This condition resulted from the maladjustment of the Gliwice Canal and the waterway as well as watergates to operate the barges. In 2017, after the water watergates were renovated on the Gliwice Canal, the first transports of coal from

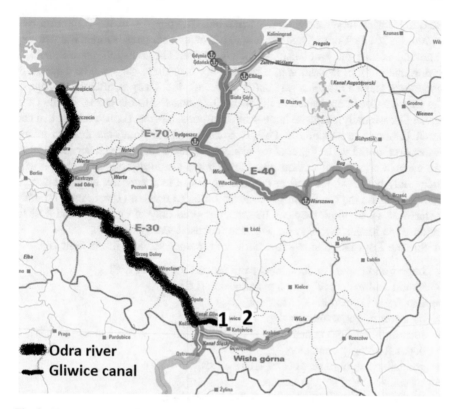

Fig. 9 The waterway of the Oder River and the Gliwice Canal against the background of waterways in Poland. 1—Logistics centre in Gliwice (SCL Gliwice), 2—Logistics centre in Sławków (Euroterminal Sławków). *Source* Own elaboration based on [15]

Fig. 10 Jib crane for bulk material handling in the Gliwice logistic centre

SCL in Gliwice to Wrocław were carried out. The purpose of the completed renovation in 2017 was to reduce the failure of water watergates, shorten the working time and reduce the amount of water consumed. The transport season starts in spring and ends in late autumn, especially for coal-filled barges. The carriage of goods is organized in such a way that first the individual barges, after the passage of the Gliwice Canal, are joined in the two sailing boats with the pusher on the Oder River. Occasionally, there are large-scale transports, e.g. steel constructions that can reach Germany or seaports in Poland, Szczecin and Świnoujście through inland waterways. Initially, the depth of the channel was 3.50 m, however, for many years the canal was operated without due care for the technical condition, that is why today only barges about 1.50 m dipping, or whose load capacity does not exceed 500 ton can use the canal. They are mostly outdated units, no longer used on other waterways. In addition, they can be sets composed only of one barge and pusher, because the longer, two-bar, do not fit in the canal watergates [10].

Silesian Logistics Centre (SCL Gliwice) provides services in the field:

- import and export of containers from around the world (FCL),
- import and export of groupage shipments (LCL),
- transports in relation door to door,
- transport of dangerous goods,
- securing (stowage) containers,
- depot for empty containers,
- specialized storage services of ADR goods,
- specialized steel storage services,
- Specialized paper and electronics storage services,
- specialized services for e-commerce,
- other services, including terminal services: reloading, sweeping, cleaning, minor repairs, power connections for refrigerated containers, assembly of flexi tanks.

Table 2 presents the handling capacity and basic information about the logistics centre SCL Gliwice in the Silesian Voivodeship.

In June 2016, the container terminal at the logistics centre SCL Gliwice was opened after modernization. The terminal is operated by PCC Intermodal dealing with intermodal transport, which has several railway and road container terminals in Poland and Germany. The entire investment–reconstruction of the container slab, purchase of 2 gantry cranes, construction of railway tracks and adaptation of the terminal to higher capacity and efficiency costed over 12 million EUR. The operational area of the terminal is 50,000 m^2, with a storage capacity of 2.9 thousand TEU. Annual terminal handling capacity is 150,000 TEU. The terminal has 4 railway lines with a length of 650 m each, 2 gantry cranes, 3 reachstackers, 50 parking spaces for trucks and 60 power connections. The terminal is connected by regular railway connections with the ports in Gdańsk and Gdynia, Central Poland— the terminal in Kutno, Lower Silesia—with the terminal in Brzeg Dolny, Frankfurt/ Oder, Hamburg, Rotterdam, Antwerp and Brześć on the Polish border with Belarus (Fig. 11). Every day the terminal organizes hundreds of road transfers to: Katowice,

Table 2 Basic characteristics of the logistics centre SCL in Gliwice

Characteristics of the container plate	Unit	
Area of logistic centre	ha	47
Area of container terminal	ha	5
Loading capacity	TEU/year	150,000
Number of storage layers	–	4
Commodities	t/year	1,600,000
Bridge cranes (load capacity 41 t)	–	2
Reachstackers	–	3
Jib crane for bulk materials (20 t)	–	1
Storage capacity	TEU	2900
Connections to isotherms	–	60
Warehouse area	m^2	28,000
Distance to the national road	km	10 to A1 10 to DK4/A4 4 to S96
Distance to the railway line	km	0 to E30/C-E30 15 to E65/C-E65
Number and length of railway tracks for loading and unloading on terminal	-/m	4/650
Total length of railway tracks (1435 mm)	m	11,000
Total length of railway tracks (1520 mm)	m	0
Number of parking spaces for trucks	–	50

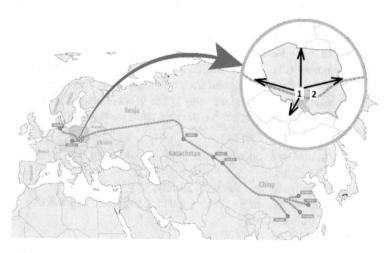

Fig. 11 Railway connections from Gliwice in the east and west direction. 1—Logistics centre in Gliwice (SCL Gliwice), 2—Logistics centre in Sławków (Euroterminal Sławków). *Source* Own elaboration based on [8]

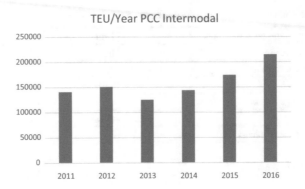

Fig. 12 Number of containers transported by the terminal (SCL Gliwice) operator—PCC Intermodal

Dąbrowa Górnicza, Opole, Kędzierzyn Koźle, Tychy, Oświęcim, Bielsko Biała, Częstochowa, Kraków and Kielce. Figure 12, according to [11], shows the number of containers transported by the terminal operator—PCC Intermodal—during the last 6 years. It should be assumed that about 1/3 of these transports has been transhipped at the terminal in Gliwice, which gives around 70,000 TEU in 2016. The upward trend for 2017 is maintained. However, also in this case, a decrease in the number of transports in 2014 and 2013 is noticeable due to the partial collapse of the eastern market.

The logistics centre in Gliwice has many specialized warehouses for storing various goods for many industry sectors. In 2016, the construction and commissioning of a new bimodal storage facility was completed. It is adapted for both road and rail transport. The warehouse has a storage area of 10,000 m². It is dedicated to handling industrial products in such industries as: paper, chemical, furniture and household appliances (Fig. 13). In addition, the logistics centre in Gliwice is particularly well-stocked in the storage of demanding goods. A special warehouse for the storage of steel products and dangerous goods in accordance with ADR should be mentioned here. The warehouse for steel products with an area of 6600 m² is equipped with:

- 3 overhead cranes controlled by means of a radio from "0" level with load capacity 32 and 25 t,
- rack system type "roll-stop" for storing coils,
- 3 gates enabling access for trucks to the hall interior (5.0 m × 4.5 m),
- 2 railway and vehicle gates,
- floor strength is 41 t/m².

The second specialist warehouse is also a class A warehouse, adapted to store dangerous goods in accordance with ADR with a storage area of 5800 pallet spaces on high storage racks. The warehouse is equipped with:

Fig. 13 The interior of the bimodal warehouse in the logistics centre in Gliwice

- monitored and regulated temperature area and humidity system,
- docks for front unloading and a gate for side unloading,
- railway ramp,
- non-dusty and tight floor topping,
- floor covered with anti-electrostatic layer,
- tank for possible leakages or water from sprinkler installation,
- ventilation enabling the exchange of the entire air from the warehouse within one hour,
- smoke flaps.

As part of services for the e-commerce sector, the logistics centre SCL Gliwice offers [8]:

- connection to e-commerce platforms,
- 14,000 pallet places,
- delivery service and storage of products,
- receiving and completing orders,
- labelling and marking,
- control of product features and parameters,
- fiscalization of sales on behalf of the store,
- distribution of parcels,
- returns handling,
- inventory of resources.

The area of SCL Gliwice is almost entirely developed and it is no longer possible to expand further within this logistic centre in the port of Gliwice. Further development would have to involve the acquisition of land in a different location.

3.3 Comparison of Handling Machines in Silesian Logistics Centres

Both terminals in Silesian logistics centres in Gliwice and Sławków have been thoroughly modernized in recent years. In Sławków a containerboard was enlarged, and one gantry crane was installed, while in Gliwice a new containerboard was built, and two gantry cranes were installed Figs. 14 and 15.

Fig. 14 Main bridge crane for reloading of containers in Euroterminal Sławków logistic centre

Fig. 15 Main bridge cranes for reloading of containers in Gliwice logistic centre

Fig. 16 Construction and basic elements of container gantry crane (RMG) in logistics centre in Sławków

Although both types of gantries perform very similar tasks, i.e., managing the input and output of containers to/from a block or stack of containers, they must meet very different functional and technical requirements for this type of machinery [12, 13].

Euroterminal Sławków is equipped with two container gantry cranes moving along the track (RMG—rail mounted gantry, Fig. 16). These are typical gantry cranes equipped with 2 booms to increase the spans and allow handling also outside the main field—usually loading or unloading means of rail and road transport. The crane travels along a track on steel wheels (12 on each side) and is supplied with electrical energy by a cable drum. Manipulation of the container is possible by mounting the lifting mechanism on the winch, and the rotation of the container is caused by the rotation of the entire winch (including the operator's cab). Cranes are equipped with a spreader to handle all containers and a piggy back handle for lifting removable bodies and semi-trailers.

The PCC terminal in the Silesian Logistics Centre in Gliwice is equipped with two RTG (rubber tired gantry, Fig. 17) container cranes. RTG crane are equipped with diesel engines (600 HP), which provide the power needed for driving and lifting. All mechanisms are electric, and the internal combustion engine is only a generator. The electricity and data transfer to the RTG trolley is carried out via a system of conductive conductors. Because the gantry is located in the port its main dimensions have forced the dimensions of the quay. However, its construction, no booms, prevents reloading in the terminal-water relation, which is surprising at first glance. The explanation of such a construction used by the terminal is economics— the port cannot be used to transport containers by waterway. The operating

Fig. 17 Construction and basic elements of container gantry crane (RTG) in logistics centre in Gliwice

Table 3 Comparison of technical parameters of container cranes

Characteristics	Unit	Sławków	Gliwice
Structure type	–	Container, gantry, on railway track, with booms	Container, gantry, with rubber wheels, without booms
Span	m	14.5+51.7+22.5	35
Hoisting speed	m/min	30	26/52
Bridge speed	m/min	100	130
Winch speed	m/min	80	70
Load capacity	t	40/50	41

principle and operations on the container are the same as the RMG crane. Basic parameters are shown in Table 3.

Handling machines used in logistics centres are machines designed for specific tasks, and basic construction and operation trends (on the basis of Silesian logistics centres) can be divided into:

1. Specialization—adaptation to specific tasks and resulting design features such as span, operating speeds, gripping bodies. The emphasis is primarily on

performance, although specialization does not mean a degree of universality—in terms of the crane being analysed, it is primarily a spreader.

2. Reduction of the weight of the structure using modern working mechanisms to reduce the negative impact of dynamics (advanced control automation)—cranes are working intermittently. In the literal sense, the use of modern materials and computational methods to optimize the weight.

3. Modular construction—the use of the same structural components for both the load-carrying structure and the mechanisms, and depending on the intended use of the specific crane, only minor modifications are made. Typisation and unification affect cost reduction both during construction and operation, reducing the cost and time of possible repairs.

4. The use of modern drive control that assists the operator in reducing the load swing, which significantly shortens the loading cycle time. In many cases, the use of semiautomatic or automatic bridge cranes on large container terminals.

5. Cab operator equipment with many necessary sensors to support the loading process. Today's cranes are operating in 24-h mode, so the equipment needed for lighting and cameras has become the norm. However, there are also laser sensors supporting the distance measurement process, sensors to keep up to date information on the state of the individual components of the structure, the position of the gripper body, atmospheric conditions. A very important element is the ergonomics of the operator, position, cabin visibility, control access.

6. Modern gantry cranes are equipped with many diagnostics to monitor the operation of engines and components.

7. Ecology is becoming increasingly important both in the construction phase and primarily in the crane operation. Gantry cranes are manufactured to work continuously for a period of 20 or more years, but after this period they should be fully recycled. In the case of RTG crane, it is important to optimize the fuel consumption, and what is associated with the emission to the atmosphere of combustible substances. The use of other structural components—including tire. RMG crane does not emit harmful substances to the atmosphere at the workplace, but only the naive believe that electricity is coming from the socket. Therefore, energy efficiency is also very important in this case. These features are related not only to ecology, but also to the economics of use.

For the efficient operation of logistics centres, it is necessary to use appropriate handling machines. In Silesian logistic centres, these machines are definitely different, and their characteristic features are:

- Construction and structural features of container cranes in Silesian logistics centres vary widely and depends on the specificity of the container terminal.
- Trends in the design and use of container cranes are related to many factors. First, with efficiency, ergonomics and control, with an ecological approach to construction and operation.
- Analysed gantries differ significantly by the span and source of the drive used—in one case electric, in the second diesel.

- RMG cranes are larger and are designed to work on larger container slabs, while RTG cranes have the ability to travel also to other terminal locations and to other container slabs.

4 Summary and Conclusions

The quality of point and linear infrastructure in European countries has a significant impact on increasing the competitiveness of rail transport in relation to other modes of transport. In Poland, the technical condition of the terminals, the lack of appropriate reloading equipment and the insufficient length of loading and unloading tracks make it difficult for the railway to start regular connections. Silesian logistics centres and transhipment terminals in them located are in a much better condition, and the development and modernization carried out in recent years gives hope for rapid development.

The average speed, for intermodal transport in domestic transport in Poland, was at 28 km/h in 2016. From the point of view of customer needs, for which the delivery time is a key issue, the low transport speed in intermodal transport limits the competitiveness of rail transport towards the road [2]. A stable policy of supporting intermodal transport in the long-term will help to reduce the disproportion between Poland and the European Union. Poland should actively participate in activities for the development of intermodal transport.

The geographical location of the Silesian region with the correct political situation in the world creates the chance that the logistic centres located here can become a significant partner in trade and transit transport. The Government Strategy for Responsible Development predicts a three-year increase in intermodal transhipments of at least 30%, which exceeds current logistics capabilities in Poland.

The advantage of the Logistics centre in Sławków is the huge potential for further investments in infrastructure in the field of international and regional transport, as well as in the field of intermodal transport in intercontinental relations. Due to its unique location, on the broad-gauge (1520 mm) railway track with European standards (1435 mm), Euroterminal Sławków can participate in the transport of cargo by rail between Europe and Asia–primarily China. As part of the so-called The New Silk Road, terminal could be a transhipment point even for several hundred trains per year. PKP Linia Hutnicza Szerokotorowa became the first Polish company co-creating the International Trans-Caspian International Transport Route (TMTM), which connects China with Europe. The weight of goods transported by sea from the Asia and Oceania region to the European Union in 2015 reached almost half a billion ton. Reloading in Polish ports amounted to 70 million ton, of which approx. 10% were from Asia. Taking over a fraction of this stream of goods through the Polish rail and logistic centre in Sławków would be a great opportunity for development.

The concept of the New Silk Road, presented in autumn 2013 by the representatives of the People's Democratic Republic of China, will be the largest infrastructure investment in history. From China to Poland, trains must overcome 11,000 km. They ride from 11 to 14 days, while transport by sea takes around 40 days. The new Silk Road leads through Central Asian countries to Western Europe through Poland, and thus through Silesian logistics centres. The concept of the New Silk Road must use the potential of transit, a very favourable location of Poland through the use of the broad-gauge railway track, ultimately reaching the Logistics Centre in the south of Poland, where all major transport routes intersect, was developed as part of the infrastructure program POLAND 3.0. The most important infrastructural projects include the restoration of the navigability of Polish rivers, starting from the Oder River, construction of the Danube-Oder-Łaba and Vistula-Oder connections, joining the Central European Transport Corridor and building a broad-gauge (1520 mm) railway track from Logistics centre in Sławków to the Logistics Centre Gorzyczki-Věřňovice. This new Silk Road running through Silesia and Czech Moravia will be both a transport and commercial route.

Due to the ongoing conflict between Russia and Ukraine and the resulting problems with rail transit through Ukraine, container traffic in this direction is currently minimal. The development of the connection network of the terminal in Sławków with the countries of Western Europe is a necessity, associated with little use of connections to the east. At the end of 2017, Euroterminal Sławków is planning to start servicing two successive new intermodal connections a week.

In Gliwice, the possibilities of expanding the infrastructure in the current location (inland port) have already virtually exhausted. At the moment, the capacity of the logistics centre is used in about 50%. However, there is a very good chance to develop transport by waterway. For this purpose, it is necessary to introduce further investments, and those made until 2017 are only the first phase. By the end of 2017, a transport to Wrocław of 20,000 tonne of coal per month is planned. Barges between Gliwice and Wrocław have about 200 km to cross. The cruise requires passage through 27 watergates.

As part of the Oder Waterway in 2021–2030, it is planned to build the Koźle-Ostrawa canal, which would be the Polish part of the waterway to the Danube. However, in 2017, an initiative to extend the broad-gauge railway line from Sławków to river ports in Gliwice and Kędzierzyn-Koźle appeared. Implementation of a new track section of about 80 kilometres would enable the dispatch of goods delivered from Asia through the intermodal rail by standard gauge, land and water roads—through Oder, Gliwice Canal and the Baltic Sea. The advantageous location of Silesian logistics centres allows for shipping goods to the Szczecin-Świnoujście seaport and Germany, and further to Scandinavia and Western Europe. The investment is justified due to the fact that Silesia is the most industrialized part of Poland and at the same time it is best equipped with transport infrastructure. Considering all other places in Europe indicated as potential ends of the New Silk Road, only the port and logistics centre in Gliwice can provide such wide access to all types of transport (railway, inland, road) and in all possible directions while maintaining the lowest possible costs. The connection of the inland waterway with a broad-gauge (1520 mm) railway track from the Far East is a project of European interest, because all European and non-European countries will benefit from this connection—from Japan, through North and South Korea, China, Mongolia, Russia and Ukraine.

Bibliography

1. Fechner I (2004) Centra logistyczne. Cel-Realizacja-Przyszłość, seria Biblioteka Logistyka, ILiM, Poznań [In Polish: Logistic centers. Goal-Implementation-Future]
2. Analiza kolejowych przewozów intermodalnych w Polsce (2016) Urząd Transportu Kolejowego. Warszawa [In Polish: Analysis of railway intermodal transport in Poland]
3. Transport intermodalny w Polsce (2016) Głowny Urząd Statystyczny [In Polish: Intermodal transport in Poland]
4. Transport – wyniki działalności (2016) Główny Urząd Statystyczny. Warszawa [In Polish: Transport—results of operations (2016) Central Statistical Office. Warsaw]
5. Euroterminal Sławków (2017) http://www.euterminal.pl/pl/10:Firma
6. Majszyk K (2017) Konflikt w Donbasie: Jak polskie firmy transportowe poradziły sobie z problemem? http://serwisy.gazetaprawna.pl/transport/artykuly/1026693,konflikt-w-donbasie-jak-polskie-firmy-transportowe-poradzily-sobie-z-problemem.html [In Polish: Conflict in Donbass: How did Polish transport companies handle the problem?]
7. Strategia Rozwoju Systemu Transportu Województwa Śląskiego. Diagnoza system transportu województwa śląskiego (2012) Urząd Marszałkowski województwa śląskiego. Katowice [In Polish: Strategy for the Development of the Transport System of the Śląskie Voivodeship. Diagnosis of the transport system of the Silesian Voivodeship]
8. Silesian Logistics Centre (SCL Gliwice) (2017) http://scl.com.pl/o-nas/
9. Raport o stanie Górnośląsko-Zagłebiowskiej Metropolii Silesia (2009) Górnośląski Związek Metropolitalny. [In Polish: Report on the state of the Silesia Metropolis (2009) Upper Silesian Metropolitan Union]
10. Bartosiewicz S (2014) Usługi outsourcingowe świadczone dla Śląskiego Centrum Logistyki S.A. w Gliwicach. Systemy Logistyczne Wojsk. No. 41 [In Polish: Outsourcing services provided for the Silesian Logistics Center S.A. in Gliwice. Forces Logistics Systems]
11. Periodic reports of PCC Intermodal (2017) http://www.pccintermodal.pl/en/investor-relations/periodic-reports/
12. Gąska D, Haniszewski T, Margielewicz J (2013) The product safety issues at the design and use of cranes. In: Sładkowski A (ed) Some actual issues of traffic and vehicle safety. Faculty of Transport, Silesian University of Technology, Gliwice, pp 243–270
13. Krośnicka K (2014) Nowoczesne terminale kontenerowe w porcie Rotterdam. Zeszyty naukowe Akademii Morskiej w Gdynii. No. 87, pp 139–153 [In Polish: Modern container terminals at the port of Rotterdam. Scientific notebooks of the Maritime University of Gdynia]
14. Antonowicz M, Syryjczyk T, Faryna P (ed) (2015) Biała Księga. Kolejowy transport towarowy. Warszawa [In Polish: White Book. Railway freight transport]
15. Żegluga śródlądowa (2017) https://www.zegluga.wroclaw.pl/images/news/161013maparzekimgm.jpg [In Polish: Inland waterway transport]
16. Gąska D, Margielewicz J (2017) Trends in the construction and operation of container cranes. Transport problems 2017. In: IX International scientific conference Silesian University of Technology. Faculty of Transport, pp 189–193
17. Bocheński T (2017) Transport intermodalny w przewozach rozproszonych w Polsce. Seminarium Technologie Transportowe – Rozwój, Bezpieczeństwo, Finansowanie. Dąbrowa Górnicza 9 marca 2017 [In Polish: Intermodal transport in dispersed transport in Poland. Seminar: Transport Technologies—Development, Safety, Financing]

18. Bocheński T (2016) Przemiany towarowego transportu kolejowego w Polsce na przełomie XX i XXI wieku. Rozprawy i Studia. Szczecin: Wydawnictwo Naukowe Uniwersytetu Szczecińskiego. Vol. 938 [In Polish: Transformations of rail freight transport in Poland at the turn of the 20th and 21st century. Dissertations and Studies. Szczecin: Scientific Publisher of the University of Szczecin]
19. Bocheński T (2017) The Operation of Handling Areas of 1435 mm and 1520 mm Gauge Railways in Europe. Studies of the Industrial Geography Commission of the Polish Geographical Society 31(3):80–94
20. Diagnoza system transportu Województwa Śląskiego (2013) Urząd Marszałkowski Województwa ŚląskiegoKatowice, p 234 [In Polish: Diagnosis of the transport system of the Śląskie Voivodeship]
21. Strategia rozwoju system transportu województwa śląskiego (2014) Katowice: Sejmik Województwa Śląskiego [In Polish: Development strategy for the transport system of the Śląskie Voivodeship (2014), Silesian Regional Assembly, Katowice]
22. Strategia rozwoju województwa śląskiego. Śląskie 2020+(2013) Katowice: Urząd Marszałkowski Województwa Śląskiego [In Polish: Strategy for the development of the Śląskie Voivodeship. Silesia 2020+(2013), Office of the Marshal of the Silesian Voivodeship, Katowice]
23. Szczepński M (ed) (2010) Strategia Rozwoju Górnośląsko-Zagłębiowskiej Metropolii „Silesia" do 2025 r. Górnośląski Związek Metropolitalny [In Polish: Strategy for the Development of the Upper Silesia and Zagłębie Metropolis "Silesia" until 2025. Upper Silesian Metropolitan Union]

Perfection of Technical Characteristics of the Railway Transport System Europe-Caucasus-Asia (TRACECA)

George Tumanishvili, Tamaz Natriashvili and Tengiz Nadiradze

Abstract The segment of the TRACECA main line passing on the territory of the former SU is made according to Russian standard, which is characterized by the low technical characteristics. Influence of the third body on the wheel and rail damage types, conditions of its formation and destruction are considered in the work taking into account the boundary layers stability and a thermal load of the contact zone. The influence of properties of the third body and degree of its destruction on the character of variation of the friction coefficient (negative, neutral and positive) and wear types (mild, sever and catastrophic) is ascertained. The conditions of destruction of the third body are developed and new ecologically compatible friction modifiers for the wheel and rail steering and tread surfaces are tested in the laboratory conditions. The wheel-sets with the separate modification of the tread and steering surfaces and increased durability of the wheel-set axle and the rail fastening device ensuring the constant rail cant are also developed.

Keywords Railway vehicle · Friction · Wear · Wheel · Rail · Tread and steering surfaces

1 TRACECA—Alternative Road

The International Transport Corridor TRACECA (Transport Corridor Europe-Caucasus-Asia) is a complex multi-modal transport system. It includes the European Union and 14-member States of the Eastern European, Caucasian and

G. Tumanishvili (✉) · T. Natriashvili · T. Nadiradze
Institute of Machine Mechanics, 10 Mindeli Str., 0186 Tbilisi, Georgia
e-mail: ge.tumanishvili@gmail.com

T. Natriashvili
e-mail: t_natriashvili@yahoo.com

T. Nadiradze
e-mail: tengiz_nadiradze@yahoo.com

© Springer International Publishing AG, part of Springer Nature 2018
A. Sładkowski (ed.), *Transport Systems and Delivery of Cargo on East–West Routes*, Studies in Systems, Decision and Control 155,
https://doi.org/10.1007/978-3-319-78295-9_8

Central Asian regions and has a significant contribution to the revival of one of the most famous historical route of the Silk Way, one of the ancient routes in the world.

The TRACECA main line, and also Baku-Tbilisi-Ceyhan oil pipeline and Baku-Tbilisi-Erzurum gas pipeline are already operating within the limits of new Silk Way and they significantly changed the economic reality in this region, created new conditions for development and enhanced energetic safety not only of the countries of this region but also of Europe.

The growing energy prices and several environmental factors have promoted the expansion and development of railway transport in the last few decades. For short and medium distances, the modern high-speed trains are able to compete with air transportation, having the advantage of presenting better energy efficiency and causing less pollution. For larger distances, the railway system is still the most economical means for transportation of goods and starts to have some competitive edges in the passenger transportation.

In order to improve the competitiveness of the railway systems, railway vehicle manufacturers are investing large resources in the research and development activities. These research activities contribute decisively to the development of new design concepts by using recent simulation techniques, modern production methods and innovative optimization procedures. The purpose is to develop sophisticated railway vehicles that answer to the increasing demands for faster, safer and more comfortable vehicles.

The main mode of overland transportation and delivery of great volumes of bulk goods over medium and long distances is still the railway but for pipelines. The pan-European Conference held in 1997 in Helsinki gave a priority to the transport corridor passing through South Caucasus. This initiative aims to develop economic relations, trade and transport communications in the regions of Europe, the Black Sea, the Caucasus, the Caspian Sea and Asia. It was officially launched in 1998 by the signature of the "Basic Multilateral Agreement on International Transport for Development of the Europe–the Caucasus–Asia Corridor". The route via Turkey forms an alternative to a rail ferry service connecting the Port of Varna with the Port of Poti. The tunnel under the Bosporus (Marmaray project: 77 km of railway linking Europe with Asia), as well as several parts of rail (e.g. between Ankara and Sivas) and the rail link between Kars in Turkey and Kartsakhi in Georgia (total of 105 km) are in completion phase. The TRACECA nations are investing heavily in the rail and port infrastructure. Azerbaijan is modernising its railway from Baku on the Caspian Sea to the Georgian border and building a 2 billion USD port south of Baku. The energy-rich Turkmenistan—which sits on the World's fourth-largest natural gas reserves and is estimated 12 billion tonnes of oil is building a 5.2 billion USD port at Turkmenbashi on the Caspian sea. According to the Reuters, the new port will see freight traffic grow from 10 million tonnes today to 2 million tonnes by 2020.

At present the competitors of TRACECA are Trans-Siberian Railway, North-South and South corridors.

The main routes between Europe and Asia are:

- North-South route from Azerbaijan to Baltic Sea;
- Trans-Siberian Railway from Kazakhstan, Mongolia and Vladivostok to Baltic Sea via Russia;
- TRACECA (Transport Corridor Europe-Caucasus-Asia) from the countries of middle Asia to the Black Sea and Turkey via the countries of Southern Caucasus;
- South Corridor from Turkmenistan to Istanbul via Iran.

The main part (about 80%) of the goods transported by Georgian Railway is transit. The wagons from the Caspian Sea shore move towards Batumi and Poti ports via Azerbaijan and Georgia and then by means of rail ferry to the West.

The TRACECA railway Corridor originates in Eastern Europe (Bulgaria, Romania, Moldova and Ukraine) and Turkey. The route then passes through the Black Sea to the ports of Poti and Batumi (Georgia). From Azerbaijan, using the Caspian Sea rail-ferries (Baku-Turkmenbashi, Baku-Aktau) the TRACECA route is connected to the railway systems of Turkmenistan and Kazakhstan, whose transport systems are connected to Uzbekistan, Kyrgyzstan and Tajikistan railways and extend to the borders of China and Afghanistan. The railway network of Turkey, which is linked directly with the railway line Kars-Tbilisi will provide uninterrupted access to the European railway network via the "Marmaray" tunnel. The railway transportation of wagons in the South Caucasus countries of TRACECA corridor and a map of TRACECA are shown in Figs. 1 and 2.

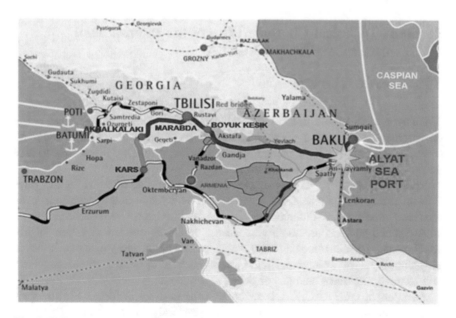

Fig. 1 Railway transportation of wagons in the South Caucasus part of TRACECA corridor

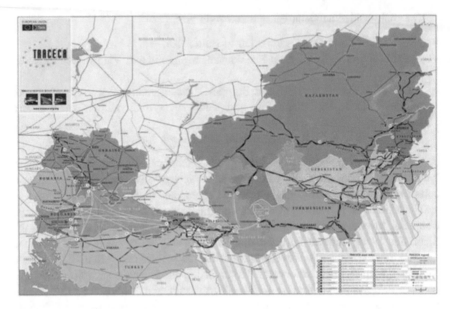

Fig. 2 Map of TRACECA

Some technical characteristics of the TRACECA European and Russian standard railways and ways of their perfection are shown below.

2   Some Technical Characteristics of the International Transport Corridor TRACECA

2.1   General Characteristics

There is a considerable number of problems that railway transport faces in its development in the region. As long as rail transport does not overcome the main obstacles it faces, which are numerous in the region, it will not be possible to take advantage of its full potential and capacity.

These are caused by expensive and time-consuming activities related to poor infrastructure and equipment, the low levels of safety and security; obsolete locomotives; existence of different legislations and transport requirements across countries and poor interoperability between different transport modes, compulsory warehousing, trans-loading/shipment activities, change of rail gauge and informal payments, lack of synchronization between border agencies, out of sequence notifications, excessive amounts of documents and etc.

East part of the TRACECA is situated on the territory of former SU and their railways correspond to Russian standard. The rail freight traffic declined with the

Table 1 Some railway infrastructure characteristics of Azerbaijan and Georgia

Country	Total rail length (km)	Double track (km)	Electrified (km)	Electrification system	Gauge (mm)
Azerbaijan	2929.4	804.7	1271.4	3 kV DC	1520
Georgia	1582	293	1523	3 kV DC	1520

fall of SU. The main reasons were a dramatic decrease in rail-based industries, an improving road network and strong competition from the trucking industry. Freight train speeds vary from 60 to 80 km/h, with restrictions of 20–40 km/h on some sections within particular countries. Currently all TRACECA member states are taking steps towards reforming their railway sector, some being already quite far into the process. Some railway infrastructure characteristics of Azerbaijan and Georgia are summarized in Table 1.

In most of the countries, the current fleet has a service life exceeding the target average of 30 years. One immediate consequence of this situation is greatly increased maintenance costs.

The characteristic and rather conflicting objectives for the modern railways are increase of speeds, loads on the axles and traffic safety, decrease of damageability of the rail track and running gear of the rolling stock and pollution of the environment by vibrations, noise and friction modifiers. Their realization calls for new approaches to solution of the raised problems. But during more than half a century the perfections were limited to the small changes of the existent constructions, for example bogies ЦНИИ-Х3, Y25, Barber; rail fastening devices and etc.

Avoidance of derailment and ensuring of durability of wheel-sets, rails, brake shoes etc., are vital for railways, both for safety and economic reasons [1]. The prevalent case of derailment is the climbing of a wheel on the rail, which is influenced by the main parameters such as: the friction forces, the flange angle, vertical and lateral forces, angle of attack etc. There are many works devoted to these phenomena [2–12] but the mentioned problems are not solved yet properly.

Some geometric parameters of the wheel and rail and their positional relationship for the straight segment and at symmetrical disposition of the wheel-set (according to Russian standard) are shown in Fig. 3.

The working profiles of the wheel and rail consist of the rolling surfaces and steering surfaces but there is no distinct boundary between them. It should be noted that a radius of curvature of new (not worn out) flange root 13.5 mm is less than radius of curvature of the rail angle (15 mm) that prevents their contact at the initial period of work (till conform contact when they are worn out). In the course of work a point contact of the wheel and rail changes into linear contact due to the wear. Under such conditions the wheel flange root and rail corner come in direct contact and they perform the functions of both, the rolling and steering simultaneously.

As it is known there are Russian standard railways in the countries of the former SU whose rail gauge (distance between rails) was decreased from 1524 to 1520 mm (leaving the distance between wheels unchanged) in 1970–1980. So, at present

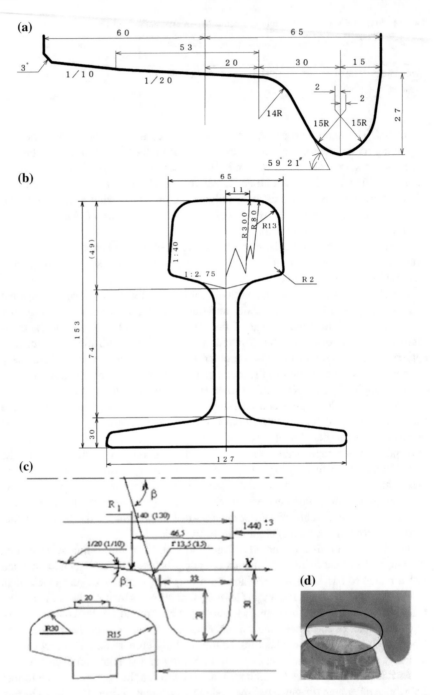

Fig. 3 Geometric parameters of the wheel (**a**) and rail (**b**), their mutual disposition (**c**) and tread (green) and steering (red) surfaces (**d**)

there is a difference between rail gauges of the European standard and Russian standard railways which are correspondingly 1435 and 1520 mm.

Besides, there are differences as well in speeds, loads on the axle, tolerances of the constructive elements, allowable values of damage, maintenance and exploitation expenses and etc. All this is negatively expressed on the effectiveness and coordination of joint work of these two railways.

In spite of many attempts (improvement of the wheels and rails mechanical and geometrical parameters, use of various kinds of lubricants and devices and etc.) a lot of questions are still left un-answered: why does the derailment of the known good rolling stock takes place from the known good rail and often this is explained by doubtful reasons? Why the allowable load on the axle is 23.5 t for the Russian standard railway despite tenfold factor of safety and for the European standard railway—32.5 t? Why the speed of our trains is lot smaller than that of the advanced countries? To answer these and many other questions and improve the situation it is necessary to take urgent measures.

2.2 The Evolution of Some Parameters of the Russian Standard Wheel-Sets and Rail Track and Their Wear

2.2.1 The Evolution of Some Parameters of the Russian Standard Rail Track

As it was noted above there is a Russian standard railway in Georgia as well as in other countries of the former SU whose normative-technical documents (ЦП/2023, ЦП/2913, ЦП-515) have been varied during the years. This was mostly expressed in widening of the allowable deviations of the rail-track normative-technical parameters. This concerns the allowable deviations of deflections of the curved rails, estimation of the rail-track state by grades and technical data obtained as a result of deciphering of the rail-track meter recordings.

Thus, for example since 1936 for the curved rails with radius of curvature less than 650 m allowable deviation of deflections measured on the two adjacent 20 m length chord was 8 mm. It should be noted here that for the curves with greater radius of curvature allowable deviation of deflections is less than for the curves with smaller radius of curvature. For instance, for neighboring group of curves with radius of curvature greater than 650 m allowable deviation of deflections is 6 mm. The explanation of these illogical normative documents is even more illogical —"correction of the curves with smaller radius of curvature is more difficult than that of the curves with greater radius of curvature".

According to the next instructions (1959, 1972) the above mentioned allowable deviation (8 mm) was increased to 10 mm, then it was not regulated at all and besides it no longer appeared in the estimation of the rail-track state by grades. But it is known that irregularities of the rail head lateral surfaces have a great influence on the steering lateral forces acting between the wheel flange and rail head lateral surface and wear rate of these surfaces. Therefore, it is evident that a main reason for sharp increase of the wheels and rails wear rate in the railways of countries of the former SU since this period (1972) is the roughening of the rail-track normative-technical parameters. The rail-track gauge 1524 mm was replaced by the 1520 mm gauge (so called "unified gauge") in the same period. Besides, if allowable lateral wear of the curved rail with radius of curvature 450 m at 1524 mm gauge was 11 mm and at 1520 mm gauge it became 26 mm. From the numerous experimental, theoretical and operational researches is seen that the increase of the rail wear rate began just from this period. Measures to narrow the track gauge to the norm of 1520 mm were a technical error [12], which occurred mainly because of the insufficient knowledge of the interaction of the track and the rolling stock.

2.2.2 The Wear of the Rail Head Lateral Surfaces and Wheel Flanges

For the railway to operate effectively, it is important to provide minimal power and thermal loading during the wheel/rail interaction and to strictly maintain necessary, economically justified parameters. It is generally known that the wear rate is highly sensitive to the load, relative sliding speed and friction path [13–15]. This becomes especially noticeable when the train passes the curved paths, where the rise of the rail curvature increases the shearing stress, relative sliding speed and friction path and reduces operational resource of the rails.

The dependences of the rail wear rate on the rail-track gauge obtained on the base of experiments carried out on the East Siberia Railway are shown in Fig. 4.

As it is seen from the graphs at increase of the rail-track gauge from 1513 to 1528 mm the rail wear rate decreases in all the three cases and at further increase of the gauge due to the wear the rail wear rate increases. In 1981 estimation of the

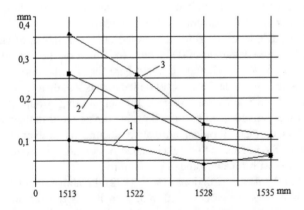

Fig. 4 Dependences of the rail wear rate on the rail-track gauge: 1—At rail-track gauge increased due to the rail wear; 2—at rail-track gauge increased due to deformatinns of sleepers and rail-pads; 3—at rail-track gauge increased due to the rail wear and deformatinns of sleepers and rail-pads

rail-track state by the technical data obtained as a result of deciphering of the rail-track meter recordings became even rougher. If before unsatisfactory state of the rail-track was estimated by 251 points, since 1981 it is estimated by 500 points. Because of the rail head lateral surface wear 3500 rails were replaced annually in Russia till 1981 while in 1995 were replaced 47631 rails. The wheel flange wear rate was also increased with increasing the rail wear rate.

For the purpose of decrease of probability of contact of the flange and rail head lateral surface the wheel repair profiles are developed and standardized whose flange thickness is reduced from 33 to 27 and 30 mm (GOST 903688). The tests of the wagon wheel-sets with the flange thicknesses 33 and 30 mm have shown [16] that the flanges of both wheel-sets were worn out to 25 mm after travelling 27,000 km. The average wear intensity of the wheel-sets with the flange thickness 33 mm turned out to be 1.6 times greater than that of the wheel-sets with the flange thickness 30 mm.

But as it was mentioned above the rail wear intensity is proportional to the flange wear intensity, i.e. the wheel with flange thickness of 30 mm causes the rail wear with 1.6 times less intensity while the service life of this flange is equal to that of the flange with thickness 33 mm. The graphs of variation of the flange wear as well as flange thickness due to the flange wear in function of the travelled path are shown in Fig. 5.

After travelling 27,000 km the flange with 33 mm thickness worn out by 8.1 mm and with 30 mm thickness—by 4.8 mm, i.e. the difference in the wear of these flanges reached 3.2 mm. But as it was mentioned above the rail is also worn out in parallel to the flange. From this point of view the use of flanges with 30 mm thickness must be considered more expedient.

With increase of the path travelled by the wheel and flange wear the clearance between the flange and rail head lateral surface increases also and the friction path decreases that causes reduction of the wear intensity. The graphs of variation of the wagon wheel flange wear intensity depending on the path travelled by the wheels

Fig. 5 The graphs of variation of the flange wear (k) and flange thickness (h) in function of the travelled path (graphs 3,1 and 4,2 for the flanges with thicknesses correspondingly 33 and 30 mm)

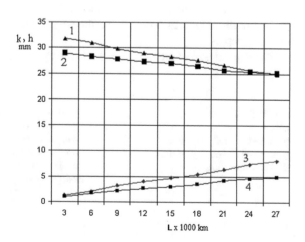

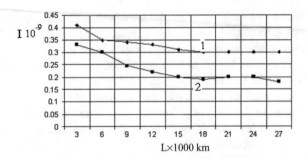

Fig. 6 The graphs of variation of the wagon wheel flange wear intensity depending on the path travelled by the wheels for one of the segments of the Siberian Railway (graph 1—for the flange with 33 mm thickness, graph 2—for the flange with 30 mm thickness)

are shown in Fig. 6 (the flange wear intensity is calculated according to the path travelled by the wheel).

It should be noted that at re-machining of wheels their diameters are often determined according to the tire thickness. After removal of the wheels from the locomotives and wagons their diameters are not measured neither. They usually measure the tread thickness in the depots that can't give precise information about the wheel diameter.

For the purpose of increase of the wear resistance of wheels and rails new profiles of wheels were also developed (GOST 11018-87) for the rail track gauge 1520 mm with maximum 29 mm and minimum 24 mm thickness of the wheel flange.

The researches have shown the high sensitivity of the wear rate on the relative sliding velocity [17–21]. The wear intensity in the curves is 2–3 times greater than in the straight segments of the rail-track that is a serious problem of the railway transport. The number of the treads re-machining and their replacements due to the flange utmost wear and limiting thickness of tires is raised sharply. In trying to overcome these phenomena new profiles of tires were developed and a technology of the tire restoration by welding and their further re-machining were implemented in many depots.

The dependence of the rail resource on the rail track curvature is shown in Fig. 7 [22].

Fig. 7 Dependence of the rail resource on the rail track curvature: S—rail life (million gross tons); R—radius of the rail track curvature, m

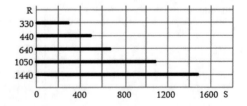

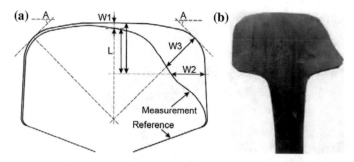

Fig. 8 Wear magnitudes of different parts of the wheel working surface: W1—vertical wear; W2 —horizontal wear; W3—450 wear (**a**) and worn rail head (**b**)

As it is seen from the graphs with decrease of the radius of the rail track curvature the rail resource decreases sharply. In Fig. 8 shown the wear magnitudes of different parts of the rail working surfaces.

The wear depths W2 and W3 (which are approximately equal) at the rail gauge surface are always greater than that at the rail head W1 [23–25]. For example, for unlubricated UIC 900A grade rail W2 is roughly 10 times greater than W1.

To the greater wear depth correspond the greater friction forces and the magnitude of the total friction force depends on the size of the contact zone (or quantity of actual interacting places). In most cases the contact of the wheel and rail is realized by the pairs of their surfaces: wheel tread-rail head and wheel flange-rail gauge [26]. The tread surface gradually passes into flange surface via flange root with gradually increasing relative sliding thereat and these two latter commonly carry out the functions of both, the tread and steering surfaces. As it is seen from Fig. 8 the maximum wear (W3) takes place at the rail corner and a similar picture we have at the flange root. It is natural that on the greater wear the greater friction energy is spent and for reduction of this latter it expedient to avoid interaction of the rail corner and flange root.

Not only at two-point contact or conformal contact but even at one-point contact the interacting points in the contact zone are practically located on the various diameters. Therefore, the increased relative sliding, especially in the case of conformal contact, can become a reason of the raised thermal loading and shearing stresses, destruction of the third body and adhesive wear process and scuffing. Due to immediate vicinity of the tread, flange root and flange surfaces the friction modifiers meant for tread surfaces and steering surfaces can be mixed and their characteristics can be changed (for more detailed information about the wear rate see Chap. 6).

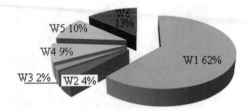

Fig. 9 The damage types of the Russian standard freight wagons wheel-sets. W1—wear of the flanges; W2—wear of the rolling surfaces; W3—wear due to brakeage; W4—wear due to differences in the wheel diameters; W5—wear due to plastic deformations of the flanges, W6—other damage

2.2.3 Damage Types of the Wagon Wheels

Friction plays a key role in the wheel and rail wear, rolling contact fatigue and other related railway maintenance and operational problems. Increase of loads on the axles and train speeds have a negative influence on the traffic safety, ecology and maintenance cost. Since the southern Caucasus railway is underlined in the work we will consider mainly Russian standard railway. The damage types of the Russian standard freight wagons wheel-sets are shown in Fig. 9.

- Till 1970 a dominant type of the wheel-set damage was fatigue damage of their rolling surfaces (pitting, layering, appearance and development of cracks);
- After 1970—wear of the flanges (adhesive wear, scuffing).

The other changes were also realized in 1970s: rails R50 were replaced by rails R65; reinforced concrete sleepers with new rail fastening devices and boxes with roller bearings were implemented. The wear rate of flanges in the Russian depot (Smolianovo) reached 2.5 mm/1000 km (in the locomotives—5 mm/1000 km) and average resource of the wheel-sets decreased 6 times; the average mileage between the re-machinings became 30,000 km and resource of the flanges—120,000 km. Because of this some depots were running out of time to do repairing works and were practically paralyzed [27]. The characterizing wear of wheel profiles till 1960 and after 1986 is shown in Fig. 10 [28].

In the late 1990s the wear rate of the flanges somewhat decreased.

The widespread types of the wheel damage are: change of the tread surface profile, its deviation from the circularity, appearance of cracks and pits on it, change of the flange geometry and etc. The value of the tread surface groove-like (hollow)

Fig. 10 The characterizing wear of the wheel profiles till 1960 and after 1986

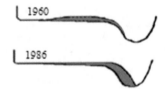

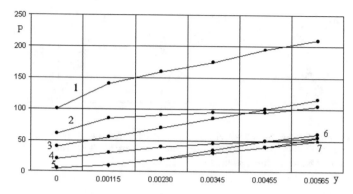

Fig. 11 Influence of value of the wheel hollow wear and rail-track curvature on the wheel rolling resistance. P—the wheel rolling resistance (N/t); Y—curvature of the rail-track (1/m); 1—hollow wear value 6.2 mm; 2—5.7 mm; 3—3.7 mm; 4—2.9 mm; 5—1.9 mm; 6—1 mm; 7—0 mm

wear is regulated and according to the norms of the former SU for the freight wagons is allowable 9 mm, for passenger wagons—6 mm and for high-speed wagons—1.2 to 1.5 mm.

The factual conicity measurements of the operational cargo wagon wheels have shown that no more than 10% of the wheelsets have 1/20 conicity. About 30% of the wheelsets have practically cylindrical surface and even negative conicity [28]. The influence of value of the wheel hollow wear and rail-track curvature on the wheel rolling resistance is shown in Fig. 11 [29].

As it is seen from the graphs hollow wear less than 2 mm has a little influence on the wheel rolling resistance. At wear value 2.9 mm and more the wheel rolling resistance increases sharply. On the base of this the railways of North America, as well as Canada and South America consider re-machining of the wheels expedient if the groove-like wear exceeds 2 mm.

3 Dependence of the Wheel and Rail Friction Path on Various Parameters

3.1 Dependence of the Wheel Flange Friction Path on the Relative Sliding Velocity of the Tread Surfaces

The friction path, sliding velocity of the flange and location of the instantaneous center of rotation of the wheel depend on the relative sliding of the tread surface at lateral displacement of the wheel relative to the rail.

A wheel with radius of the tread circumference r is shown in Fig. 12. The wheel-set axis is perpendicular to the rail-track axis.

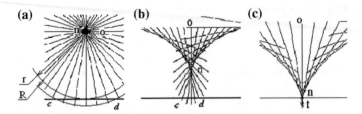

Fig. 12 The arc of contact of the wheel flange and lateral surface of the rail head at various ratios of velocities of the rolling and sliding: **a** sliding without rolling; **b** rolling with sliding of the rolling surface; **c** rolling without sliding of the rolling surface

The points of contact of the flange and lateral surface of the rail head are located on the circumference of radius R and length of the contact line is arc cd. For estimation of influence of the wheel movement on the value of the contact zone of the flange and rail lateral surface the wheel is divided into 40 equal radial parts. At rotation of the wheel about its geometric center O and sliding of the rolling circumference on the rail without rolling (Fig. 12a) the geometric center O and instantaneous center of rotation n coincide and length of the contact zone is maximal and equals to arc cd.

At rolling of the wheel with sliding (Fig. 12b) the instantaneous center of rotation no longer coincides with geometric center O. It is shifted towards the rail and length of the arc of contact cd is decreased. At rolling of the wheel without sliding of the rolling circumference (Fig. 12c) the instantaneous center of rotation "n" is located on the rail rolling surface and length of the arc of contact is minimal and direction of the creep force is close to the lateral direction on the steering surface of the rail head (that is noted also in [30]). When the wheel rolls on the rail without sliding (Fig. 12c) the instantaneous center of rotation "n" is located on the rail rolling surface and the length of the arc of contact is minimal and the direction of the friction force is near to lateral direction on the gauge surface (Fig. 13).

As it is seen from the figures with decrease of the friction (sliding) path of the wheel rolling surface the friction path of the flange on the rail lateral surface is decreased and at pure rolling of the rolling surface the sliding of the flange reaches a minimal value.

3.2 Influence of the Wheel Disposition Relative to the Rail on the Flange Relative Sliding and Friction Path

The angle of inclination of the tangent to the wheel flange root at the contact point can be expressed as follows

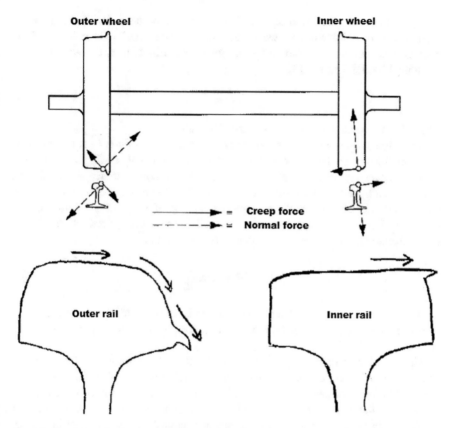

Fig. 13 Examples of rail wear for outer and inner rails [30]

$$\beta = arctg \frac{x}{\sqrt{r^2 - x^2}}, \tag{1}$$

where r is a radius of curvature of the flange root, x—coordinate of the flange root contact point on the X axis (Fig. 3).

The angle of inclination at the contact point is different from the angle of inclination of the rolling surface that can significantly increase loads on the wheel-sets and worsen dynamical characteristics of the movement [31]. Relation between the coordinate of the transition curve (flange root) contact point and corresponding radius of the wheel can be presented in the form

$$R = R_1 + k = R_1 + r - \sqrt{r^2 - x}^2, \tag{2}$$

where R_1 is the wheel radius at the onset of the transition curve; x—coordinate of the contact point on the X axis; k—increment of the wheel radius.

Estimation of the friction path is very important for prediction of thermo-mechanical loads and wear rate of the wheels and rails contact. If the average wear intensity is known for the given conditions, then the wear value is calculated by the formula [32]

$$h = IL, \tag{3}$$

where I is the wear intensity; L—the friction path.

The rolling without sliding is only possible on one circumference at two-point or conform (linear) contact. The rolling will be realized with sliding on the other circumferences due to their different radii. The flange point contacting the rail will roll on it with presence of sliding. With increase of the rolling velocity the rolling radius and friction path will decrease. At sliding of the wheel rolling surface without rolling, if its center does not move relative to the rail, the friction path for one revolution of the wheel is calculated by the formula

$$L = 2R \arccos \frac{R_1}{R}. \tag{4}$$

As it is seen from (4) the friction path depends on the ratio of radii of the wheel contact points, which in their turn are the functions of the contact point coordinate according to (2).

Thus, a friction path of the flange exceeds that of the rolling surface. Sliding velocity and friction path of the flange are minimal at rolling of the rolling surface without sliding. Decrease of the rail-track gauge causes reduction of the clearance between the flange and rail-head lateral surface and passage of the contact point on the flange. In the case of the two-point or linear contact, the increase of the wheel radius on the straight rail-track segment leads to the rise of the flange relative sliding, friction path, wear rate and intensity at passage of the contact point on the flange root. With increase of the flange and rail relative sliding (at acceleration/braking of the locomotive, two-point or conform contact due to the increase of sliding velocity) the friction path, time of dwelling of the flange and rail separate points in the contact zone and thermal and mechanical impact and wear intensity increase.

The wheel lateral displacement on the rail is restricted by the conic form of the wheel profile. The inclination of the wheel profile for the rapid wagons is 1/40 and for other wagons 1/20. But in the process of operation the wheels tread surfaces are worn out (whose allowable value is quite great) and angle of inclination of the tread surface changes.

3.3 The Features of the Wheel-Set Movement on the Rail-Track and Estimation of the Sliding Distance

3.3.1 The Features of the Wheelset Movement on the Rail-Track

Movement of the wheel-set is characterized by the following features. In the straight a wheel-set performs a zigzag movement close to the sinusoid which is accompanied by creeping. At moving of the locomotive wheel-set in the curve the same rotation is transmitted to its both wheels from the electric drive and they pass the equal distances on the outer and inner rails because of which the wheel-set axle will deviate from the radial position. At moving of the wagon wheel-set in the curve its outer and inner wheels will travel the different distances (at pure rolling of the wheels) but due to the fact that at rotation through the given angle the difference between the lengths of the outer and inner rails is different for the curves with different radii of curvature and inclination of the wheel tread surface (1:20) is constant here also the wheel-set axle will deviate from the radial position and will be twisted at the same time (Fig. 14).

The contact of the wheel flanges of both, the front and rear wheel-sets with the rail leads to increase of the angle of attack, lateral force and rolling resistance of the outer front wheel of the bogie, climbing of the wheel-flange on the rail and derailment, squealing noise, flange wear, gauge face wear of the high-rail and corrugations basically of the low-rail. In such conditions, to return the wheel-set into radial position it is necessary that one of the wheel of the wheel-set slide on the rail in the longitudinal direction [30]. It should be noted that there are a lot of attempts to maintain more or less radial position of the wheel-sets [33, 34].

The intermittent slipping of one of the wheel of the wheel-set is accompanied by torsional vibrations of the wheel-set shaft and the longitudinal vibrations of the vehicle (that have been identified as flange noise [35]) and the respective wear of rails—rail corrugation, which occurs preferentially on the low rail in curves [36]. Despite the multiple attempts there is no reliable solution to the corrugation problem yet except the expensive technology—grinding.

The movement of a wheel-set in the curve and a corrugated inner rail are shown in Fig. 15.

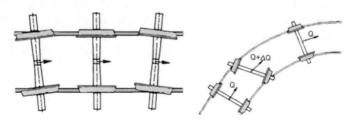

Fig. 14 Movement of the wheel-set in the straight and curve

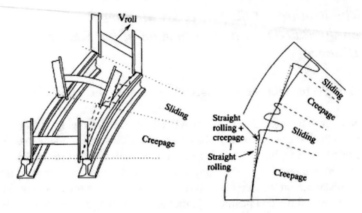

Fig. 15 Movement of a wheel-set in the curve and a corrugated inner rail

The similar picture would be at difference of diameters of the tread surfaces or at various brake efforts of the wheelset. The wheel–rail contact is a simultaneous rolling and sliding contact, which can be divided into stick (no slip) and slip zones. The slip zone of the contact of the tread surface is related to the traction force, creep, and geometry of the contact. The slip rate increases at movement in the curves, braking, and acceleration.

The tread surface of the wheel is conical which passes gradually into the flange surface through its root. Therefore, the differences between diameters of interacting surfaces inside of the contact zone, relative sliding, contact stresses, deformations and temperatures towards the flange are greater. The sliding distances for the curves and straights and for elliptic wheels are determined and the numerical calculations are given.

3.3.2 Estimation of the Distance Between Worn Out Segments and Friction Path in the Curve

We estimate parameters of the rail corrugation, distances between worn-out segments of the rail, lengths of these segments and depth of the worn-out layers caused by the three different reasons. We determine first distances between worn out segments and friction paths for the rail.

At pure rolling (without sliding) of the wagon wheel-set with velocity V in the curve of radius R through angle α, the inner wheel will rotate relative to the outer wheel in the counterclockwise direction if it is seen from axial direction A (Fig. 16) and it will twist the wheel-set axle through angle φ. This latter will be equal to the ratio of the difference Δl of the outer and inner arcs (or friction path for the rail) to the radius D/2 of the wheel tread surface:

$$\varphi = 2\Delta l/D.$$

Fig. 16 Movement of the
wheel-set in the curve

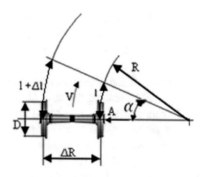

From the drawing $\alpha = 1/R = (1+\Delta1)/(R+\Delta R) = \Delta1/\Delta R$ from where

$$\Delta1 = 1\Delta R/R, \tag{5}$$

and therefore

$$\varphi = 21\Delta R/DR. \tag{6}$$

The maximum angle of twist of the wheel-set axle φ_{max} depends on the friction force

$$F = fQ \tag{7}$$

and is calculated by the formula

$$\varphi_{max} = ML/IpG, \tag{8}$$

where M is a torque caused by the friction force

$$M = FD/2 = fQD/2; \tag{9}$$

f—friction coefficient; Q—thrust load of the wheel on the rail; L—length of the wheel-set axle; Ip—polar moment of inertia of the wheel-set axle cross section; G—modulus of rigidity of the axle material.

We determine path 1 necessary for the axle to be twisted on the maximum angle φ_{max} (or distance between worn out segments of the rail since at travelling this path the wheels are rolling on the rail without sliding) from (6) replacing φ by φ_{max}:

$$1 = DR \, \varphi_{max}/2\Delta R = MLDR/2I_pG\Delta R \tag{10}$$

and putting the found 1 into (5) we obtain correspondingly the friction path for the rail

$$\Delta1 = MLD/2I_pG. \tag{11}$$

3.3.3 Estimation of the Distance Between Worn Out Segments and Friction Path in the Straight Track for the Case of Wheels of Different Diameters

At rolling of the wheel-set with the wheels of different diameters D and D + ΔD in the straight track, in the case of the same angle of rotation these wheels will pass correspondingly the distances l and l + Δl (Fig. 17a).

At rolling of the smaller wheel through distance l it will rotate relative to the bigger wheel in the counterclockwise direction if it is seen from axial direction A and it will twist the wheel-set axle through angle $\varphi = 2\Delta l/(D + \Delta D)$, from where, considering the formula (9) we obtain the value of Δl corresponding to the maximum angle of twist φ_{max} of the axle

$$\Delta l = \varphi_{max}(D + \Delta D)/2 = ML(D + \Delta D)/2IpG. \tag{12}$$

The following proportion can be written from the drawing $(l + \Delta l)/l = (D + \Delta D)/D$ or $\Delta l/l = \Delta D/D$, from which we obtain distance l (distance between the worn out segments) at passing of which the wheel-set axle will be twisted through angle φ_{max}

$$l = \Delta lD/\Delta D = ML(D + \Delta D)D/2IpG\Delta D. \tag{13}$$

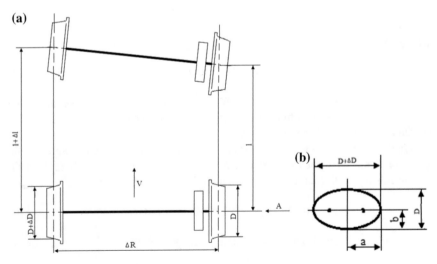

Fig. 17 Movement of the wheel-set in the straight track: **a** with the wheels of different diameters or with one wheel having ellipticity; **b** parameters of ellipticity

3.3.4 Estimation of the Distance Between Worn Out Segments and Friction Path in the Straight Track for the Case of One Elliptical Wheel

Consider movement of such wheelset in the straight track whose one wheel has a tread surface of diameter D and the other wheel has ellipticity with the small D and bigger D + ΔD diameters (Fig. 17a).

At one revolution, these wheels will pass the different distances, correspondingly l and l + Δl, because of which a wheel with diameter D will rotate relative to the elliptical wheel in the counterclockwise direction when it is seen from axial direction A and it will twist the wheel-set axle through angle

$$\varphi = 2\Delta l/D_{av}, \tag{14}$$

where average diameter of the elliptical wheel

$$D_{av} = D + \Delta D/2 . \tag{15}$$

The difference of distances passed by the wheels at one revolution is Δl = L − πD, where the length of the elliptical tread surface

$$L = \pi \left[3(a+b) - \sqrt{(3a+b)(a+3b)} \right] , \tag{16}$$

or

$$\Delta l = \pi \left[3(a+b) - \sqrt{(3a+b)(a+3b)} \right] - \pi D . \tag{17}$$

The value of the friction path Δl corresponding to maximum angle of twist φ_{max} of the axle is obtained from (13)

$$\Delta l^l = \varphi_{max} D_{av}/2 = ML(D + \Delta D/2)/2IpG. \tag{18}$$

The distance l at passing of which the wheel-set axle will be twisted on the angle φ_{max}

$$l = \pi D \Delta l^l/\Delta l. \tag{19}$$

3.3.5 Estimation of the Rolling-Sliding Distances, Wave Lengths, Relative Sliding Velocities and Depth of the Worn-Out Layer

In all the three cases considered above at removing or decrease of the torque M acting on the wheel that takes place at its vertical vibrations when the friction force F decreases, the angle of twist of the axle will start to decrease. Suppose φ_{max} falls

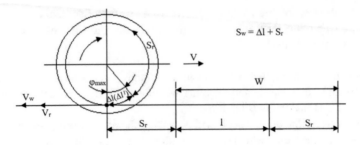

Fig. 18 The rolling and sliding distances on the rail and wheel

down to zero during time t. This will take place at rotation of the inner wheel in the clockwise direction relative to the outer wheel on the angle φ_{max} since the flange of the outer wheel is pressed on the rail and its movement is additionally restricted by the friction force arisen between the flange and rail. Obviously, during this time t the inner wheel will roll and slide simultaneously on the rail and the rolling and sliding distance on the rail will be

$$S_r = Vt. \tag{20}$$

We note that the rolling and sliding distance on the wheel tread surface is

$$S_w = \Delta l + S_r \text{ or } S_w = \Delta l^I + S_r \tag{21}$$

for the variant C, where Δl or Δl^I is a sliding friction path and the wave length of the worn-out rail (Fig. 18)

$$W = l + S_r. \tag{22}$$

This value of the wave length assumes that at release of the inner wheel the friction force acting on it from the rail is zero. When the friction force differs from zero the wave length will be less since its both components will be decreased and its value depends on the friction force magnitude.

To determine time t we present the wheel-set as a one-mass torsion vibratory system (Fig. 18a), where C is the torsion rigidity of the wheel-set axle, I—total moment of inertia of the inner wheel and gear wheel located near it. Then, angle of twist φ_{max} will fall down to zero in conformity with a law of free vibrations of this vibratory system during the period P/4 (Fig. 19b).

At that, period of free vibrations

$$P = 2\pi\sqrt{I/C} \tag{23}$$

and consequently time t will be

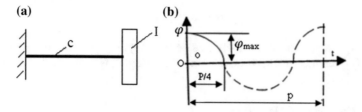

Fig. 19 a One-mass torsion vibratory system; **b** graph of the system free vibrations

$$t = P/4 = \frac{\pi}{2}\sqrt{I/C} . \tag{24}$$

The average velocity of the wheel contact point relative to the wheel center (Fig. 18)

$$V_w = -\frac{D\varphi max}{2t} + V_r , \tag{25}$$

where $V_r = -V$ is a velocity of the rail contact point relative to the wheel center.

We note that maximum velocity of the wheel contact point relative to the wheel center

$$V_w^I = -\frac{A\omega D}{2} + V_r = -\varphi_{max}\sqrt{\frac{C}{I}} \times \frac{D}{2} + V_r \tag{25I}$$

where $A = \varphi_{max}$ is an amplitude of the wheel-set shaft torsion vibrations;

$\omega = \sqrt{C/I}$—cyclic frequency of vibrations.

Sliding velocity

$$V_{sl} = V_w - V_r \tag{26}$$

Relative sliding velocities for the rail and wheel

$$K_r = \frac{V_{sl}}{V_r} \times 100\% \text{ and } K_w = \frac{V_{sl}}{V_w} \times 100\% \tag{27}$$

The depth of the worn-out layer a year of the rail segment S_r

$$h = i\Delta lN, \tag{28}$$

where i is the wear intensity and N—number of cycles which is determined as follows

$$N = N_1 N_2 N_3 N_4, \tag{29}$$

where N_1 is a number of the trains passing by a day; N_2—number of wagons in the train; N_3—number of wheels on one side of the wagon; N_4—number of days a year.

3.3.6 Numerical Calculation

Numerical calculation is carried out for the locomotive wheel-set with the following parameters:

$D = 1250$ mm; $\Delta D = 3$ mm; $L =$ mm; $Ip = 1.57 \times 10^{-4}$ m^4; $G = 75$ GPa; $Q = 115$ kN;

$I = 176$ kg m^2; $C = 7450$ kN m; $f = 0.4$; $\Delta R = 1546$ mm.

The torque $M = fQD/2 = 0.4 \times 115 \times 1.25/2 = 28.75$ kN m;

Maximum angle of twist of the axle

$$\varphi_{max} = ML/IpG = (28.75 \times 10^3 \times 1.58)/(1.57 \times 10^{-4} \times 75 \times 10^9)$$
$$= 0.00386 \, \text{rad} = 0.221°;$$

Determine first distances l between the worn out segments and friction path Δl for the considered cases:

(a) On the base of (6) we have:

For $R = 200$ m; $l = DR \, \varphi_{max}/2\Delta R = (1.25 \times 200 \times 0.00386)/(2 \times 1.546) = 0.312$ m $= 312$ mm;

$R = 400$ m; $l = 624$ mm;

$R = 600$ m; $l = 936$ mm.

Putting the found l into (1) we obtain the same result for all the R

$\Delta l = l\Delta R/R = (0.312 \times 1.546)/200 = 0.00241$ m $= 2.412$ mm;

(b) On the base of (8) we have

$\Delta l = ML(D + \Delta D)/2IpG = (28.75 \times 10^3 \times 1.58)(1.25 + 0.003)/2 \times 1.57 \times 10^{-4} \times 75 \times 10^9 = 2.417$ mm.

Distance l on the base of (9) will be $l = \Delta lD/\Delta D = 2.417 \times 1250/3 = 1007.1$ mm;

(c) Semi-axes of the ellipse on the base of Fig. 4b will be

$a = (D + \Delta D)/2 = (1250 + 3)/2 = 626.5$ mm; $b = D/2 = 1250/2 = 625$ mm.

Difference of the paths travelled by the wheels at one revolution on the base of (13) will be

$$\Delta l = \pi \left[3(a+b) - \sqrt{(3a+b)(a+3b)} \right] - \pi D$$
$$= 3.14 \left[3(626.5+625) - \sqrt{(3 \times 626.5+625)(626.5+3 \times 625)} \right]$$
$$- 3.14 \times 1250 = 4.17 \, \text{mm}.$$

The value of the friction path Δl^I corresponding to the angle of twist φ_{max} of the axle on the base of (14) will be

$$\Delta l^I = ML(D + \Delta D/2)/2IpG$$
$$= \left(28.75 \times 10^3 \times 1.58 \right)(1.25 + 0.003/2)/2 \times 1.57$$
$$\times 10 \times 75 \times 10^9 = 2.414 \, \text{mm}.$$

Distance 1 on the base of (15) will be

$$1 = \pi D \Delta l^I / \Delta l = 3.14 \times 1250 \times 2.414/4.71 = 2012.7 \, \text{mm}.$$

Period of the free vibrations for all the three cases considered above on the base of (19) will be

$$P = 2\pi \sqrt{\frac{I}{C}} = 2 \times 3.14 \sqrt{\frac{176}{7450 \times 10^3}} = 0.03 \, \text{s}.$$

Time of the simultaneous rolling and sliding according to (20) will be

$$t = P/4 = 0.03/4 = 0.0075 \, \text{s}.$$

The parameters S_r, S_w, V_w, V_w^I, V_{sl}, K_r and K_w calculated according to expressions (16), (17), (21)–(23) correspondingly are the same for all the three variants (a), (b) and (c) since Δl is practically the same for these variants and the results for the three different velocities V are given in Table 2:

Table 2 The parameters S_r, S_w, V_w, V_w^I, V_{sl}, K_r and K_w calculated three different velocities

Velocity (V)	S_r (mm)	S_w (mm)	V_w (m/s)	V_w^I (m/s)	V_{sl} (m/s)	K_r (%)	K_w (%)
30 km/h = 8.33 m/s	62.5	64.9	−8.65	−8.83	−0.32	3.84	3.70
50 km/h = 13.88 m/s	104.1	106.5	−14.2	−14.38	−0.32	2.30	2.25
80 km/h = 22.22 m/s	166.6	169.0	−22.54	−22.72	−0.32	1.44	1.40

As for the wave length W its values calculated according to (18) under supposition that to radii of curvature R = 200, 400, 600 m of variant (a) correspond successively velocities V = 30. 50 and 80 km/h and to variants (b) and (c) corresponds velocity V = 80 km/h, are given below:

(a) For R = 200 m; W = 1 + Sr = 312 + 62.5 = 374.5 mm;
    R = 400 m; W = 624 + 104.1 = 728.1 mm;
    R = 600 m; W = 936 + 166.6 = 1102.6 mm.
(b) W = 1007.1 + 166.6 = 1173.7 mm;
(c) W = 2012.7 + 166.6 = 2179.3 mm.

Depth of the worn-out layer on the rail segment S_r according to (24) and (25) will be $h = i\Delta l N_1 N_2 N_3 N_4 = 10^{-6} \times 2.4 \times 30 \times 20 \times 4 \times 365 = 2.1$ mm, where $i = 10^{-6}$; $\Delta l = 2.4$ mm; $N_1 = 30$ trains; $N_2 = 20$ wagons; $N_3 = 4$ wheels; $N_4 = 365$ days.

Though real parameters of the rail corrugation, length of the worn-out rail segment S_r, distance l between these worn out segments and depth of the worn-out layer on the rail segment S_r caused by the multiple movement with different velocities of the wheel-sets of various types, their vertical vibrations, torsion vibrations of the wheel-set axles and etc., differ more or less from the obtained numerical values, these latter display a general tendency of the wear of such type. It should also be taken into account that in reality the three considered reasons of the rail corrugation take place simultaneously with various combinations.

4 Analysis of Thermal Loadings of the Wheel and Rail Contact Zone

4.1 Formulation of the Problem

Problems relating to the determination of the friction generated heat, temperature distribution and heat partition factor in the wheel-rail contact are well known, and a great number in studies of the fields of railway and tribology deal with them. Those problems are related to rolling contact fatigue (RCF) and wear of the wheel-rail system. Relative slip at the wheel-rail interface leads to an increase in temperature causing thermal softening of the steel that can even lead to its metallurgical transformation. Furthermore, thermal stresses cause occurrence of residual stresses which contribute to the propagation of cracks in materials [34].

When contacting surfaces are moving at friction, energy emitted by asperities in contacting surfaces transforms into heat. The problem of friction contact has been of high interest for a long time and a large number of scientific works have been devoted to various kinds of technical and mathematical aspects of this problem (see. e.g. [37–39] and references therein). Frictional contact is susceptible to several factors. The main one is the thermal and power loading of contact zone, which negatively impacts on working conditions and durability of the interacting surfaces.

For example, the temperature of wheel-rail contact zone can exceed 800 °C [40], and cause different kinds of surface damage and other negative phenomena. The primary source of surface damage is arising in discrete parts of contact zone. The measurement of the ambient temperature, body temperature, surface temperature near the contact and the mean surface temperature is determined experimentally. The [13] deals with the solution of the one-dimensional problem of micro roughness temperature, but it does not provide for the influence of the environment of the micro roughness.

Under the particular combinations including the contact pressure, properties of the third body, speed and friction, a critical temperature (that collapses the third body) is reached in the separate areas of the contact zone with local welding or adhesion of the contacting surfaces can appear. As result of local adhesion, the friction increases and causes a still higher temperature, and then the wear process can become catastrophic. There is little agreement in the literature on what scuffing is, what causes it, or what appearance is. This has resulted in significant confusion because various researchers have point out very different physical processes into the general heading of scuffing. Scuffing involves the sudden collapse of the third body and is generally regarded as resulting from thermal phenomena. In this connection, it is useful to describe thermal process in discrete zones of contact—on micro roughness with regard to heat transfer in bodies and in environment of the micro asperities. Although in recent years a lot of works has been carried out on the temperature rise at the contact between rough rubbing bodies due to its significance in the friction and wear behavior of sliding bodies, a clear correlation between the contact temperature rise and the surface topographic parameters is lacking. The direct precise measurement of temperature in contact zone due to limited sizes of contact zone, especially in micro asperities is technically difficult. Theoretical researches often include the different assumptions and not always take into account the influence of properties of body outside ambient of the micro asperities. Therefore, the solution of engineering problems of the contact loading and load-carrying capability of interacting bodies usually are considered in isothermal statement which limits the area of their application. However, despite the fact that these theories are commonly in use and their results often differ from practical observations.

In this study, the effects of contact area on the surface temperature on local contact area were investigated based on the mathematical model of the three-dimensional steady state temperature field. The numerical simulation of the model indicates that the dimensions and the thermo-physical characteristics of the environment of the local contact roughness has an essential influence on the surface temperature and the maximal contact temperature; the mathematic model of local contact temperature field can be used to predict the critical sliding velocity and allowable rate of energy absorption.

In the present work, the heat distribution in a asperity is mathematically modelled by boundary value problem, describing heat transfer in cylinder (Fig. 20).

In the present work, the heat distribution in the micro-asperity is mathematically modeled by boundary value problem, describing heat transfer in the cylinder.

Fig. 20 Model of the
micro-asperity

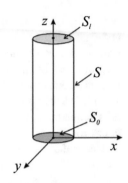

Let $K = \{(x, y, z) : x^2 + y^2 < R^2, \ 0 < z < l\}$ be a cylinder of length l and radius R in coordinate system $Oxyz$. Denote by S it's lateral surface

$$S = \{(x, y, z) : x^2 + y^2 = R^2, \quad 0 \le z \le l\}$$

and the lower surface by S_0

$$S_0 = \{(x, y, z) : x^2 + y^2 \le R^2, \quad z = 0\}$$

and its upper surface by S_1

$$S_l = \{(x, y, z) : x^2 + y^2 \le R^2, \quad z = l\}$$

it's lower and upper bases respectively.

In the cylinder K consider the steady state temperature field $u = u(x, y, z)$ where the temperature on the lower base is described by the given function $\vartheta_0(x, y)$ and on the lateral surface by the function $\vartheta_S(x, y)$, whereas on the upper base the heat flux function $\vartheta_l(x, y)$ is given.

The above physical problem can be reformulated as the following boundary value problem.

In the cylinder K find the thermal field described by the function $u = u(x, y, z)$ which solves steady state thermal field equation (coinciding to Laplace equation)

$$\Delta u(x, y, z) = 0, \quad (x, y, z) \in K, \tag{30}$$

(where $a^2 = \frac{k}{c\rho}$, k is the heat conductivity coefficient, c is the heat capacity and ρ is the density of the medium), and satisfy the following conditions:

Dirichlet conditions on the lower base S_0 and on the lateral surface S_R

$$u(x, y, 0) = \vartheta_0(x, y) \quad x^2 + y^2 \le R^2, \tag{31}$$

$$u(x, y, z) = \vartheta_S(z), \quad x^2 + y^2 = R^2, \quad 0 \leq z \leq l, \tag{32}$$

and Neumann condition on the upper base S_l

$$\partial_z u(x, y, l) = \vartheta_l(x, y) \tag{33}$$

Additionally, we assume that thermal field $u(x, y, z)$ and the boundary data $\vartheta_0(x, y)$ and $\vartheta_l(x, y)$ are bounded functions

$$|u(x, y, z)| < M, \quad (x, y, z) \in K, \tag{34}$$

$$|\vartheta_0(x, y)| < M, \quad |\vartheta_l(x, y)| < M, \quad x^2 + y^2 \leq R^2. \tag{35}$$

Consider the case, when thermal field, as well as the boundary data is axially symmetric with respect to the z-axis

$$u(x, y, z) = u(r, z), \ \vartheta_0(x, y) = \vartheta_0(r), \ \vartheta_l(x, y) = \vartheta_l(r), \ \vartheta_S(x, y, z) = \vartheta_S(r, z), \ r = \sqrt{x^2 + y^2},$$

and introduce cylindrical coordinates (r, φ, z), where

$$x = r \cos \varphi, \quad y = r \sin \varphi$$

The boundary value problem in the new coordinates (r, φ, z) reads as

$$\partial_{rr}^2 u + \frac{1}{r} \partial_r u + \partial_{zz}^2 u = 0, \quad (r, \varphi, z) \in K, \tag{36}$$

$$u(r, 0) = \vartheta_0(r), \quad 0 \leq r \leq R, \tag{37}$$

$$\partial_z u(r, l) = \vartheta_l(r), \quad 0 \leq r \leq R, \tag{38}$$

$$u(R, z) = \vartheta_S(z). \tag{39}$$

4.2　Solution of the Problem

To solve the problem, we decompose its solution u into the sum

$$u = u_1 + u_2 + u_3, \tag{40}$$

where u_1 is the solution of the boundary value problem

$$\begin{cases} \partial_{rr}^2 u_1 + \frac{1}{r}\partial_r u_1 + \partial_{zz}^2 u_1 = 0, & (r,\varphi,z) \in K, \\ u_1(r,0) = 0, & 0 \le r \le R \\ \partial_z u_1(r,l) = \vartheta_l(r), & 0 \le r \le R, \\ u_1(R,z) = 0, & 0 \le z \le l, \end{cases} \tag{41}$$

the function u_2 is the solution of the boundary value problem

$$\begin{cases} \partial_{rr}^2 u_2 + \frac{1}{r}\partial_r u_2 + \partial_{zz}^2 u_2 = 0, & (r,\varphi,z) \in K, \\ u_2(r,0) = \vartheta_0(r), & 0 \le r \le R \\ \partial_z u_2(r,l) = 0, & 0 \le r \le R, \\ u_2(R,z) = 0, & 0 \le z \le l, \end{cases} \tag{42}$$

whereas the function u_3 solves the boundary value problem

$$\begin{cases} \partial_{rr}^2 u_3 + \frac{1}{r}\partial_r u_3 + \partial_{zz}^2 u_3 = 0, & (r,\varphi,z) \in K, \\ u_3(r,0) = 0, & 0 \le r \le R \\ \partial_z u_3(r,l) = 0, & 0 \le r \le R, \\ u_3(R,z) = \vartheta_s(z), & 0 \le z \le l. \end{cases} \tag{43}$$

We solve all three auxiliary boundary value problems by applying variable separation method [41]. For u_1 we obtain expansion

$$u_1(r,z) = \sum_{k=1}^{\infty} C_k^{(1)} J_0\left(z_{0,k}r/R\right) \sinh\left(z_{0,k}z/R\right), \tag{44}$$

where

$$C_k^{(1)} = \frac{2}{Rz_{0,k}\cosh(z_{0,k}l/R)J_1^2(z_{0,k})} \int_0^R J_0(z_{0,k}r/R)r\vartheta_l(r)dr, \quad k = 1,2,\ldots, \tag{45}$$

Functions J_ν are the ν-th order Bessel functions of the first kind, $z_{0,k}$ is the k-th zero of J_0, and sinh and cosh are the hyperbolic sine and the hyperbolic cosine, respectively.

Similarly, for u_2 we have

$$u_2(r,z) = \sum_{k=1}^{\infty} C_k^{(2)} J_0\left(z_{0,k}r/R\right) \cosh\left(z_{0,k}(z-l)/R\right), \tag{46}$$

where

$$C_k^{(2)} = \frac{2}{R^2 \cosh(z_{0,k}\, l/R) J_1^2(z_{0,k})} \int\limits_0^R J_0(z_{0,k}\, r/R) r\vartheta_0(r)dr, \quad k = 1, 2, \ldots \quad (47)$$

The expansion for u_3 reads as follows

$$u_3(r, z) = \sum_{k=1}^{\infty} C_k^{(3)} I_0\left(\left(k + \frac{1}{2}\right)\frac{\pi r}{l}\right) \sin\left(\left(k + \frac{1}{2}\right)\frac{\pi z}{l}\right), \quad (48)$$

where

$$C_k^{(3)} = \frac{2}{I_0\left(\left(k + \frac{1}{2}\right)\frac{\pi R}{l}\right) l} \int\limits_0^l \sin\left(\left(k + \frac{1}{2}\right)\frac{\pi z}{l}\right)\vartheta_S(z)dz, \quad k = 1, 2, \ldots, \quad (49)$$

and I_0 is the zero-order modified Bessel function of the first kind.

Obtained expressions for the solution of the boundary value problem allow to analyze the behavior of the steady temperature field in the asperity as well as to calculate solution numerically.

Figure 21 shows the results of solution of the problem for different ambient materials—for water and for oil lubricant: intensity of the temperature reduction toward the base of the micro-asperity is higher for water than for oil lubricant.

As it is seen from the figure the raised heat capacity and thermal conductivity of water contributes to better cooling conditions of the micro-asperity and at increasing distance from the heat source the temperature for water is lower than for oil lubricant.

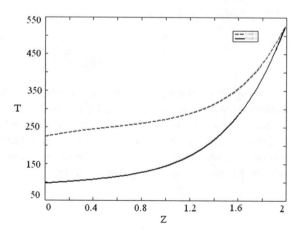

Fig. 21 Dependence of the temperature on the distance from the micro-asperity base for various ambient: solidline water dashedline oil lubricant

The separate parts of interacting surfaces can be considered as moving heat sources with capacity $q = fpv$, where f is the friction factor, p is the contact pressure and v is the sliding velocity. For considering distribution of the heat flow the method, given in the work [42] will be used. Let us admit that whole heat is spent on heating of bodies. To estimate the temperature of interacting surfaces it is possible to use the linear differential equation of the thermal conductivity [43]

$$\frac{\partial \theta}{\partial t} = a \frac{\partial^2 \theta}{\partial x^2}. \tag{50}$$

With boundary conditions:

$$\theta_1(x,0) = \theta_2(x,0) = \theta_0 = const;$$
$$\frac{\partial \theta_1(\infty,t)}{\partial x} = \frac{\partial \theta_2(-\infty,t)}{\partial x} = 0;$$
$$\theta_1(+0,t) = \theta_2(-0,t);$$
$$\lambda_1 \frac{\partial \theta_1(0,t)}{\partial x} + \lambda_2 \frac{\partial \theta_2(0,t)}{\partial x} + q = 0.$$

Solution of the equation of heat conductivity has a form

$$\theta_1(x,t) = \theta_0 + \frac{2q\sqrt{a_1 t}}{\lambda_1} \frac{K_\varepsilon}{1+K_\varepsilon} ierfc \frac{x}{2\sqrt{a_1 t}};$$
$$\theta_2(x,t) = \theta_0 + \frac{2q\sqrt{a_2 t}}{\lambda_2} \frac{1}{1+K_\varepsilon} ierfc \frac{x}{2\sqrt{a_2 t}}. \tag{51}$$

where $K_\varepsilon = K_\lambda/K_a$; $K_\lambda = \lambda_1/\lambda_2$, is the criterion of relative heat conductivity of bodies, $K_a = a_1/a_2$, the criterion which is characteristic for the thermo-inertial conditions of the two bodies. Solution of this equation with the use of method of sources, has the form [44]:

$$\theta(R,t) = \frac{Q}{c\gamma(4\pi a t)^{3/2}} \exp\left(-\frac{R^2}{4at}\right) \tag{52}$$

Here Q is heat quantity, c—heat capacity, ε—specific weight, R—distance between any point of the body and heat source.

Let us consider interaction of the brake shoe and wheel as a sliding of one cylinder on another with constant speed and temperature change for lot of instant sources of a heat during time t. The temperature change for lot of instant sources of heat in the given point during time t will be

$$\theta(x, y, z, t) = \int\limits_0^t \int\limits_s \frac{2qds dt_2}{c\gamma(4\pi a t_2)^{\frac{3}{2}}} \exp\left[-\frac{(x+vt_2)+y^2+z^2}{4at_2}\right], \tag{53}$$

where s is the nominal contact area. Designate $u_2 = R_2/4at_2$, where $R^2 = x^2 + y^2 + z^2$, then, after transformation we will have

$$\theta(R, x, t) = \frac{qbr\alpha}{\pi R\lambda} \exp\left(-\frac{vx}{2a} - \frac{vR}{2a}\right) erfc \frac{1}{2\sqrt{F_0}}, \tag{54}$$

where q is the power of source, r—is average radius of the brake shoe, b—width of the sample, l—average length of the contact zone, α—angle of contact, F_o—Fourier criterion.

The obtained results show dependence of temperature on the thermo-physical characteristics of the micro-asperity environment; temperature change in various points of the interacting zones depends on the friction power, duration of action of the heat sources, geometric and kinematic parameters of the contact, a distance between the heat source and any point of the contact zone and thermo-physical properties of bodies. These results are useful for more precise definition of the scuffing temperature criteria [45, 46]. The main problem in determining of thermal loading of the contact zone is ascertainment of the friction power and its conformity with law of the friction factor variation. But for estimation of the conditions of destruction of the third body and friction factor it is necessary to carry out special experimental researches.

5 The Friction of a Wheel on a Rail and the Third Body

5.1 The Friction of a Wheel on a Rail

The complex physical, mechanical and tribo-chemical phenomena proceeding in the contact zone of the wheel and rail at the direct impact of the environmental conditions raise the problems to be solved for many-sided study of these processes. There are many works devoted to these phenomena [47–51] but the mentioned problems are not solved yet properly. Namely, prediction of the friction coefficient in the contact zone, its control and character of influence of many parameters on its variation are still problematic. This hinders from avoidance of the wheel derailment and decrease of the energy consumed, damage intensity, vibrations, noise and maintenance expenses.

The wheel-rail interaction occurs on the: tread surfaces during rolling, traction and braking; steering surfaces (flange and rail lateral head side) in curves; flange root and rail corner at rolling, traction, braking or steering. The friction coefficient for wheel-rail interaction can vary in the range 0.05–0.8. The values of the friction

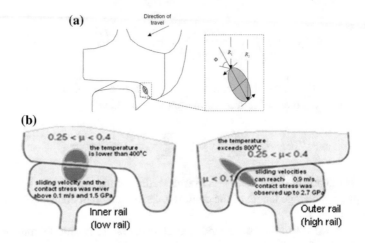

Fig. 22 The ideal values of the friction coefficients and the thermal loading, stress distribution and sliding velocities of profiles (**a**) and schematic view of various radii inside the contact zone of the wheel flange (**b**)

coefficient for the tread and steering surfaces must be correspondingly in the ranges 0.25–0.4 and <0.1 [1]. Optimal value of the friction factor for tread surfaces is 0.35 [11]. This is one of the required conditions for normal interaction of wheels and rails which are not properly realized yet.

The creep is characteristic for wheel and rail interaction. Different parts of interacting surfaces of wheels and rails have different functions; different creep and therefore they need to have different properties. The friction factor for wheel flange and rail gauge face should be as low as possible—less than 0.1. Excessively high friction and torque on tread surfaces leads to the plastic flow and fatigue and low friction can cause poor traction and braking and causes severe wear at traction and at braking. In Fig. 22a is shown a schematic view of various radii inside of the contact zone [5], desirable values of friction coefficients [47] of the contact zone and the thermal loading.

The existent profiles of wheels and rails can be divided into the tread surfaces (which take part in the "free" rolling, traction and braking) and steering surfaces (the wheel flange and rail gauge, which take part in the steering mainly in curves and prevent the wheel-set from derailment). The flange root can roll on the rail corner, and it can take part in traction, braking and steering. But traction (braking) and steering require mutually excluding properties: the relatively high value of the friction factor at traction and braking and the relatively low value of the friction factor at steering. And the "ideal" value of the friction coefficient ($\mu < 0.1$) in the contact zone of the flange root and the rail corner is not acceptable for both cases. At gradual displacement of the interacting points of the wheel and rail from tread surface towards the flange root and rail corner and then towards the flange and rail gauge, the relative velocity, sliding distance, the scuffing probability, vibrations, noise and wear rate increase. Besides, immediate vicinity of the tread, flange root and flange surfaces promote mixing of the friction modifiers for tread and flange surfaces. Therefore, the tread and the steering surfaces must be separated and modified by the friction modifiers with corresponding properties.

The processes accompanying the interaction of profiles especially at increase of creeping (at increase of the traction force or at moving of interacting places towards flange) increase the probability of destruction of the third body and interacting surfaces too. But increase of the relative slipping of wheels causes rise of thermal and power loads in the contact of superficial layers, generating vibrations, typical noise and the most dangerous type of wear—scuffing. Therefore, the preservation of the third body between interacting surfaces has a crucial importance.

The above-mentioned problems are of current importance for any railways and their solutions require the special experimental and theoretical researches.

It is known that working conditions of the wheel-sets are very hard in the curves, especially for the first wheel-set. The weakest part of the wheel at rolling in the curves is a flange. This issue became burning especially in the last 40 years and many works appeared devoted to enhancing of stability of the wheel flanges against the operational impacts.

As it is seen from Fig. 19 the power and the thermal loading of tread surfaces are relatively low. At working of wheels in regimes of traction and braking, at lateral movement, at rotation around vertical axis and at skidding, the value of the sliding velocity and sliding distance will grow. The flange root and gauge face in the contact zone have considerably high level of creeping, shearing stress and temperatures. They perform the role of both, the tread and steering surfaces and the value of the friction factor 0.1 is not acceptable for both cases. The flange root and rail corner participate in traction, braking and steering of the wheel that demands the mutually excluding properties: the relatively high value of the friction factor at the traction and braking and the low value of the friction factor at steering.

Not for only the two-point contact or conformal contact, but for the one-point contact as well the interacting points will be located on various diameters in the contact zone (Fig. 19a). Therefore, the increased relative sliding in all cases can become a reason for raised thermal loading, shearing stresses, destruction of the third body, adhesive wear process and scuffing. Even more, inadequate and high friction on the wheel flange-rail gauge contact can cause the wheel climb on the rail head and derailment [52]. Due to immediate vicinity of tread, flange root and flange surfaces the friction modifiers for tread and flange surfaces can be mixed and their characteristics can be changed.

5.2   Influence of the Ambient Conditions on the Wheel and Rail Interaction

A lot of the complexity of the wheel-rail contact is brought about by the open nature of the system and the constantly varying environmental conditions, for instance temperature, humidity, plants, dust etc. Compared to oil, water has a high cooling capacity but its low viscosity and corrosive properties make it unacceptable for the tribological applications. In order to adjust the performance and improve the properties of the water-based lubricants, high quality additives are used, such as surface/interface active molecules [48]. Water has many unique properties, such as the polarity of the molecule, which makes that aqueous lubrication technology distinctively differs from oil-based lubrication. However, water-based lubricants also have specific disadvantages, such as corrosiveness, the risk of vaporization and low viscosity [49].

It is known that in spite of the low lubricant properties of water the existence of the moisture in the contact of the wheel and rail promotes the decrease of the friction coefficient and prevents the wheel from climbing on the rail. As it is seen from the graphs the increase of the number wheel-sets removed from the loco-motives due to the worn-out flanges is observed in the relatively warm and dry period of the year. The existence of water on the rail in the contact zone promotes the decrease of the friction coefficient and thermal loading in the contact zone due to its high thermal capacity and evaporation power. Thus, it has a significant influence on the wear resistance of the wheels and rails, it decreases the friction coefficient between the flange and rail head lateral surface, noise and wear intensity.

Increase of the wheel flanges wear intensity is observed in the summer period that is explained by increase of the friction coefficient and thermal load. Increase of the average daily precipitation leads to decrease of the wear rate because of cooling of the contact zone and transition of the water in the contact zone and generation of the third body [53, 54].

On the graphs (Fig. 23) are shown the number of: removed wheel-sets from locomotives because of the extreme wear of flanges by months in Tbilisi

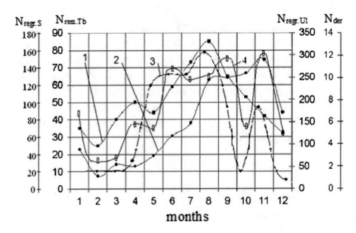

months

Fig. 23 The number of: re-machined wheel-sets in the locomotive depot Smolyanovo ($N_{rem.S}$, curve 4); wheelsets removed due to extreme wear of the flanges in the Tbilisi locomotive depot ($N_{rem.Tb}$, curve 1); re-machined wheel-sets due to extreme wear in Ulan-Bator railway ($N_{rem.Ul}$, curve 2) and derailed empty wagons ($N_{der.}$, curve 3)

Locomotive Depot (curve 1), re-machined wheel-sets because of the worn flanges on the Ulan-Bator Railway (the Russian Railway, curve 2) [55], in the Locomotive Depot Smolianovo (the Russian Railway, curve 4) [56] and the number of derailed empty wagons (curve 3) [57].

As it is seen from graphs the number of removal of the wheel-sets and their re-machining is especially great from June to November. From December to May the precipitation and cold weather predominate and the atmospheric moisture appears on the rails. Consequently, the water, having low viscosity and bearing capacity, gets into the contact zone separating partially the interacting surfaces. But due to the high thermal capacity it promotes improvement of the thermal removal from the contact zone, decrease of thickness of the heated-up layer predisposed to destruction, decrease of the friction coefficient, its stabilization and decrease of the number of removed and re-machined wheel-sets and derailed wagons.

6 Modification of the Interacting Surfaces of Wheels and Rails and Stability of the Third Body

6.1 Modification of the Interacting Surfaces and Formation of Friction Forces

Wheels and rails are characterized by poor conditions of modification of interacting surfaces and they are subjected to direct influence of the environment (brake shoes interacting surfaces commonly are not modified; they are subjected to direct influence of the environment although as well). All these factors worsen working conditions of wheels, brake pads and rails; they have a negative influence on the transport safety and operational expenses. In addition, the lubricants and friction modifiers, vibrations and noise pollute the environment. Damage accumulation because of structural change of interacting surfaces, wear, fatigue and plastic deformation significantly change profiles of wheels and rails and reduce their service life.

For thermomechanical stability of the third body it is necessary to take special measures to provide their desirable properties: thermal stability, relatively low and stable friction factor for steering surfaces (wheel flange and rail gauge surfaces) and relatively high and stable friction factor for tread surfaces, modification of which is also necessary. The movement of the wheel-set in the curves is accompanied by advancing creep of the inner wheel causing periodical torsion deformations of the wheel-set shaft. In this case, various kinds of damage of surfaces can appear: corrugation because of plastic deformations and adhesive wear, fatigue and etc.

At the common operational conditions, interacting surfaces are covered by various types of boundary films—products of interaction with the environment that prevent the direct contact of rubbing surfaces. Depending on the friction conditions, environment and surfaces properties, these layers may have various tribological properties that will have the great influence on the boundary friction [58–61]. This is confirmed by the results of the experimental research in an inert gas environment and in a vacuum, which excludes the possibility of oxidation in friction. Under such conditions the unhindered seizure and intensive wear rate is observed.

To prevent the aforementioned undesirable phenomena, it is important to provide presence of the third body in the contact zone with appropriate properties, control the friction factor and protect it from destruction. But until recently, despite of considerable quantity of works, devoted to investigations of dependences between wheel/rail and wheel/brake shoe friction forces and their durability, expected results are not obtained yet.

Control over friction and wear during sliding is needed to ensure the proper functioning of the tribological system. One of the approaches is to use well selected and carefully formulated lubricants, as lubricants can provide a low shear strength layer separating the interacting surfaces. Thus, lubrication can reduce the mechanical damages that arise from accommodating the velocity difference in sliding contacts. They have begun to lubricate the steering and tread surfaces in 1990, with the use of appropriate devices.

The lubricant film can be a solid, (commonly graphite or MoS2), a solid/liquid dispersion, a liquid, or, exceptionally, a gas [62]. Most of the conventional lubricants are liquids. Wheels are generally lubricated with liquid lubricants because of their outstanding merits over solid lubricants such as excellent torque characteristics, easiness to carry the applied load and to manage and replenish. Many liquids have been used as lubricants to minimize the friction, heat, and wear between mechanical parts in contact with each other. Typically, lubricants are oil based or water based. Oil lubrication is traditionally used for mechanical applications, and lubrication theory and technology were developed based on using oil in these systems. Most of the commercial additives are for oil-based lubrication [63]. However, oil-based lubricants have some major drawbacks such as poor biodegradability and high toxicity for humans.

Usually the surfaces are covered with layers of various origin which at interaction represent the third body. At separation of interacting surfaces by the third body the friction forces depend mainly on the third body rheological properties—internal friction forces (viscosity) of the third body, velocity gradient and area of the contact. Thereat, the viscosity commonly depends on the contact temperature and pressure. The neglecting of the properties and state (continue or discontinue) of the third body in the contact zone leads to inadequate description of processes and inadequacy of results of theoretical and experimental researches. It is known about positive influence of the third body with necessary properties on the adhesive wear, scuffing and fatigue damage of interacting surfaces. The adhesion approach of friction implies three stages: interpenetrating of asperities in the contact zone, intimate contact (at absence of the third body) and adhesive junction of micro-asperities and disruption of these junctions. This is accompanied by the thermal effects that have a direct influence on the surface structure and physical and mechanical characteristics of the surfaces and wear process.

When the continuous third body separates the interacting surfaces, their movement resistance (the friction forces) mainly depends on the rheological properties of the third body. The friction coefficient at the wheel/rail interface is strongly related to the rheological properties of the third-body constituents [57–61]. Under such conditions the movement resistance depends on the inside friction forces between the layers of the third body (viscosity) and the area of those layers. $F = f(\eta, \frac{\Delta v}{\Delta x}, S)$ $F = \eta \frac{\Delta V}{\Delta X} S$, where S—area of contact of layers, η—viscosity, $\frac{\Delta v}{\Delta x}$—velocity gradient. It is known that the viscosity depends on the properties of the third body, contact temperature and pressure, and various lubricants are characterized by different properties in this regard.

The friction force Ff between two direct interacting bodies (at absence of the third body on the actual kontact area) is a function of the actual area of microscopic contact, $F_f = \psi(\sum \tau A asp)$ [59], where τ is effective shearing strength of the contacting bodies and Aasp is an actual area of the microscopic contact. It also depends on the thermal loading of the contact zone and the environment of separate micro-asperities.

Thus, the friction forces depend on the contact area in both cases—when the interacting surfaces are separated from each other fully and partly.

The boundary layers in the contact zone, and the self-generated organic films among them, would provide 30 times the capacity of the back-to-back endurance (of loading cycles) when the acting stress was exceeding twice the endurance limit [13].

6.2 The Theory of Elasto-hydrodynamic Lubrication and Calculation of Lubrication Film Thickness

The EHD theory of lubrication is considered as most perfecting mathematical model among the components of third body—lubricant film of the heavy loaded frictional contact.

The main purpose of modification is to control of the friction between the interacting surfaces in relative motion by introducing a friction modifier (lubricant) between them. The third body in the contact zone usually consists from lubricant, boundary films and wear products. Depending upon the thickness of lubricant film presented between the interacting surfaces, four well defined lubrication regimes are identified such as hydrodynamic or elasto-hydrodynamic (EHD), mixed and boundary lubrication regimes and they are clearly from the Stribeck/Hersey curve.

The kinematic, geometric, mechanical and thermos-physical parameters of heavy loaded frictional contact that have the great influence on the lubrication film and the third body parameters are shown in Fig. 24.

The EHD theory is the theoretic basis of the lubrication of the heavily loaded interacting surfaces. Below are given the integro-differencial equations of EHD

Fig. 24 The kinematic, geometric, mechanical and thermo-physical parameters of heavy loaded frictional contact

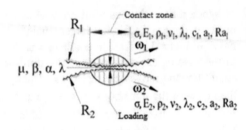

theory of lubrication with the consideration of the thermal processes which take place in the lubricant film and on the surface boundary:

$$\frac{dp}{dx} = 6\mu(V_1 + V_2)\frac{h - h_0}{h^3}, \quad \text{when } x = -\infty, p = 0 \text{ and } x = x_0, p = \frac{dp}{dx} = 0;$$

$$h = h_0 + \frac{x^2 - x_0^2}{2R} + \frac{2}{\pi}\left(\frac{1 - \nu_1^2}{E_1} + \frac{1 - \nu_2^2}{E_2}\right) \int p(\xi) \ln\left|\frac{\xi - x}{\xi - x_0}\right| d\xi;$$

$$\rho c V \frac{\partial t}{\partial x} = \lambda \frac{\partial^2 t}{\partial y^2} + \mu\left(\frac{\partial V}{\partial y}\right)^2, \quad \text{when } x = -\infty, t = t_0;$$

$$t(x, 0) = \left(\frac{1}{\pi \rho_1 c_1 \lambda_1 V_1}\right)^{0.5} \int_{-\infty}^{x} \lambda \frac{\partial t}{\partial y}\Big|_{y=0} \frac{\partial \varepsilon}{(x - \varepsilon)^{0.5}} + t_0;$$

$$t(x, h) = \left(\frac{1}{\pi \rho_2 c_2 \lambda_2 V_2}\right)^{0.5} \int_{-\infty}^{x} -\lambda \frac{\partial t}{\partial y}\Big|_{y=h} \frac{\partial \varepsilon}{(x - \varepsilon)^{0.5}} + t_0;$$

$$\mu = \mu_0 \exp(\beta p - \alpha \Delta t)$$

$$(55)$$

where P_1—linear load; V_1 and V_2—peripheral speeds; μ—Dynamic viscosity of lubricant oil in normal conditions; P_1—linear load; h—clearance; h_0—minimum clearance; R—radius of curvature; E_1 and E_2—modulus of elasticity; ν—Poisson's ratio of body materials; t—temperature; ρ, c, λ, ρ_1, c_1, λ_1, ρ_2, c_2, λ_2—correspondingly density, specific heat capacity and thermal conductivity of lubricant and interacting surfaces; β—piezo coefficient of lubricant viscosity; λ—lubricant thermal conductivity; $\acute{\alpha}$—thermal coefficient of lubricant viscosity; ξ, ε—complementary variables; x_0—abscissa in the place of lubricant outlet from the gap.

The most interesting practical aspect of the EHD lubrication theory is the determination of lubricant film thickness which separates the bodies. Calculation of the oil film thickness is the main problem of the EHD lubrication theory and there are a numerous literature sources about it. In EHD lubrication, the load is carried by the elastic deformation together with the hydrodynamic action of the lubricant. There are various formulas for isothermic and anisothermic solutions for EHD problems describing the behavior of oil film thickness with various accuracy. The formulas for calculation of oil film thickness obtained by different authors and the results obtained by their use, and the results of experimental research of oil film thickness are given in the table. There are number of formulas that meet isothermal or an-isothermal tasks of EHD theory of lubrication, which describe the lubricant film thickness with various accuracy. In Table 3 you can see the oil film thicknesses (μm) calculated using the formulas by various authors and the relevant results of experimental researches in the conditions of frictional rolling.

The formulas of the first four authors given in the Table do not take into account thermal phenomena proceeding in the contact zone and they give the results with the same precision on the low rolling velocities. At high rolling velocities (10–20 m/s) the oil viscosity, velocity and other parameters stipulate the thermal processes in the oil layer that first decreases intensity of the oil layer growth and then decreases it. Operation of the interacting bodies in the conditions of rolling-sliding activates significantly the thermal phenomena and isothermal solution of the EHD theory equations is inacceptable in this case.

At very thin oil layer when the thermal processes proceeding on the boundary of the interacting surfaces are essentially important the formula for calculation of the oil layer thickness has the form [64]:

$$\frac{h}{R} = K \left(\frac{\mu V_{\Xi K}}{P_n}\right)^{0.7} \cdot \left(\frac{P_n \beta}{R}\right)^{0.6} \cdot \left(\frac{\lambda}{\alpha \mu V_{CK}^2 P_{el,2}^2}\right)^e \tag{56}$$

where $Pe_{1,2}$ is average value of Pecle's number ($Pe = bV/a$) in the contact zone: $Pe_{1,2} = (Pe_1 + Pe_2)/2$, where b is a half-width of the area of Hertz, V—velocity. As it is known the third body consists of the oil and boundary layers for the modified (lubricated) contact zone. The latter are especially important for the modern alloyed lubricants that reveal the quite different properties for the oils with the same viscosity due to the tribo-chemical interaction of their active components with the surface. These properties are revealed in formula (27) by the values of coefficients K and e that are ascertained on the base of results of the experimental researches.

6.3 Experimental Researches into the Third Body

6.3.1 Experimental Researches on the High-Speed Twin Disk Machine

The experimental researches into the third body properties were carried out on the high-speed twin disk machine (Fig. 25) with the use of the traditional lubricants and ecologically compatible and heat resistant frictional and anti-frictional friction modifiers developed by us in the frame of the European Union project. The tests were carried out at variation of the acting parameters in the wide range.

Table 3 Oil film thicknesses

Author	Formula	Sum rolling speed (m/c)						
		1	5	10	20	30	50	70
Ertel and Grubin [62]	$\frac{h}{R} = 1.19 \left[\frac{\mu_0 V_\Sigma \beta}{P}\right]^{8/11} \left[\frac{ER}{P}\right]^{1/11}$	0.44	1.41	2.34	3.89	5.23	7.6	9.7
Petrusevich [63]	$h = \dfrac{\left(\frac{V_c}{1000100}\right)^{2/3} \left(\frac{g}{E}\right)^{0.25}}{\left(\frac{a_0}{250}\right)^{0.5} \left(\frac{\sigma_g}{10^4}\right)^{0.4h}}$	0.34	0.98	1.56	2.47	3.24	4556	5.7
Dayson and Higginson [79]	$\frac{h}{R} = 1.6 \frac{(\beta E)^{0.6} (\mu V_\Sigma / ER)^{0.7}}{(P/ER)^{0.13}}$	0.55	1.7	2.74	4.45	5.9	8.45	10.7
Kodnir [80]	$h = 2.224 \frac{(\mu_0 V_\Sigma)^{0.75} \beta^{0.6} R^{0.4}}{P^{0.15}}$	0.24	0.82	1.385	2.33	3.16	4.3	6.0
Murch and Wilson [81]	$\frac{h}{R} = 1.19 \left(\frac{\mu_0 V_\Sigma}{P}\right)^{8/11} \left(\frac{ER}{P}\right)^{1/11} \left(\frac{3.94}{3.94 + V^{0.88}}\right)$	0.44	1.36	2.146	3.2	3.87	4.578	4.85
Drozdov, Tumnishvili [82]	$\frac{h}{R} = 1.57 \left(\frac{\mu_0 V_\Sigma}{P}\right)^{0.7} \left(\frac{P\beta}{R}\right)^{0.6} \left(\frac{\lambda}{z\mu_0 V^2}\right)^{0.36}$	0.5	1.4	2.4	2.36	2.32	2.25	2.2
Experiments of Tumanishvili		0.6	1.7	2.6	2.7	2.65	2.6	2.6

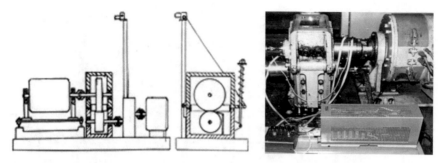

Fig. 25 High speed twin disk machine

Fig. 26 Experimental pieces (**a**) and a working surface of the roller with traces of scuffing (**b**) at total speed of rolling 7 m/s, sliding speeds of 3 m/s, linear load 100 N/m

In Fig. 26 are shown experimental pieces and a working surface of the roller with traces of scuffing.

The experiments have been devoted to the character of the wear process of working surfaces, research into influence of various parameters on the lubricant film thickness and the factor of friction at the use of popular mineral lubricants. The researches were executed with the use of the roller machine with independent drive of rollers.

Conditions of the experiments and measured sizes:

- Rolling speed—up to 70 m/s;
- Diameters of rollers 183 and 143.3 mm; width of rollers 12 and 17 mm;
- Sliding velocity—up to 35 m/s;
- Contact pressure—$5 * 10^5$ to $2 * 10^6$ N/m;
- Dynamic viscosity 49–140 mN s/m^2;

Mean temperature of operational surfaces of the rollers and the fillet-bound lubricant 50–1000 C.

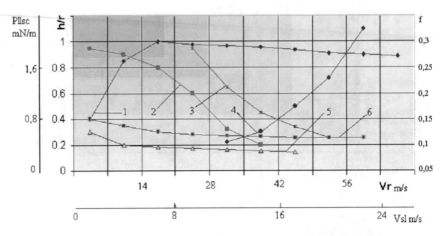

Fig. 27 Dependence of relative lubricant film thickness (h/R), linear scuffing load (P_{llsc}) and coefficient of friction (f) from rolling speed (Vr) and sliding velocity (Vs) at various viscosities (v) of lubricants

During experiments were measured: friction torque, sliding velocity and lubricant film thickness. For measurement of speeds of rotation was utilized magnetic pickups, for measurement of friction torque was utilized the strain gage transducer and contactless skate. The lubricant film thickness was measured by capacitance method. The beginning of occurrence of scuffing was defined on splashes of the moment of friction. Development of process was accompanied by sharp rise of temperature and characteristic noise. Results of experimental research are shown in Fig. 27.

1. $h/R = \varphi(V_r)$; $P_{ll} = 10^6$ H/m; $v = 157$ cSt;
2. $h/R = \varphi(V_{sl})$; $P_{ll} = 2 \times 10^6$ H/m; $v = 157$ cSt; $V_r = 50$ m/s;
3. $P_{llsc} = \varphi(V_{sl})$; $v = 49$ cSt; $V_r = 50$ m/s;
4. $P_{llsc} = \varphi(V_r)$; $v = 157$ cSt; $V_{sl} = 22$ m/s;
5. $f = \varphi(V_{sl})$; $P_{ll} = 1.5 \times 10^6$ H/m; $v = 49$ cSt; $V_r = 50$ m/s;
6. $f = \varphi(V_r)$; $P_{ll} = 10^6$ H/m; $v = 157$ cSt

The researches have shown that a dominating kind of deterioration of working surfaces for the given conditions is the scuffing. Researches have shown also high sensitivity and the limited stability of used lubricants to thermal influence.

For rolling with sliding heavy-loaded bodies the most dangerous (dominating) kind of deterioration is scuffing. It arises at destruction of a lubricant film (the third body) and direct contact of bodies in extreme friction conditions. For prevention of this phenomenon they try to minimize a sliding distance (structures of wheels and rails, dynamic characteristics of the track and a rolling stock, etc. must be improved) and to improve the tribological characteristics of the contact (improves properties of contacting surfaces, applied lubrication). Necessary condition of scuffing is destruction of the third body and direct contacting of surfaces. Liquid

Fig. 28 The interacting elements of machines with various types of damage: **a** train wheel with fatigue damage of the rolling surface and damage of the flange due to adhesive wear (scuffing); **b** gear wheel with the traces of scuffing on the tooth face; **c** inner ring of the rolling bearing with the traces of fatigue damage

lubricants are frequently used for formation of the third body. Consistent and solid lubricants are also used.

Our experimental researches have shown that at the same friction modifier variation of the friction coefficient depends on the degree of destruction of the third body: negative friction takes place at continuous third body; neutral friction—at the partly destructed and restorable third body and positive friction—at the progressive destructed discontinuous and the non-restorable third body.

It should be noted that various types of damage take place simultaneously and with various intensity in the heavy loaded contact and visually they are seen as dominant types of damage. The experimental researches have shown that the type of degradation of interacting surfaces mainly depends on the combination of the relative sliding velocity and shearing stress. For example, the main type of damage at low relative sliding is fatigue wear [13] (generation of cracks, plastic deformations and exfoliation), though adhesive wear takes place in parallel. Such phenomenon takes place on the rolling surface of the train wheel, near the pitch point of the gear drive, in rolling bearings and etc. The portion of adhesive wear increases at increase of the relative sliding and it becomes dominant type of damage. For example, on the steering surfaces of the train wheel, in the remote places of the gear drive line of action (Fig. 28), in the cam mechanisms and etc.

6.3.2 Experimental Studies on the Twin Disk Machine MT-1

The neglecting of the properties and state (continuous or discontinuous) of the third body in the contact zone leads to inadequate description of processes and inadequacy of results of theoretical and experimental researches. Because of this we performed the experimental researches on the twin disk machine MT-1 (Fig. 29) in the broad range of operational factors with the use of existing lubricants and ecologically friendly friction modifiers, developed by us.

The tests were performed at single application of the friction modifier on the rolling surface of the roller. After certain number of revolutions, a thin layer of the

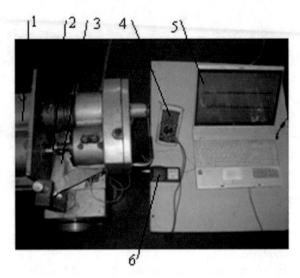

Fig. 29 The twin disk machine model MT1 and measuring means: 1—twin disk machine, 2—tribo-elements, 3—the wear products, 4—tester, 5—personal computer, 6—vibrometer

(a)   **(b)**   **(c)**

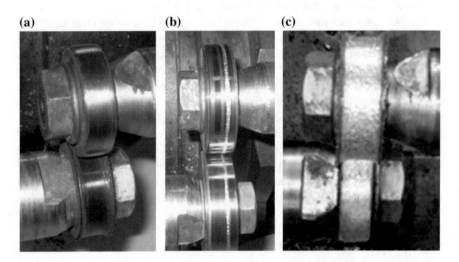

Fig. 30 The stages of damage of the interacting surfaces: **a** damage in the separate points; **b** damage in the form of the narrow strip; **c** damage on the whole area of the contacting surfaces

friction modifier was destroyed that was revealed by sharp increase of the friction moment and initial signs of scuffing on the surfaces. Without repeated feeding the friction modifier the damage process was progressed. The rollers with various degree of damage are shown in Fig. 30: (a) with initial signs of damage; (b) damage in the form of a narrow strip; (c) damage on the whole contacting area.

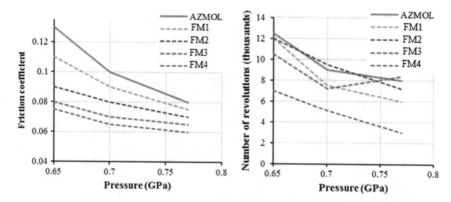

Fig. 31 Dependences of friction coefficients and numbers of revolutions till appearance of the first signs of scuffing on the contact stress for initial linear contact of disks and different anti-frictional friction modifiers

To assess capabilities of the friction modifiers (FM), the tests were carried out on the twin-disc machine. Experimental research was performed at rolling of discs with up to 20% of sliding. The rollers imitating the wheel and rail had diameters of 40 mm and widths of 10 and 12 mm. A benchmark was provided by tests with the use of lubricant "AZMOL" made in Ukraine. For these heavy loaded rolling/sliding surfaces the intimate contact of the interacting surfaces resulted in scuffing.

The graphs of dependences of the friction coefficient and number of revolutions of rollers till appearance of the first signs of scuffing at the contact stress for initial linear contact of disks and anti-frictional friction modifiers are shown in Fig. 31. It is seen from these graphs that for the initial linear contact, when the contact stress is in the range of 0.65–0.77 GPa (flange contact) increase of the contact stress leads to decrease of the friction coefficient. This is characteristic for the solid modifiers [13]. It can also be seen that increase of the contact stress leads to decrease of number of revolutions till the destruction of the third body and onset of scuffing. The FM-s developed by us in the frame of SUSTRAI project provided lower friction coefficients than AZMOL and some of them endured greater number of revolutions before scuffing at specific contact pressures.

The graphs of dependences of the friction coefficient and number of revolutions of rollers till appearance of the first signs of scuffing at the contact stress for initial point contact of disks and anti-frictional friction modifiers are shown in Fig. 32.

When the contact stress is in the range of 2.42–3.96 GPa the friction coefficient increases with increase of the contact stress. It can also be seen that increase of the contact stress leads to decrease of the number of revolutions till the destruction of the third body and onset of scuffing more intensive than in the previous case.

Dependences of friction coefficients and number of revolutions till appearance of the first signs of scuffing at the contact stress for initial linear contact of disks and for three different frictional FM-s are shown in Fig. 33. As shown from the graphs

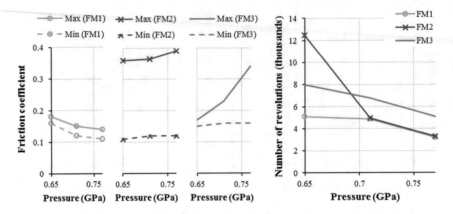

Fig. 32 Dependences of friction coefficients and numbers of revolutions till appearance of the first signs of scuffing at the contact stress for initial point contact of disks and three different FM-s

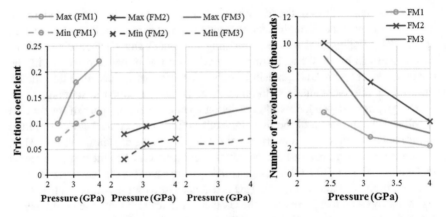

Fig. 33 Dependences of friction coefficients and numbers of revolutions till appearance of the first signs of destruction of the third body (first signs of appearance of the first signs of scuffing) on the contact stress for initial linear contact of disks and three different frictional FM-s

the friction coefficients for such FM-s is too high and in some cases, they are characterized by significant dispersion.

The carried-out researches have shown that a third body of various generations and properties always exists in the contact zone and the grade of their coating has a great influence on the tribological properties of the interacting surfaces.

At existence of the continuous film of the third body in the contact zone the stable friction force and low intensive surface damage take place. But at destruction and restoration of the third body, as the experimental researches have shown there exist the conflicting processes: at operation in the steady regime when destruction and restoration of the third body take place the separate faint impulses occurred that can be explained by destruction of the seizures generated in the individual places of destruction of the third body. At continuation of operation in such conditions because of destruction of the third body, the processes of the destruction of the third body and scuffing were progressed and expanded. This had a great influence on the friction moment—its constant and especially variable components (Fig. 33), wear rate, vibrations and noise were increased. Restoration of the third body in the contact zone by additional feeding the friction modifier promoted regaining of the normal working conditions—the friction moment has stabilized.

Destruction of the third body in separate parts of the contact zone of wheels and rails was followed by instantaneous seizures, scuffing and rise of the friction moment. At low numbers of revolutions this caused low frequency vibrations and at high revolutions—a squeal like noise was rising up to 120 dB and temperature on the actual contact area of interacting surfaces could reach the temperature of metal fusion.

A particular instability of the friction coefficient was observed on the low speeds and high loads: intensive impulses of low frequency were marked and the scuffing marks of significant sizes—scratches and pits were noticed on the surfaces of rollers. With increase of the speed the time of interaction of the surfaces and thermal impact in the real contact zone, values of the impulses and pits and amplitude of the friction force variable component decrease, the frequency increases and the separate impulses turn into noise. With further increase of the speed the process is progressed, tonality of the noise rises and it turns into whistle. At further increase of its frequency (above 20 kHz) it becomes imperceptible for man.

6.4   Dependences of Tribological Characteristics of the Contact Zone on the Relative Sliding Velocity

The adhesive wear and scuffing are the most common damage types at destruction of the third body [64]. In Fig. 34 shown the stages of damage of the interacting surfaces at conditions of rolling with sliding.

To various degrees of destruction of the third body corresponded the various types of variations of the friction moment: to continuous third body corresponded the stable friction moment which varied smoothly at increase of the thermal load.

A great number of the scientific works are devoted to ascertainment of the laws of variation of the friction coefficients as well as to other problems of tribology and with the perfection of machines its actuality increases. Usually the friction process proceeds at presence of the continuous or discontinuous (restorable or progressively destructible) third body that, as our experiments have shown, stipulates the character of variation of the friction force (tractive force). The researches have shown that stability of the third body is very sensitive to the relative sliding. The friction/creep relationship is shown in Fig. 35 [65, 66].

As it is noted in [13], the variation of the friction coefficient is mainly caused by the changing composition of the interfacial layer between the wheel and rail

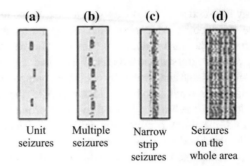

Fig. 34 The stages of damage of the interacting surfaces due to seizures and scuffing of the surfaces: **a** unit seizures; **b** multiple seizures; **c** seizures in the form of the narrow strip and **d** on the whole area of the roller

Fig. 35 Friction/creep relationship

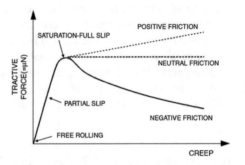

(the "Third Body"). As it were shown our experimental researches, to the various types of variations of the friction moment (and destruction rate of surfaces) correspond the various degrees of destruction of the third body: to continuous third body correspond the stable (or smooth varied) friction moment and low destruction rate of surfaces. As the third body destructed, the friction moment and destruction rate of surfaces were increase. Experimental researches have shown that for the given friction modifier the variation of the friction coefficient mainly depends on degree of destruction of the third body. At presence of the continuous third body increase of the relative sliding velocity leads to increase of the friction power and the contact temperature; decrease of the lubricant viscosity, film thickness and friction force ("negative friction").

Worsening of the working conditions caused by the partial, non-progressive damage of the third body in the separate unit places that corresponds to the separate small impulses of the friction moment. Destruction of the third body in the multiple places leads to the multiple damage of the third body, multiple adhesive junction of micro-asperities, disruption of these junctions, and comparatively certain increased impulses of the friction moment and to "neutral friction".

The progressing of the third body destruction leads to spacious discontinuous third body, adhesive junction of micro-asperities, disruption of these junctions and increase of the friction forces ("positive friction").

As it is seen from Fig. 31 negative, neutral and positive behaviors of the friction forces are stipulated by the degree of destruction of the third body (the continuous, unit, multiple, narrow strip and vast).

The researches have revealed the processes of the initial destruction of the third body and following restoration at steady conditions of operation: restoration of the third body in the places of destruction and destruction in other places that was seen visually. After certain number of revolutions if a new dose of the friction modifier was not applied the process of scuffing progressed. This resulted in increase of the friction moment, especially its variable component and also the wear rate, vibrations and noise. After repeated feed of the friction modifier into the contact zone the normal working conditions was restored.

Therefore, for ensuring the high wear resistance and stable friction force in the contact zone of the wheels and rails it is necessary to provide continuous film of the third body with due properties between interacting surfaces. Therefore, the condition for the destruction of a third body can be used for the basics of estimating the coefficient of friction and the damages for the given peculiarities of surfaces materials. Damage map according to the stress state, wheel steel wear rates and regimes and dependences of tribo-technical characteristics of the contact zone of wheels and rails on the sliding velocity are shown in Fig. 36.

Dependences of the damage type on the contact stress and sliding velocity are shown in Fig. 36a [67]: at small sliding velocity, the dominant type of damage is fatigue and at increase of the sliding velocity—wear and thermal damage.

Dependence of the wear rate on the relative sliding velocity (in percent) is shown in Fig. 36b [47]: at low sliding velocity, the wear rate is low. The middle and sever wear rates are not greatly differ from each other and when the sliding velocity exceeds 0.14 high intensive catastrophic wear begins.

The various dominant damage types, wear rate and friction coefficient are characteristic for various relative sliding. Three zones can be distinguished in Fig. 36c: in the zone 1 and at the beginning of zone 2 deformations of the subsurface layers reach the maximal values and the interacting surfaces undergo cyclic deformations (at kinematic or friction rolling). The zone "no damage" in Fig. 36a corresponds to zone 1 in Fig. 36c. With the rise of relative sliding the contact stress gradually increases towards the surface [65] and the friction factor decreases reaching the minimum value. The zone "mild" in Fig. 36b corresponds to zone 2 till noticeable increase of the friction moment variable component in Fig. 36c. In the case of moderate shearing stresses, the principal types of destruction are fatigue failure of the surfaces—generation and propagation of cracks, shelling and pitting, plastic deformation, and also adhesive wear and beginning of scuffing. These types of damage take place in the following zones: "RCF (spalling and head checks)" in Fig. 36a, zone "sever" in Fig. 36b and zone "2" in Fig. 36c. In the zone 3 (Fig. 36c) scuffing is the main type of damage but plastic deformations and fatigue

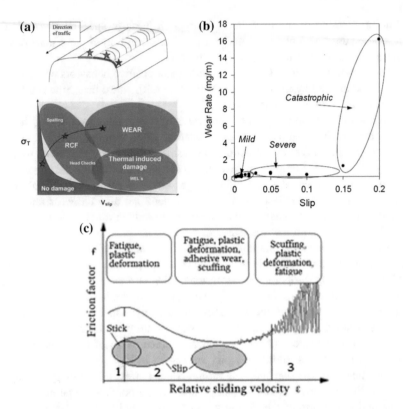

Fig. 36 Dependences of tribo-technical characteristics of the wheel and rail contact zone on the sliding velocity: **a** damage map according to the stress state and slip speed [67]; **b** R8T wheel steel wear rates and regimes [47] and **c** dependence of the coefficient of friction (f) on the relative sliding (ε) and expected kind of surface damage

damage can also take place. The zones "wear and thermal induced damage" in Fig. 36a and the zone "catastrophic" in Fig. 36b correspond to these damage types.

At full separation of the interacting surfaces by the third body the tribo-technical properties of the contact zone mainly are depend on the properties of the third body and they provide high wear resistance of the interacting surfaces and relatively stable friction coefficient. Destruction of the third body in the separate unit places corresponded the separate small impulses of the friction moment and adhesive wear of the mean intensity and the balance of destruction and restoration of the third body was observed, that corresponds to the "mild" wear type [47]. At destruction of the third body in the multiple places corresponded the multiple, comparatively increased impulses of the friction moment and a heavy form of the adhesive wear—scuffing; separate seizures of the micro-asperities (or unit seizures located in one or several narrow strips) and that corresponds to the "sever" wear type. But then, without feeding or insufficient renovation of the third body, continuation of work leads to extension of destruction of the third body and appearance of multiple seizures (Fig. 34b). Then the friction moment and its variable component are rapidly growing and the unit and multiple seizures (causing scuffing) are becoming uninterrupted and then they propagate on the whole width of the interacting surfaces (Fig. 34c) and the corresponding wear becomes "catastrophic".

7 Main Damage Problems of Wheels and Rails and Prevention of Catastrophic Wear Rate

7.1 General Overview

For the purpose of increase of durability of the wheels and rails contacting surfaces they usually improve their volumetric and superficial properties and working conditions: they use various lubricants and friction modifiers; establish wear types (mild, sever and catastrophic [47]), try to estimate the parameters having influence on the wear [1–3, 9], friction coefficient between the wheels and rails, that has a great influence on the behaviour of the railway vehicle. But some physical and chemical aspects of the friction process are not sufficiently studied yet and the corresponding mathematical models do not exist. The proper friction control in the contact zone of the wheel and rail can play an important role in reduction of the environment pollution, vibrations and noise and the maintenance cost.

For the wheels and rails are characteristic various types of damage: various forms of the adhesive wear (mild and sever adhesive wear, scuffing, scoring, galling, spaling, squats and etc.), fatigue, fretting-corrosion, plastic deformations, abrasive wear and etc., whose mechanisms are different for their early reveal a proper approach is needed. These types of damage often take place simultaneously and are revealed by the visually similar signs though the mechanisms of their generation and development and methods of their prevention are also different. At present the methods and means of the damage prevention are not properly developed that complicates their identification.

It should be noted that fatigue damage in contrast to other types of damage depends on the number the loading cycles and value of the load. The third body promotes the even distribution of the load on the surface and in some cases, it can increase the number of cycles 30 times and more [13]. The third body promotes also prevention of the adhesive and catastrophic wear.

7.2 Methods of Estimation of the Adhesive Wear

It is known several approaches to prediction of the wear rate of wheels and rails by numerical simulation. Fries and D'avila [68] compare different wear modelling techniques, the material loss being dependent on the friction work on the contact area, normal load and sliding distance, contact pressure, and normal load.

In the 80s of the last century in 31 laboratories of 7 countries the tests were carried out on the similar samples and the obtained results show great dispersion of the friction coefficients. The authors explain this by the rigidity difference of the testing devices [69].

At present, there is not a fundamental method for estimation of the wear intensity of the contacting bodies and existing methods are characterized by the low information value and precision [68, 70]. As it is noted in [71] about 100 variables are included in more than 180 models but none of these variables describes the wear process unambiguously.

A reliable wear model can be used for estimation of qualitative and quantitative influence of the acting parameters. Many important research works regarding the wear models based on both, the global and local approaches to wear estimation can be found in the literature. It is known several approaches to prediction of the wear rate of wheels and rails by numerical simulation. Fries and D'avila [68] compare different wear modelling techniques, the material loss being dependent on the friction work on the contact area, normal load and sliding distance, contact pressure, and normal load. Kalker [72] calculates the wheel tread wear for a metro vehicle assuming a material loss proportional to the dissipated energy and defining probability functions for the lateral contact position. In the 80s of the last century in 31 laboratories of 7 countries the tests were carried out on the similar samples and the obtained results show great dispersion of the friction coefficients. The authors explain this by the rigidity difference of the testing devices [69].

One of the most popular formulas for calculation on wear appertains to Archard [73, 74] and has the form:

$$Q = \frac{KWL}{H}$$

where Q is the total volume of the wear debris produced; K—dimensionless constant; W—total normal load; L—sliding distance; H—hardness of the softest contacting surface.

As it is seen from the formula it contains the insufficient number of parameters affecting the wear and formation or destruction of the third body, heat processes in the contact zone and etc. All this is reflected negatively on the precision of the calculation results.

7.3 Development of a New Concept of the Wear Mechanism

The researches have shown the two quite different damage periods and characteristic processes: damage at existence of the continuous third body between the interacting surfaces and at destructions of various degreases (types) of the third body.

The researches have also shown that separation of the interacting surfaces by the third body is a universal method of providing the contact zone with needed properties: at presence of the continuous third body tribologic properties of the contact zone depend mainly on the properties of the third body and for ensuring desired tribologic properties it is necessary to choose proper friction modifier, materials and methods of modification of the interacting surfaces. At destruction of the third body the various undesired phenomena (high and unstable friction forces, neutral and positive friction coefficient, heavy and catastrophic wear intensity, vibrations and noise during friction and etc.) can be explained by the properties of the third body and degree of destruction. Thus, ensuring of desired tribology properties of the heavy loaded contact and friction control must be based on the third body properties and inadmissible degree of its destruction, which can cause its progressive destruction and catastrophic wear, must be avoided.

Materials of the interacting surfaces and environment with heterogeneous physical and chemical properties participate in the process of formation and destruction of the third body in the contact zone. Further destruction of the adherent parts in the places of destruction of the third body takes place also at variable power and thermal loadings. This causes various undesirable phenomena (significant instability of the friction forces, high and catastrophic wear intensity, vibrations, noise and etc.) many aspects of which are not studied sufficiently. Therefore, for rational solution of the problem it is necessary to combine the existent theoretical fundamentals and results of the proper experimental researches that will allow correction of results of the theoretical researches and generalization of the obtained results.

7.4 Estimation of Stability of the Third Body on the Base of EHD Theory of Lubrication

The quality of mineral oil is determined by the chemical composition and physical properties of the hydrocarbon-based substances. The properties are viscosity, viscosity index, low temperature properties, high temperature properties, density, demulsification, foam characteristics, pressure/viscosity characteristics, thermal conductivity, electrical properties and surface tension.

The effectiveness of the EHD theory of lubrication is described by ratio λ or film parameter [75], which is the ratio of central film thickness at the Hertzian contact zone to the r.m.s. of the rolling element surface finish:

$$\lambda = \frac{h_{\min}}{\sqrt{(S_r^2 - S_l^2)}} , \qquad (57)$$

where S_r and S_t are the r.m.s of the surface finish.

The EHD regime is characterized by ratio λ between 3 and 10, which corresponds to the film thickness between 0.1 and 1 μm. It has been pointed out that a full film can be obtained with no asperity contact only when $\lambda > 3$. If $\lambda < 3$, it will lead to mixed lubrication with some asperity contacts (Hamrock and Dowson [76]).

Till recently, the third body was only considered from mechanical point of view and only the oil layer of hydrodynamic or elasto-hydrodynamic generation was thought to protect the contacting surfaces against the direct contact. Elasto-hydrodynamic theory of lubrication only considers the mechanical phenomena proceeding in the contact zone. The modern friction modifiers and products of tribo-chemical interaction of the medium and surfaces have a great influence on the operational properties of the frictional contact and like the oil layer are ingredients of the third body.

Chang has proposed the first approximate solution of the system of integral and differential equations of the EHD theory of lubrication (considering thermal processes and dependence of the oil viscosity on the pressure and temperature) [77], though for prognosis of adhesive and other types of wear his results turned out not to be satisfactory. Solution of the tribological problems must be based on the fundamental laws of physics, chemistry, mechanics of rigid bodies, thermo-dynamics, materials science [5], but elastic and hydro-dynamical theory of lubrication only takes into account mechanical phenomena proceeding in the contact zone. Such a conclusion is made due to ignoring remaining components of the third body that have a significant influence on the interacting contact exploitation properties.

The tribological properties of the interacting details greatly depend on the properties of their contacting surfaces (which differ from their volumetric properties) and the third body [1], existing in the contact zone. The third body separates the interacting surfaces fully or partly and it is formed in the contact zone naturally (because of interaction of the environment and surfaces and wear products) or artificially (because of participation of the lubricants or friction modifiers along with natural third body). The boundary layers of the contact zone including the self-generated organic films increase significantly the contact endurance limit of the materials [2]. Destruction of the third body stipulates the direct contact of the interacting surfaces and the heaviest form of the wear—scuffing at heavy working conditions [3].

The thickness of the rough surface boundary layers cannot be measured with the use of the modern methods of measurement of the oil layer thickness. Information about destruction of the boundary layers (and about onset of scuffing as well) can be obtained by the onset of the friction moment sharp increase on the oscillogram.

The criterion of destruction of the third body (or scuffing) according to formula (57) has the form:

Table 4 Exponents of formula (58) parameters

a	b	c	d	l	f	g	h	i	j	n
0.37–0.7	(−0.36) to (−1.32)	(−0.15) to (−0.265)	0.04–0.52	0.25–0.36	−1	0.6	0.18–0.66	(−0.18)–(−0.66)	0.09–0.33	0.045–0.165

$$K\left(\frac{R}{\sqrt{R_{a1}^2 + R_{a2}^2}}\right) \cdot \left(\frac{\mu V_{\Xi K}}{P_n}\right)^{0.7} \cdot \left(\frac{P_n \beta}{R}\right)^{0.6} \cdot \left(\frac{\lambda}{\alpha \mu V_{CK}^2 P_{el,2}^2}\right)^e \leq 1, \quad (58)$$

As it follows from the formula (58), conditions of destruction of the third body depends on the mechanical and thermo-physical characteristics of interacting surfaces, geometric and kinematic parameters, thermo physical and tribotechnical parameters of the third body; researches have shown also the special sensitivity of the third body on thermal loading and relative sliding, which must be taken into account to improve working conditions of the electric locomotive running gear.

The criterion of the third body destruction which is developed on the base of elasto-hydrodynamic theory of lubrication and results of experimental researches, considering stability of the boundary layers, has the form:

$$K \cdot V_{\Sigma k}^a \cdot V_{sl}^b \cdot P_{ll}^c \cdot \mu_0^d \cdot R^l \cdot \left(\sqrt{R_{a1}^2 + R_{a2}^2}\right)^f \cdot \beta^g \cdot \lambda^h \cdot \alpha^i \cdot a^j \cdot E^n \leq 1 \quad (59)$$

where $V_{\Sigma k}$ is a total rolling velocity; V_{sl}—sliding velocity; P_{ll}—linear load; μ—dynamic viscosity of the lubricant; R—reduced radius of curvature of the surfaces; R_{a1} and R_{a2}—average standard deviation of the interacting surfaces; β—piezo coefficient of the lubricant viscosity; λ—the lubricant thermal conductivity; α—thermal coefficient of the lubricant viscosity; a—thermal diffusivity; The exponents a, b, c, ..., n and coefficient K are specified on the base of the experimental data obtained by T. I. Fowle, Y. N. Drozdov, Vellawer, G. Niemann, A. I. Petrusevich, I. I. Sokolov, K. Shawerhammer, G. Tumanishvili and are given in Table 4.

As it is seen from Table 4, destruction of the third body is especially sensitive to the degree b of sliding velocity. It follows from formulas (58) and (59) that with increase of the rolling velocity, viscosity, radius of curvature, piezocoefficient of viscosity, heat conductivity factor, thermal diffusivity and coefficient of elasticity the stability of the third body increases and with increase of the sliding velocity, linear loading, roughness of surfaces and thermal coefficient of viscosity it decreases.

8 Design Methods for Perfection of the Technical Features of Running Gears

8.1 The Wagon Wheelset

Figure 37 shows the wagon wheel-set with the independent rotating wheels with a new profile of the wheel and compound shaft.

For reduction of the un-sprung mass, stress concentrations on places of mounting of wheels on the shaft and damping of the bending and torsional deformations, the shaft consists of two embedded bushings 3, 4. The wheel 2 is mounted on the outer bushing 3 motionlessly and the wheel 6 is mounted on the outer bushing 3 with rotation possibility or motionlessly. The motionless wheel (2) provides rotation of the shaft in axle boxes, and the second wheel (6) that can be mounted with rotation possibility, in case of lagging of the outer wheel in curves will give the possibility of additional rotation around outer bushing 3 avoiding slipping of inner wheel on a rail. Outer bushing 3 of the shaft is mounted on inner bushing 4 motionlessly in the place of action of the minimum bending moment (in the middle part 8), and in other parts—with the clearance fit, that unloads the inner bushing from torque. The shafts of wheels are un-sprung details and their hollow form, as the preliminary estimation shows, will allow to reduce a mass of the shaft, retaining the required bending strength.

The device is working in the following way: one wheel it is established motionless on an axis and another with rotation possibility. A rotary movement is transferred from rails 2, 8 to wheels 3, 7, from which wheel 3 rotates bushing 4 and the axle 5 fixed on it rigidly and rotating in the bearings 1, 10 of the units of boxes. A low friction coefficient is provided between bushing 4 and wheel 7. At the zigzag movement of the wheelset in the straight section of the track or at the different distances rolled by them in the curves, wheel 7, whose friction coefficient with the axle is lower than one with the rail, it will rotate (slide) relative to the bushing and not to the rail. The tread surfaces of wheels and rails, as well as the steering surfaces [pressure gauge and flange root (flange)] are separated by a groove [78], and its modification is carried out separately. Use of the hollow shaft allow to decrease of un-sprung masses more than 1 kN.

Fig. 37 Wagon wheel-set with the independent rotating wheels with a new profile and assembled shaft

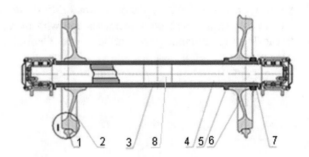

8.2 The Rail Fastening Device

The problem consists in working out of a rail fastening which will capable of damping the lateral fluctuations and maintain a constant rail cant. But the damping is performed by deflection of the rail in the external direction (Fig. 38), promoting displacement of the contact point on the flange downwards. It increases the difference of diameters of interacting surfaces, the sliding distance and causes other undesirable phenomena.

To maintain a constant rail cant and ensure damping of lateral fluctuation and constant distance between heads of the rails, a new design of fastening of the rail was developed (Fig. 39).

The rail fastening device consists the sleeper 1, immovable rail pad 2, elastic member 3, rails 4, movable rail pad 5 and controlling beam with treaded ends 6. The base plate consists of the movable 5 and immobile 2 arts connected with each other by sliding kinematic pair. The rail 4 is fastened on the movable part 5 with optimal angle of inclination 20° and the immobile part 2 is fastened on the sleeper 1. An elastic element—compression spring 5 is placed between movable 5 and immobile 2 parts with possibility of preliminary compression by screw pair 6.

The rail fastening device provides damping of the wheelsets by equidistant movement of the rails and preserves the constant inclination of the rails.

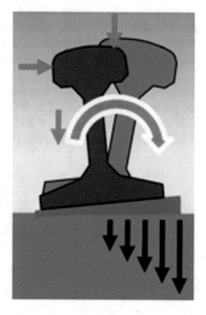

Fig. 38 Variation in rail cant

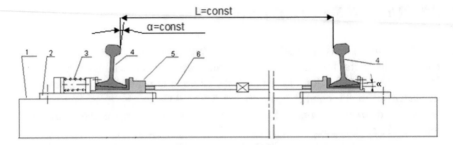

Fig. 39 The rail fastening device: 1—sleeper; 2—immovable rail pad; 3—elastic member; 4—rail; 5—movable rails pad; 6—controlling beam with treaded ends

8.3 The Electric Locomotive Wheelset

Supporting elements of the freight locomotive traction engine are the unsprang wheel-set and the sprung part of the bogie or engine (Fig. 40). Partial leaning of the traction engine on the wheel-set increases its un-sprang mass and contact loading of the wheel and rail.

Therefore, it is necessary to elaborate such running gear of the freight locomotive, which will ensure to decrease the unsprung masses of the freight locomotive running gear, the lateral forces and relative slipping in the contact zone.

Such scheme is shown in Fig. 41, where the traction engine is completely sprung and when the lateral force exceeds the allowable value in the curves, the wheel moving on the inner rail will be automatically disengaged from the running gear kinematic chain and at decrease of the lateral force the wheel will again be engaged in the kinematic chain.

At movement in the curve the flange of the outer wheel 8 will contact with outer rail (not shown) gauge and produce the axial force. This force will be transmitted to the bushing 12, which will move in the direction of this force together with the wheels 8, 9. The teeth of half-coupling 11 will enter deeper in the teeth of the half

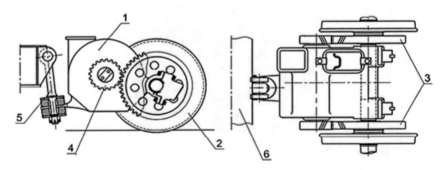

Fig. 40 The freight locomotive running gear 1—traction engine; 2—wheel-set; 3—gear wheel; 4—pinion; 5—suspension of the traction engine; 6—transversal beam of the bogie

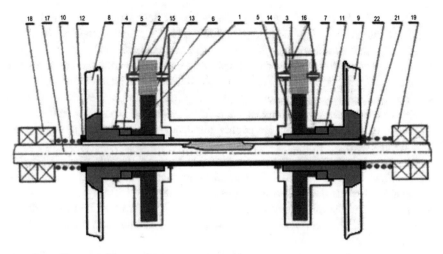

Fig. 41 A scheme of the electric locomotive running part

coupling of gear wheel 5 and the teeth of the half coupling of wheel 8 will be disengaged from the teeth of half-coupling 13 of gear wheel 1. Springs 10 and 21 return wheels 8, 9 in the initial position. This will cause movement of the axle 17 towards its initial position, decrease of the angle of attack, axial force, and rolling resistance and wear rate of the wheel flange and rail gauge.

Teeth gears 1, 5 together with wheels 8, 9 and shaft 17 will roll up on driver pinions 2, 3 (mounted on the engine shaft) at action of the vertical loading from rails (it is not shown on the scheme), so this loading will not be transmitted directly to the engine 1, sprung on the bogie or body. It will considerably decrease the un-sprung mass (approximately by 2.5 tons).

The suggested locomotive running gear allows to decrease the un-sprung masses and the relative sliding velocity.

9 Conclusions

- The wheels and rails are the most vulnerable parts of railways of both Russian and European standard and one of the ways to increase competitiveness of the TRACECA railway is perfection of the technical characteristics of wheels and rails interaction by improving their tribological properties;
- Heavily loaded interacting surfaces can be separated by the continuous, partially destructed, but recoverable or progressively destructible third body, that leads respectively to the mild, severe or catastrophic wear;
- To ensure high wear resistance and stable friction force of the wheels and rails, it is necessary to provide the contact zone by continuous or restorable film of the third body with due properties;

- For heavily loaded surfaces, at certain degree of destruction of the third body, direct interaction of juvenile parts of the surfaces leads to their seizure and scuffing, that stipulates the vibrations, noise and catastrophic wear;
- The degree of destruction of the third body has a great influence on the wear rate, value and stability of the friction coefficient, vibrations and noise;
- The traction-creepage relationship (the negative, neutral and positive friction) and dependences of tribological characteristics of the wheel and rail contact zone on the sliding velocity are mainly stipulated by the degree of destruction of the third body;
- The researches in laboratory conditions of the ecologically friendly friction modifiers for wheels and rails steering and tread surfaces and brake shoes, developed by us, have shown the positive results;
- The reason of the rail corrugation is periodic sliding of a wheel on the rail due to movement in the curves, ellipticity of a wheel and difference in diameters of the tread surfaces;
- The formula for the computation of the thermal field for the widespread model of micro asperity takes into account thermal characteristics of the micro asperity as well as its surroundings;
- For the modified surfaces, the friction coefficient depends on the degree of the third body destruction which stipulates the character of variation of the friction forces;
- The criterion of destruction of the third body reflects the factors taken into account in the EHD theory of lubrication. The stability of the boundary layers is stipulated by the coefficients determined experimentally and reflecting the third body properties;
- The wagon and locomotive wheelsets and rail fastening devices developed by us decrease the wheel and rail relative sliding velocity and accompanying undesirable phenomena.

Bibliography

1. Magel EE (2011) Rolling contact fatigue: a comprehensive review. Prescribed by ANSI Std. 239-18 298-102 DOT/FA/ORD-11/24, U.S. Department of Transportation, Office of Railroad Policy and Development Washington, DC 20590. 118 p
2. Tunna J, Shu X (2006) Regional fast rail project: turnout safety analysis. P-06-019 for Victoria's Department of Infrastructure. Regional fast rail project. Transportation Technology Center, Inc. A subsidiary of the Association of American Railroads. Pueblo, Colorado USA
3. Shust WC, Elkins JA, Kalay S, El-Sibaie M (1997) Wheel-climb derailment tests using AAR track loading vehicle, Association of American Railroads Report R-910
4. Weinstock H (1984) Wheel climb derailment criteria for evaluation of rail vehicle safety. In: Proceedings of ASME Winter Annual Meeting, 84-WA/RT-1, New Orleans, LA

5. Thompson DJ, Honk-Steel AD, Jones CJ, Allen PD, Hsu SS, Iwnicki SD (2003) Project A3— railway noise: curve squeal, roughness growth, friction and wear. MMU

6. Wu H, Elkins J (1999) Investigation of wheel flange climb derailment criteria, Association of American Railroads report R-931

7. Cristol-Bulth'e AL, Desplanques Y, Degallaix G (2007) Coupling between friction physical mechanisms and transient thermal phenomena involved in pad–disc contact during railway braking. Laboratoire de Mrecanique de Lille (CNRS UMR 8107), Ecole Centrale de Lille, BP 48, F-59651 Villeneuve d'Ascq Cedex, France

8. Elkins J, Wu H (2000) New criteria for flange climb derailment, IEEE/ASME joint railroad. Conference, Newark, NJ

9. Rhee SK, Jacko MG, Tsang PH (1991) The role of friction film in friction, wear and noise of automotive brakes. Wear 146(1):89–97

10. Gubenko S, Sladkowski A (2005) The influence of the contact stress on the structural changes of railway wheel stell. In: A. Sladkowski (ed) Rail vehicle dynamics and associated problems. Gliwice, p 135–160

11. Vasic G, Franklin FJ, Kapoor A (2003) Prepared for the railway safety and standards board. University of Sheffield. report: RRUK/A2/1

12. Митрохин АН (2008) Кто и почему « шьёт » колею шириной 1512 мм? Локомотив. No. 4. pp 5–8. [In Russian: Mitrokhin AN (2008) Who and why "sews" a gauge width of 1512 mm? Locomotive]

13. Дроздов ЮН, Павлов ВГ, Пучков ВН (1986) Трение и износ в экстремальных условиях. Москва: Машиностроение. 224 p. [In Russian: Drozdov YN, Pavlov VG, Puchkov VN (1986) Friction and wear in the extreme conditions. Moscow. Mashinostroenie]

14. Olofsson U, Lewis R (2006) Tribology of the wheel—rail contact. In: Simon I (ed) Handbook of railway vehicle dynamics. Manchester, Taylor & Francis Group, LLC, 526 p

15. Arias-Cuevas O, Li Z, Lewis R, Gallardo-Hernandez EA (2010) Rolling-sliding laboratory tests of friction modifiers in dry and wet wheel-rail contacts. Wear 268(3–4):543–551

16. Новиков ВВ, Иванов СГ, Шавшишвили ИД (2000) Ремонтные профили поверхности катания колеса. Путь и путевое хозяйство. No. 1. pp 56–57. [In Russian: Novikov VV, Ivanov SG, Shavishishvili ID (2000) Repair profiles for the tread surface of the wheel. Track and railway equipment]

17. Bolton PJ, Clayton P (1984) Rolling-sliding wear damage in rail and tyre steels. Wear 93:145–165

18. Lewis R, Dwyer-Joyce RS, Olofsson U, Hallam RL (2004) Wheel material wear mechanisms and transitions. In: 14th international wheelset congress, Orlando, USA

19. Lewis R, Dwyer-Joyce RS, Olofsson U, Pombo J, Ambrósio J, Pereira M, Ariaudo C, Kuka N (2010) Mapping railway wheel material wear mechanisms and transitions. In: Proceedings of ImechE, vol 224. Part F: J. Rail and rapid transit, JRRT328V

20. Кононов ВЕ (2005) Пути снижения износа гребней колёсных пар локомотивов. Локомотив. No. 2. pp 34–36. [In Russian: Kononov VE (2005) The ways of the wear decrease of wheel-set flanges of locomotives. Locomotive]

21. Ликратов ЮН (2004) Ездим на гребнях колёс. Железнодорожный транспорт. No. 12. pp 52–54. [In Russian: Likratov YN (2004) We ride on the wheel flanges. Railway transport]

22. Stone DH, Sawlay K, Kelly D, Shust W (1999) Wheel/rail materials and interaction: North American heavy haul practices. IHHA'99 STS—Conference, Session 3. Moscow, pp 155–168

23. Lundmark J, Hoglund E, Prakash B (2006) Running-in behavior of rail and wheel contacting surfaces. In: International conference on tribology, Parma, Italy, pp 20–22

24. Lewis R, Olofsson U (2004) Mapping rail wear regimes and transitions. Wear 257(7–8): 721–729

25. Olofsson U, Telliskivi T (2003) Wear, plastic deformation and friction of two rail steels a full-scale test and a laboratory study. Wear 254:80–93

26. Müller B, Jansen E, de Beer F (2003) UIC curve squeal project WP3. Swiss Federal Railways, Rail Environmental Center

27. Брюнчуков ГИ, Сухов АВ, Разумов АС, Тараканов ВЮ (2012) Результаты испытаний бандажей повышенной износостойкости. Бюллетень ВНИИЖТ. No. 2. pp 34–37. [In Russian: Brunchukov GI, Sukhov AV, Razumov AS, Tarkanov VY (2012) The results of tests of wheels with increased wear resistance. Bull VNIIJT]

28. Буйносов АП, Клинский ВС (1992) Об износе бандажей локомотивных колёс. Железнодорожный транспорт. No. 5. pp 45–46. [In Russian: Buinosov AP, Klinski VC (1992) About wear of the electric locomotive wheel treads. Railway transport]

29. Лысюк ВС (2004) О причинах схода вагонов и износа рельсов в кривых. Железнодорожный транспорт. No. 12. pp 50–52. [In Russian: Lysiuk VS (2004) On the causes of derailment and wear of rails in curves. Railway transport]

30. Sowlay KJ, Clark SL (1999) Engineering and economic implications of hollow worn wheels on wheel and rail asset life and fuel consumption. In: IHHA'99 STS-conference, Session 4. Moscow, pp 299–305

31. Николаев НИ (1962) Динамика локомотивов. Москва: Трансжелдориздат. 317 p. [In Russian: Nikolaev NI (1962) Dynamics of locomotives. Transzheldorizdat, Moscow]

32. Когаев ВП, Дроздов ЮН (1991) Прочность и износостойкость деталей машин. Москва: Высшая школа. 320 p. [In Russian: Kogaev VP, Drozdov YN (1991) Strength and durability of machine parts. Higher school, Moscow]

33. Swenson CA (1999) Locomotive radial steering bogie experience in heavy haul service. In: IHHA'99, pp 79–86

34. Yang C, Li F, Huang Y, Wang K, He B (2012) Comparative study on wheel-rail dynamic interactions of side-frame cross-bracing bogie and sub-frame radial bogie. Springerlink.com, Chengdu, pp 1–8

35. Grassie SL (2009) Review paper 1. Rail corrugation: characteristics, causes, and treatments. In: Proceedings of IMechE, vol 223. Part F: J. Rail and rapid transit. 16 p

36. Colette C, Horodinca M, Preumont A (2008) Rotational vibration absorber for the mitigation of rail rutting corrugation. Veh Syst Dynam 47(6):641–659

37. Hertz H (1882) Uberdie beruhung fester, elastischer Korper. Journal fur die und angewandte Mathematik 92:174 p

38. Kalker JJ (1990) Tree-dimensional elastic bodies in rolling contact. Kluwer academic Publishers, Dordrecht

39. Ohyama T (1991) Tribological studies on adhesion phenomena between wheel and rail at high speeds. Wear 114:263–275

40. Cassidy PD (2001) Wrought materials may prolong wheel life. Int Railw J 12:40–41

41. Courant R, Hilbert D (2004) Methods of mathematical physics, vol 1. WILEY-VCH Verlag GmbH & Co KGaA, Weinheim, p 560

42. Щедров ВС (1955) Температура на скользящем контакте. In: Трение и износ в машинах. Москва: Издательство АН СССР. No. 10. pp 155–296. [In Russian: Schedrov VS (1955) Temperature on the sliding contact. In: Friction and wear in machines. Academy of Sciences of SU, Moscow]

43. Лыков АВ (1967) Теория теплопроводности. Москва: Высшая школа. 441 p. [In Russian: Lykov AV (1967) Theory of heat conduction. Higher School, Moscow]

44. Рыкалин НН (1967) Расчёт тепловых процессов. Москва: Машгиз. 250 p. [In Russian: Rykalin NN (1967) Calculations of thermal processes. Mashgiz, Moscow]

45. Leach EF, Kelly BW (1065) Temperature, the key of lubricant capacity. Trans ASLE 9 (9):271–285

46. Дроздов ЮН (1971) Уточненный метод расчёта на задир пар трения в тяжелонагруженных механизмах. Вестник машиностроения. No. 4. pp 25–39. [Drozdov YN (1971) The more precise calculating method on scuffing of the friction pear in heavy loaded mechanisms. Vestnik mashinostroenia]

47. Lewis R, Dwyer-Joyce RS (2004) Wear mechanisms and transitions in railway wheel steels. Proc Inst Mech Eng Part J J Eng Tribol 218(6):467–478

48. Ghanbarzadeh A, Wilson M, Morina A, Dowson D, Neville A (2014) Development of a new mechano-chemical model in boundary lubrication, Leeds/Lyon conference

49. Wang W, Liu XJ, Liu K (2012) FEM analysis on multibody interaction process in three body friction geometry with rough surface. Tribology 6(2):59–66
50. Dwyer-Joyce RS, Lewis R, Gao N, Grieve DG (2003) Wear and fatigue of railway track caused by contamination, sanding and surface damage. In: 6th International conference on contact mechanics and wear of rail/wheel systems (CM2003) in Gothenburg, Sweden
51. Kalousek J, Magel E (1997) Modifying and managing friction. In: Railway track & structures, pp 5–6
52. Elkins J, Wu H (2000) New criteria for flange climb derailment. In: IEEE/ASME joint railroad conference, Newark, NJ
53. Tomala A, Karpinskab A, Wernera WSM, Olverb A, Störi H (2010) Tribological properties of additives for water-based lubricants. Wear 269:804–810
54. Wang J, Li T, Peng J, Chen Z (2000) Water lubrication of mechanical frictional pairs—current research and future development trends. In: Su DZ (ed) Proceedings of the international conference on gearing, transmissions, and mechanical systems, pp 761–768
55. Мелентьев ЛП (1992) Взаимодействие колёс и рельсов и их износ. Путь и путевое хозяйство. No. 5, pp 6–15. [In Russian: Melentev LP (1992) Interaction of wheels and rails and their wear. Track and railway equipment]
56. Минин СИ (1991) Причины интенсивного износа колёсных пар и рельсов. Железнодорожный транспорт. No. 1. pp 47–50. [In Russian: Minin SI (1991) Causes of intense wear of wheel sets and rails. Railway transport]
57. Ермаков ВМ, Певзнер ВО (2002) О сходах порожных вагонов. Железнодорожный транспорт. No. 3. pp 29–33. [In Russian: Ermakov VM, Pevzner VO (2002) About derailment of empty cars. Railway transport]
58. Ahmed NS, Nassar AM (2013) Lubrication and lubricants. Tribol Fundam Adv, pp 55–76
59. Yifei M, Turner KT, Szlufarska I (2009) Friction laws at the nanoscale. Nature 457, 26. https://doi.org/10.1038/07748
60. Rudnick LR (2010) Lubricant additives: chemistry and applications. CRC Press
61. Hou K, Kalousek J, Magel E (1997) Rheological model of solid layer in rolling contact. Wear 211:134–140
62. Грубин АИ (1949) Основы гидродинамической теории смазки тяжело нагруженных цилиндрических поверхностей. In: Исследование контакта деталей машин. ЦНИИТМАШ. No. 30. 219p. [In Russian: Grubin AI (1949) Fundamentals of the hydrodynamic theory of lubrication of heavily stressed cylindrical surfaces. In: Research into the contact of machine parts]
63. Петрусевич АИ, Данилов ВД, Фомичев ВТ (1975) Исследование влияния скорости скольжения на толщину масляной пленки в контакте цилиндрических роликов. In: Исследования по триботехнике. Москва: Научно-исследовательский институт информации по машиностроению. pp 158–164. [In Russian: Petrusevich AI, Danilov VD, Fomichev VT (1975) Research into influence of sliding velocity on the oil film thickness in the contact of cylindrical rollers. In: Researches into the tribotechnics. Research Institute of Information on Mechanical Engineering, Moscow]
64. Дроздов ЮН, Туманишвили ГИ (1978) Толщина смазочного слоя перед заеданием трущихся тел. Вестник машиностроения. No. 2. pp 8–10. [In Russian: Drozdov YN, Tumanishvili GI (1978) The lubricant film thickness before scuffing. Vestnik mashinostroenia]
65. Eadie DT, Kalousek J, Chiddik KC (2002) The role of high positive friction (HPF) modifier in the control of short pitch corrugations and related phenomena. Wear 253:185–192
66. Eadie DT, Bovey ED, Kalousek JOE (2002) The role of friction control in effective management of the wheel/rail interface. Presented at the railway technology conference at railtex, Birmingham, UK
67. Vidaud M, Zwanenburg WJ (2009) Current situation on rolling contact fatigue—a rail wear phenomena. In: Swiss transport research conference
68. Fries RH, D'avila CG (1985) Analytical methods for wheel and rail wear prediction. In: Proceedings of the ninth IAVSD symposium, Linkoping

69. Li S, Li Z, Núñez A, Dollevoet R (2017) New insights into the short pitch corrugation enigma based on 3D-FE coupled dynamic vehicle-track modeling of frictional rolling contact. Appl Sci 7(8):807, 22 p

70. Denape J (2014) Third body concept and wear particle behavior in dray friction sliding conditions. In: Tribological aspects in modern aircraft industry. Trans Tech Publications, pp 1–12

71. Meng HC, Ludema KC (1995). Wear models and predictive equations: their form and content. Wear, 181–183 (Part 2): pp 443–457

72. Kalker JJ (1991) Simulation of the development of a railway wheel profile through wear. Wear 150:355–365

73. Archard JF (1953) Contact and rubbing of flat surfaces. J Appl Phys 24(8):981–988

74. Archard JF, Hirst W (1956) The wear of metals under unlubricated conditions. Proc R Soc Lond A 236(1206):397–410

75. Gohar R (2001) Elastohydrodynamics, 2nd ed. World Scientific

76. Hamrock BJ, Dowson D (1986). Ball bearing lubrication: the elastohydrodynamics of elliptical contacts. Wiley, NJ

77. Chang HS, Orcutt FK (1965) A correlation between the theoretical and experimental results on the elasto-hydrodynamic lubrication of rolling and sliding contact. Pross I Mech E Lond 180, 158 (Pt. 3B)

78. Tumanishvili G, Nadiradze T, Tumanishvili I (2014) Improving of operating ability of wheels and rail tracks. Transp Prob 9(3):99–105

79. Dyson D, Higginson GR (1961) New roller-bearing lubrication formula. Eng Lond 192 (4972):158–159

80. Murch U (1975) Analysis of the input zone of elasto-hydrodynamic contact with allowance for thermal effects. Problems of friction and wear, № 2, pp 76–81

81. Коднир ДС (1976) Контактная гидродинамика смазки деталей машин. Москва: Машиностроение. 303 p. [In Russian: Kodnir DS (1976) Contact hydrodynamics of lubrication of machine parts. Mechanical engineering, Moscow]

82. Дроздов ЮН, Туманишвили ГИ (1982) Расчёт на заедание по предельной толщине смазочного слоя. Вестник машиностроения. No. 4. pp 19–29. [In Russian: Drozdov YN, Tumanishvili GI (1982) Calculation on scuffing by the extreme thickness of the lubricating layer. Vestnik mashinostroenia]

83. Knothe K, Liebelt S (1995) Determination of temperatures for sliding contact with applications for wheel-rail systems. Wear 189(1–2):91–99

84. Blok H (1937) Les temperatures de surface dans conditions de graissage sous extreme pression. Congr Mondial du petrole Paris 3:471–486

85. Дроздов ЮН, Арчегов ВГ, Смирнов ВИ (1981) Противозадирная стойкость трущихся тел. Москва: Наука. 140p. [In Russian: Drozdov YN, Archegov VG, Smirnov VI (1981) Scuffing resistance of friction bodies. Nauka, Moscow]

Potential and Problems of the Development of Speed Traffic on the Railways of Uzbekistan

Saidburkhan Djabbarov, Makhamadjan Mirakhmedov and Aleksander Sładkowski

Abstract The chapter deals with the features of the organization of high-speed train traffic on existing Uzbek railways. The possibility of preserving and increasing the carrying capacity of high-speed railways with the combined movement, of freight and passenger trains, as well as the influence of speed traffic on the construction and technological parameters of individual parts and structural elements infrastructure of the railway, from the position of ensuring traffic safety at high-speed railways of Uzbekistan is analyzed in the article. The results of the research can be useful in studying, designing and operating a railway track of high-speed roads under similar conditions.

Keywords Railway track · Train speed · High-speed train traffic
Modeling · Construction and technological parameters · Aerodynamic velocity
field

1 Basics of the Development of High-Speed Traffic in Uzbekistan

In the conditions of constant and rapid growth of population mobility, the possibility of their more active and unhindered movement, the problem of choosing the most reliable, convenient and economical means of transport arises.

S. Djabbarov (✉) · M. Mirakhmedov
Tashkent Railway Engineering Institute, Adylhodjaev 1, 100167 Tashkent, Uzbekistan
e-mail: saidhon_@inbox.ru

M. Mirakhmedov
e-mail: mirakhmedovm@mail.ru

A. Sładkowski
Department of Logistics and Aviation Technologies, Faculty of Transport,
Silesian University of Technology, Krasińskiego 8, 40-019 Katowice, Poland
e-mail: aleksander.sladkowski@polsl.pl

Based on the criteria of safety, comfort all weather, speed or time on the way, regularity and cost of the trip, high-speed railway traffic is one of the most promising areas of passenger traffic in regions with a high population density.

In this regard, the priority direction of the development of railways in the world is the increase in the speed of passenger trains, the creation of a network of high-speed railways.

In Europe, the creation of a network of national high-speed railways that meets the requirements of the EU Directive EC 96/48 [1] is one of the promising areas for the development of passenger transport. By early 2017, high-speed rail lines have built and successfully operated in 21 countries, of the world, their total length is more than 21 thousand km [2–4]. Since 2011, Uzbekistan has been included in the list of countries with high-speed train traffic, where Talgo-250 trains began to run.

The development of integration processes in the world, the expansion of business, cultural, economic and tourist ties, is a constantly acting factor, which causes the growth of requirements for high-speed rapid pace, and, according to expert forecasts, by 2020 the length of the high-speed network will reach 25,000 km [4, 5]. The longest in the world inter-regional high-speed railway line Beijing-Shanghai length of 1320 km [6].

In the long term, China plans to provide a network of domestic high-speed rail lines to neighboring countries, namely, in the northwest of the country via Urumqi and Alashankou to the settlement of Druzhba in Kazakhstan, and also through Jinghe towards Alma-Ata, exit through Kyrgyzstan to Uzbekistan [4, 6, 7].

Thus, Uzbekistan, located in the central part of the Eurasian continent, will be at the center of the transcontinental railway communication from the Pacific to the Atlantic Ocean, which is a modern interpretation of the "Great Silk Road".

The development of railways on the territory of the Republic of Uzbekistan dates back to 1874, when a special commission recognized the need for the construction of the Orenburg-Tashkent railway line. However, later in 1880 the decision was changed. Tashkent was connecting to the eastern coast of the Caspian Sea. Five years later, the builders reached Ashgabat, and in 1886 Chorzhou. In 1888, a movement was open up to Samarkand along the wooden bridge across the Amu Darya. In 1899 the road reached Tashkent. Simultaneously, a section was built from the Ursatievskaya station (now Khavast) to the Fergana valley.

At the end of the nineteenth century, the question arose again about the construction of a road from Tashkent to Orenburg, the construction of which began in the autumn of 1900 simultaneously from Tashkent and Orenburg. In 1906 the Tashkent-Orenburg road entered service, opening a direct exit to Central Russia for Central Asia.

After the acquisition of the country's independence, the priority and the main direction of the development of railways was the creation of a unified network of railway communications. In 1993, the Government of the Republic of Uzbekistan adopted the General Program for the Electrification of Railways, according to which the electrification of the main lines was provided for in the shortest possible time, and further the electrification of the whole rail network.

The first step in the implementation of this program was the construction in 1993–2003 line Uchkuduk–Nukus, and the second step was construction in 2008–2011 line Guzar-Boysun-Kumkurgan. Construction in 2015–2016 the Angren-Pap line has completed the creation of a unified railway network of Uzbekistan's transport infrastructure, ensuring the development of the productive forces of the regions, the development of their rich natural resources.

At the beginning of the 21st century, there were major changes in the passenger traffic. In order to increase the comfort of passenger transportation and bring them to the level of world standards, since 2004, the electric train "Registon" has started to run between Tashkent and Samarkand, since 2005 on the Tashkent-Bukhara line the "Shark" high-speed train, and on the Tashkent-Karshi route the high-speed train "Nasaf" with an average speed of 120 km/h.

The introduction in August 2011 of the high-speed passenger train Afrasiyob (Talgo-250 electric train) on the Tashkent-Samarkand line, in August 2015 on the Tashkent-Karshi line and in August 2016 on the Samarkand-Navoi-Bukhara line showed, that there are technical opportunities and progressive methods of organization and management of high-speed, train traffic are developed, ensuring the safety and trouble-free operation of trains with maximum speeds of up to 250 km/h and more (Fig. 1).

At present, the construction of the Bukhara-Miskin line with the transfer by 2020 to electric traction is completed, which will allow to organize in Uzbekistan a national network of high-speed railways with a length of more than 1000 km (Fig. 2).

Fig. 1 High-speed electric train Talgo-250

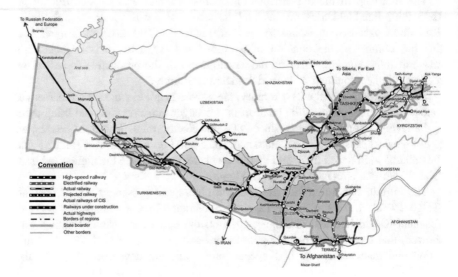

Fig. 2 The scheme of high-speed railways in Uzbekistan

The increase in the speed of passenger trains on the railways of JSC "Uzbekistan Temir Yollari" is carried out on the existing line (after their reconstruction and modernization) with the adjoining on some sections to new lines.

One of the most difficult issues in the introduction of high-speed train traffic is, first of all, bringing permanent technical means and railway infrastructure in line with the requirements imposed on them.

The problem lies in the following:

- the need to combine the movement of freight and passenger trains;
- the railway line was designed for a much lower load-carrying capacity, lower speed and weight of trains;
- the presence of a real tendency to further increase the intensity of freight trains.

Therefore, the organization of high-speed traffic on the railways of Uzbekistan means their transfer to a qualitatively higher level of transport services. This laid the foundation for a new stage of reconstruction and modernization of Uzbekistan's railway transport.

As a result of the reconstruction of railways, their reliability and quality of work has significantly increased, and the indicators of rolling stock use have been improved due to significant capital expenditures.

Modernization of railway lines for the implementation of high-speed train traffic requires significant investments in volume, and their payback is a long time. Currently, private investors are not ready for such an investment. Therefore, in the Republic of Uzbekistan the creation of the infrastructure of high-speed railways was largely undertake by the state, having formulated the following requirements for the projects of creating a national network of high-speed railways:

(1) when developing long-term plans for the development of a network of high-speed railways to improve their profitability, focus on its integration in the West through other countries of Central Asia and Russia, into a pan-European network; in the east—through Kyrgyzstan to the network of high-speed railways in China; in the south through Afghanistan, Turkmenistan, Iran to the network of high-speed railways of Turkey and Eastern Europe;

(2) take into account the experience of creating a network of national high-speed railways of the countries of the European Union, Japan, China;

(3) develop their own strategy for the implementation of projects for the reconstruction and modernization of the infrastructure of existing railways and the construction of new lines to organize the movement of high-speed trains.

It can be assumed that there are solutions that allow you to get the expected result at a much lower cost.

2 Construction and Technological Parameters of the Railway for High-Speed Train Traffic

The current regulatory documents of Uzbekistan set the maximum speed on high-speed railways within the limits of 141–200 km/h, on high-speed 201–250 km/h.

World experience shows that the highest speed of 200–350 km/h can be achieved by organizing high-speed train traffic on specialized high-speed lines. However, the construction of specialized high-speed railways and the production of specialized rolling stock for them requires large capital investments.

In cases of non-obvious positive effect of the introduction of high-speed train traffic on specialized lines, it is possible to consider the organization of high-speed passenger traffic on existing lines with combined movement of freight and passenger trains with speeds of:

- up to 120 km/h—on traditional railway lines;
- up to 160 km/h—on the lines after the road overhaul;
- up to 200 km/h—on the reconstructed lines.

The infrastructure of the railway line, where high-speed trains are used, should ensure their safe movement at the set speeds; the technical parameters of all of its facilities, track facilities, communications and computer facilities, automatics and teleautomatics, electrification and power supply and civil structures must comply with the design standards for high-speed railways. The basic requirements and standards for designing high-speed railways in Uzbekistan are given in [8, 9].

The construction and technological parameters of each structural element of the existing railway infrastructure determine the maximum level of speed of high-speed trains that must correspond to their state. A thorough study of the existing condition and construction and technological parameters of the existing rail infrastructure will allow to establish the maximum permissible speed of high-speed trains; Identify objective causes and factors that limit the speed of train traffic.

Since in Uzbekistan the organization of high-speed train traffic is carried out using the existing infrastructure, we will consider in more detail the construction and technological parameters of the railroad for high-speed railways.

The increase in the speed of train traffic on existing railways in Uzbekistan is complicated by a number of circumstances, the most important of which are the following:

- all previously constructed railways in the territory of modern Uzbekistan were designed based on the maximum train speed of no more than 120 km/h;
- for all main highways (main routes), the mixed movement of passenger and freight trains, with different levels of maximum speed, is characteristic;
- the increase in the maximum train speed to 140–160 km/h is associated with a significant amount of work on the reconstruction and modernization of the infrastructure of existing lines.

The increase in the speed of passenger trains on existing lines is hampered by a number of factors, the main of which are availability:

- numerous small-radius curves;
- inconsistencies in the design of the upper structure of the track, the contact network and traction substations, communication and signaling devices, the requirements;
- Old-style switches that do not allow the trains to follow them in the forward direction with the highest possible speed;
- the short length of the rails laid in the way, adversely affecting the characteristics of interaction between the rolling stock and the track, and also reducing the level of comfort for passengers.

Since it is not possible to carry out a radical reconstruction of all infrastructure facilities of existing lines that are part of the program for increasing the speed of trains in a relatively short time and with minimal costs, a rolling stock with a tilted body of cars designed for high-speed traffic along the existing track infrastructure is used.

To achieve the intended (given) level of speed, a number of measures and works on the modernization (reconstruction) of permanent devices and railway structures should be performed. These structures and devices include: the upper structure of the track, the roadbed, the system of signaling and communication, power supply, man-made structures, passenger platforms and other structures.

The main technical solutions that can be applied in the modernization and reconstruction of the infrastructure of existing railways are presented in Table 1.

Table 1 The main technical solutions recommended for the modernization and reconstruction of the infrastructure of existing lines

No.	Infrastructure facility	Decision taken
1	Locomotive and wagon fleet	Acquisition of electric locomotives and electric trains intended for high-speed traffic, rolling stock with inclinable car bodies
2	Electricity supply facilities	Installation of the contact suspension with the use of modern technical solutions, including materials and technologies for the production of parts and fittings of the contact network to ensure reliable current collection at high train speed; strengthening the traction power supply system, if necessary, transfer to new locations; transition to an alternating current system; replacing existing defective supports of the contact network with new ones, carrying the supports in places where the position of the route changes; change in overall dimensions
3	Earth canvas	Strengthening of the existing subgrade, dumping and pruning, cleaning of the "trails", broadening of the main site of the earth's canvas in accordance with the new norms, "treatment—diseased places" of the existing roadbed, reinforcement of embankments on weak grounds, etc.; reconstruction of drainage structures in the excavations, reconstruction of cuvettes, sub-drainage drains, reinforced concrete trays; shifting the path to a new axis; the device of transition areas on approaches to bridges, etc.
4	Permanent way	Strengthening the main section of the roadway using geomaterials; continuous track lining with rail lashes, long length of rails type R65, reinforced concrete sleepers with enlarged diagram, cleaning or replacement of crushed stone, increasing the thickness of the ballast layer, changing the size of the ballast prism; Gasket of cells with a continuous rolling surface 1/11 with rails of type P65
5	Railroad track vinyl-ready	Tracing the route to a new position, increasing the radius of the curves, the length of the transition curve, the straight insert between adjacent curves, the elevation of the outer rail
6	STSB and communication devices	Modernization of the signal system (signaling, centralization, interlocking), the use of a four-digit automatic locking system, floor and locomotive communication equipment based on the use of microprocessor technology
7	Artificial constructions	Strengthening span structures and supporting parts of bridges; lengthening of existing pipes, repair or replacement with new separate structural elements of pipes; device two-level intersection with highways

The increase in the speed of passenger trains is a stable global trend in the development of rail transport. At the same time, the first stage is the organization of high-speed passenger trains on existing lines with speeds of 160–200 (250) km/h after reconstruction and modernization of the railway infrastructure.

Instrumental research of the existing state of the railway allows the causes of the speed limits to be determined on each individual section of the railway; outline activities (or adopt a project solution) to eliminate them; determine the scope of work on the reconstruction (modernization) of certain structural elements of the railway infrastructure that limit the speed of high-speed passenger trains.

The total amount of work to eliminate both local and linear speed limits can't be reduced to a single meter. Therefore, the scope of work is expedient to determine by types of restrictions, type of devices and structures.

As a rule, the amount of work to eliminate linear restrictions is actually equal to the length of the section where high-speed traffic of passenger trains is introduced, which is confirmed by the experience of organizing high-speed train traffic on the Tashkent-Samarkand, Samarkand-Bukhara section.

The scope of work for the introduction of high-speed train traffic using the existing railway infrastructure on the same site varies significantly depending on the maximum speed level to which the speed should be increased.

For example, from the existing maximum speed of 120 km/h to 140, 160, 200 km/h. Used in practice devices for signaling, power supply, VSP structures and others are designed for different levels of maximum speed, compliance in the gradation of speed in them are absent. So, for example, SCS devices for speeds up to 140 km/h, individual elements of the power supply infrastructure up to 120 km/h, etc.

Thus, there is a functional relationship between the established maximum speed of high-speed passenger trains and the construction and technological parameters of railway infrastructure facilities. Those. The established (preset) maximum speed level of high-speed trains, determines the construction and technological parameters of railway infrastructure facilities that ensure the safe movement of trains at a specified speed. It can be said in another way that the installed construction and technological parameters of railway infrastructure facilities determine the maximum level of speed of safe movement of high-speed trains. Then the established maximum speed level of high-speed trains also determines the scope of work for the reconstruction and reconstruction of the infrastructure of existing railways.

As a rule, high-speed traffic introduced on existing railways without significant changes to the existing infrastructure. In this case, the construction and technological parameters of the route plan, which were set for small speeds, are the main obstacle to increasing the speed of passenger trains.

In some countries where existing high-speed lines have been adapted for high-speed train traffic, it has been ascertained that increasing there speed above 220–230 km/h requires costly technologies in operating and maintaining the track and entails many technical and organizational difficulties. Therefore, the construction of a dedicated high-speed railway is the only way to increase the speed of passenger trains over 200 km/h.

In the directive documents, high-speed rail transport is considered as a single technological complex (railway track, railway power supply, railway automation and telemechanic, railway telecommunications, and station buildings, structures and devices), including subsystems of the natural and technical infrastructure of high-speed rail transport and safe traffic specialized high-speed railway rolling stock with speeds of 6 than 200 km/h.

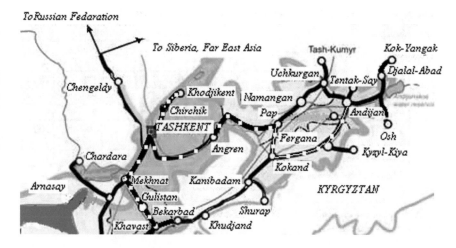

Fig. 3 The scheme of the railway Tashkent-Andijan

Thus, the adopted construction and technological parameters of all subsystems of the natural and technical infrastructure of high-speed transport should ensure the safe movement of high-speed rolling stock at speeds exceeding 200 km/h.

As an example, consider the construction and technological parameters of the section of the Tashkent-Angren-Pap-Andijan railway, which, according to experts, will eventually become an element of high-speed ground communication between Europe and Asia along the international transport corridor "Southeast Asia–Western Europe" interstate [10–13].

Presumably, the high-speed traffic of passenger trains from Tashkent to Andijan can be organized along two routes (Fig. 3); within the Tashkent region and the Kamchik pass along the single route Tashkent-Angren-Pap; in the Ferghana Valley, from the Pape station to the Andijan station, according to variants of Pap Namangan-Andijan or Pap-Kokand-Andijan, which can be conditionally designated as "northern" and "southern" options. The length of the "northern" variant is 43 km shorter than the "southern" option.

To assess the options for organizing high-speed traffic, the analysis of the parameters of the technical equipment of infrastructure devices and structures, longitudinal profile elements and the plan of the railway section was carried out. According to the construction and technological parameters of the technical equipment of the railway section from the Tashkent station to the station of Andijan can be divided into the following four components:

Tashkent-Angren (T-A); Angren-Pap (A-P); Pap-Namangan-Andijan (P-N-A) and or Pap-Kokand-Andijan (P-K-A).

As shown by the analysis, the construction and technological parameters of the sections (Table 2) of the northern and southern variants of the Fergana ring movement are identical and on the longitudinal profile the longitudinal slopes are ranked in ascending order and distributed within the slopes 0.1–6, 6.1–9, 9.1–12,

Table 2 Technical specifications section of the railroad Tashkent-Andijan

No.	Indicator	Name of sites			
		T-A[a]	A-P	P-N-A	P-K-A
1	Length [m]	113.9	122.3	138	181.8
2	Type of traction	Electric	Electric	Diesel	Diesel
3	Maximum longitudinal gradient [‰]	13	27	9	9
4	The minimum radius of the circular curves [m]	300	300	500	500
5	Communications	AB	AB	SAL	AB

[a]Here and below abbreviations are deciphered, respectively
T-A Tashkent-Angren
A-P Angren-Pap
P-N-A Pap-Namangan-Andijan
P-K-A Pap-Kokand-Andijan
T-N-A Tashkent-Namangan-Andijan
T-K-A Tashkent-Kokand-Andijan
AB automatic blocking
SAB semi-automatic blocking

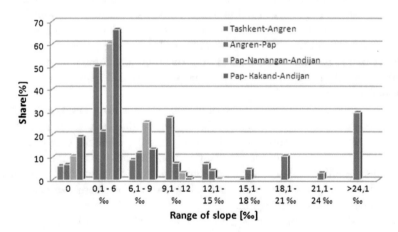

Fig. 4 Distribution of slopes of longitudinal profile elements

12.1–18, 18.1–21, 21.1–24, 24.1‰ and more. The total length of the elements corresponding to a certain range of longitudinal slopes and their percentage as a percentage of the total length of sections of the railway are determined. A graphical representation of the distribution of the slopes of the longitudinal profile in these ranges and the fraction of the total length of the elements in them along separate sections shown in Fig. 4.

Analysis of the longitudinal profile of railways shows that in the Angren-Pape section, the total length of the longitudinal profile elements designed with slopes steeper than 12‰ is 52.3%, incl. 29.7% slopes steeper than 24‰. On the sections of Tashkent-Angren, Pap-Namangan-Andijan, Pap-Kokand-Andijan, maximum longitudinal deviations do not exceed 12–15‰. Thus, only on the Angren-Pape section, the speed of passenger trains can be limit by the path profile. However, for modern locomotives and electric trains, like Uzbekistan (production of China) or Talgo-250, this factor is not limiting and subsequent calculations, this factor can be ignore.

In the same way, in increasing order, the parameters of the elements of the plan are ranked, i.e. the length of the straight sections of the path (including the straight inserts between adjacent curves) and the radii of the circular curves; their number and proportion of the total number of elements of the plan.

Graphical representation of the distribution of the number of straight inserts and straight sections of the path, as well as the radii of circular curves for individual sections is shown in Figs. 5 and 6.

The number of sections with a length of straight inserts and straight sections of a path of less than 300 m in Angren-Pap is 62%. In the other three areas, this indicator is almost two or more times less.

Analyzing the data characterizing the parameters of the elements of the plan, we can state that the main deterrent cause of the increase in the speed of passenger trains on the existing Tashkent-Andijan railway is the parameters of the elements of the route plan, and, first of all, the radii of the circular curves, and on the Angren-Pap sections length of straight inserts between adjacent curves.

Insufficient length of straight inserts does not allow, place transitional curves of longer length, accelerate to a higher speed. At the same time, the smooth running of the train along the curves is also disturbed, as a consequence, the comfort for passengers is worsened.

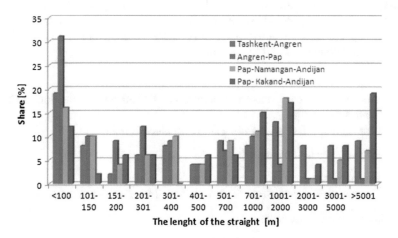

Fig. 5 The ratio of the number of straight inserts and straight sections of the path

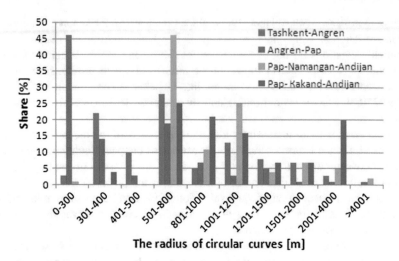

Fig. 6 Correlation of radii of circular curves

On the sections of Tashkent-Angren, Pap-Namangan-Andijan, Pap-Kokand-Andijan, more than half of the straight sections of the road allow placing the transitional curves of the calculated length and raising the outer rail to the normative value.

Almost half (46%) of the curves in the Angren-Pape plot are projected with a radius of 300 m or less. At the Tukimachi-Angren section, 22% of the curves have a radius of 301–400 m. At the Tukimachi-Angren, Pap-Namangan-Andijan, Pap-Kokand-Andizhan sections, the curves with a radius of 501–800 m, respectively, are 28, 36, 25%.

Appropriate calculations were carried out to evaluate the possibility of increasing the speed of passenger trains on existing sections of the Tashkent-Angren, Angren-Pap, Pap-Namangan-Andijan, Pap-Kokand-Andijan railroads [14] based on the following conditions and assumptions:

1. Rolling stock

   – for passenger trains electric locomotives O'zbekiston (production of China) and 4-axle wagons;
   – for high-speed trains of the Talgo-250 electric train.

2. Maximum permissible speeds

   – at the initial state of 70 km/h;
   – after the modernization of 90, 120, 140, 160, 180, 200 km/h.

3. Maximum driving speeds

   – passenger trains up to 140 km/h;
   – high-speed passenger trains 141–200 km/h.

4. Weight of formulations
   - passenger trains of 1200 ton;
   - high-speed passenger trains of 400 ton.
5. The maximum speed of freight trains with a combined movement of 70–90 km/h.
6. The highest value of the unprecedented acceleration is according to the norm 0.7 m/s².
7. The maximum elevation of the outer rail is 150 mm.
8. The rate of increase of the unoccupied acceleration in the transition curves, not more than 0.5–0.7 m/s²;
9. The speed of the passage of the switches is the same as the maximum permissible speed established in this section;
10. The whole area of Tashkent-Andijan is electrified.
11. To increase the maximum speed of passenger trains, it is planned to upgrade the permanent devices and infrastructure of the railway. At the same time, at this stage of the calculation, investments to modernize the permanent facilities and infrastructure facilities of the railway sections were not taken into account.
12. Calculations were carried out for both individual sections and for two variants of the route of the high-speed train, suggesting that trains along the whole route follow without stopping.

A distinctive feature of the single-track section of the Tashkent-Andijan railway is the combined movement of trains of all categories. Therefore, the maximum speed limit for train trains along the curves is established taking into account the safety and smoothness of movement of all the categories of trains that are circulating in this sector, including the Talgo-250 high-speed electric train with the car body tilt. Calculation of the parameters of the curve, equally satisfying the conditions for the movement of trains with maximum speeds, was performed in the same sequence as in [15]. The results of the calculations are summarized in Table 3.

Based on the analysis of the parameters of the plan elements, the longitudinal profile, the results of the traction calculations carried out, the following were established:

1. The reasons for speed limits at the level of 90, 120, 140, 160, 180 and 200 km/h;
2. Separate sections, where it is possible to set the maximum speed of passenger trains at the level of 120, 140, 160, 180 and 200 km/h;
3. Separate sections, curves or adjacent curves at which the maximum speeds of passenger trains with available track parameters should be limited to speeds of 90, 120 km/h.

Calculations show that improving the construction and technological parameters of the track, i.e. the removal of speed limits for passenger trains on all sections, with the exception of the Angren-Pap section, allows the average speed of trains to reach 110–150 km/h, reducing the time spent on the road by 3.9 h (i.e., reducing by 40%).

Table 3 Time and speed of the train along sections

№	Indicators	Maximum speed of passenger and high-speed trains (km/h)	Railroad sites				Total along the route	
			T-A	A-P	P-N-A	P-K-A	T-N-A	T-K-A
1	Length of the section (km)		114.1	122.3	141.1	180.1	377.5	416.5
2	Time of travel trains on site (min)	70	109.9	119.3	146.4	154.9	375.5	384.0
			108.7	119.7	146.8	155.4	375.2	383.8
		90	79.5	111.9	94.5	121.2	285.9	312.6
			79.3	112.3	94.5	121.5	286.1	313.1
		120	66.5	108.8	79.3	94.5	254.6	269.8
			65.8	109.6	79.3	94.7	254.7	270.1
		140	62.7	103.0	77.4	84.8	243.1	250.5
			61.5	103.4	77.0	85.2	241.9	250.1
		160	61.3	102.7	71.4	77.9	235.4	241.9
			59.4	102.9	71.2	78.0	233.5	240.3
		180	60.7	102.4	70.5	74.1	233.6	237.2
			58.4	102.5	70.2	74.2	231.1	235.1
		200	60.4	102.3	70.1	71.9	232.8	234.6
			58.0	102.4	69.9	71.9	230.3	232.3
3	Average speed over section (km/h)	70	62.0	61.0	57.0	70.0	60.3	65.1
			63.0	61.0	56.0	70.0	60.4	65.1
		90	86.0	65.0	88.0	89.0	79.2	79.9
			86.0	65.0	88.0	89.0	79.2	79.8
		120	103.0	67.0	104.0	114.0	89.0	92.6
			104.0	67.0	104.0	114.0	88.9	92.5
		140	109.0	71.0	107.0	127.0	93.2	99.8
			111.0	71.0	108.0	127.0	93.6	99.9
		160	112.0	71.0	116.0	139.0	96.2	103.3
			115.0	71.0	116.0	139.0	97.0	104.0
		180	113.0	71.0	117.0	146.0	97.0	105.4
			117.0	71.0	118.0	146.0	98.0	106.3
		200	113.0	72.0	118.0	150.0	97.3	106.5
			118.0	72.0	118.0	150.0	98.3	107.6

The most intensive reduction of train travel time and accordingly an increase in the average speed occurs within the maximum speed of 70–160 km/h (Fig. 7). In the subsequent speed range of 160–200 km/h, the increase in average speed and, correspondingly, the shorter travel time of the train, occur less intensively.

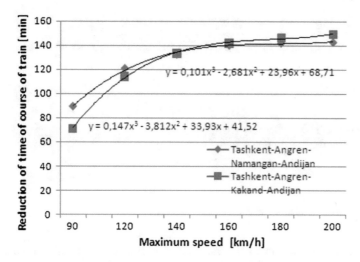

Fig. 7 Changes in train travel time

A similar analysis was carried out when studying the construction and techno-logical parameters of the Karshi-Kitab (Shakhrisabz), Karshi-Toshguzar-Boysun-Kumkurgan.

3 Features Safety for the Movement of High-Speed Trains

To increase the speed of trains in the 50–60 years of the last century, great efforts have been made to conduct research to increase the power of the power plant (engines) of diesel locomotives, electric motors of electric locomotives and electric trains. All resources, both intellectual and material, were aimed at increasing the energy component of locomotives and electric trains. At the same time, the study of air currents arising around the train, to study their physics was considered a sec-ondary task.

A further increase in the speed of trains led to a qualitative change in the physics of air flows around high-speed trains, the transformation of train energy into aerodynamic resistance, noise and vibrations (vibrations) and other phenomena adversely affecting the dynamics of a high-speed train. As the speed increased, their character changed greatly.

At relatively low speeds, many theoretical problems and technical problems in the development of high-speed trains concerning (relative to) passenger discomfort, aerodynamic drag, noise, vibrations, pulse strength, and others arising during the movement (single and towards each other in parallel way) of high-speed trains.

An essential problem of further increasing the speed of trains is the aerodynamic resistance of the train. To increase the speed of high-speed trains, the aerodynamic

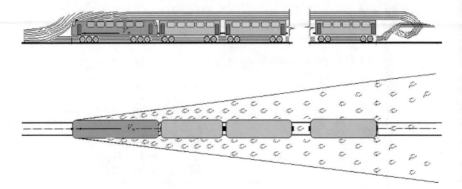

Fig. 8 Air flows around the high-speed train

drag must be kept as low as possible, which is associated with air currents arising around the moving train (Fig. 8). Depending on the place of their occurrence and the impact on the environment (objects), they can be conditionally divided into five groups:

- air flows near the nose (head car) of the train;
- air flows along the train;
- air flows behind the train;
- air flows under the train;
- air flows over the train.

In addition, complex flows arise when moving high-speed trains along parallel paths, towards each other, and also when driving in a tunnel. The origin and impact of each of these flows has its own patterns. In sum, they form a complex in nature and diverse in their impact air flow, which must be considered in a complex and integrated into a single task.

Scientists from Europe, USA, Japan, China, Korea, Russia and others are conducting experimental and theoretical studies on the study of aerodynamics of a high-speed train and related issues [16–19]. The purpose of the research is to obtain aerodynamic data on the dynamics of the movement of high-speed trains, assess their impact on other trains, as well as on people in proximity to the passing train. So successfully solve the issues of the shape of the nose of the train to reduce the resistance to movement of the train [20–24].

Among the diverse studies of aerodynamics of a high-speed train, the problem of the interaction of the aerodynamic field of a high-speed train on the environment, for example, on the sandy surface and infrastructure elements near the railway track, and people on the platform remain unresolved. First of all, it is required to theoretically investigate the velocity field around a moving high-speed train.

The problem is the peculiarities of the movement of trains, which do not allow us to use the laws of the physics of aircraft. These features are as follows:

- the length of a moving train is much greater than its width;
- the train moves along a predetermined trajectory (along the railway track);
- the train moves in close proximity (along) the railway infrastructure facilities: top track elements, booths, contact-line towers (overhead lines), overpasses, bridges, station buildings, etc.;
- the train can move in a confined space (tunnel);
- a moving train through the flow of air affects the surface of the earth, a train moving toward or along a parallel path, people on a platform.

Currently, the speed of high-speed trains is approaching 300–350 km/h. The maximum speed of high-speed trains, which are in constant operation, reached 380–400 km/h. The speed record in the "rail-wheel" system is 574.8 km/h.

Increasingly relevant for high-speed train traffic is the study of their aerodynamic impact on people, railway infrastructure objects that are located near the passing train, ensuring their safety by applying scientifically based design standards for railway infrastructure facilities.

The introduction of high-speed passenger train traffic on the Tashkent-Samarkand section and the further expansion of the high-speed train range requires the improvement of certain sections of the current design standards for high-speed and high-speed railways in Uzbekistan [8, 9, 25, 26]. This, first of all, concerns the establishment (normalization) of the minimum permissible distance of the location of individual objects of the infrastructure of railways and people (passengers, railway workers) from the axis of the high-speed railway.

To establish the location (determine the minimum permissible distance) of railway infrastructure objects and people (passengers, railway workers), the effect of aerodynamic pressure on them, theoretical studies of the propagation of the velocity field around a high-speed train were carried out.

Below we analyze the distribution of the aerodynamic field around a high-speed train as an axisymmetric body with a head and tail in the form of a cone or come to life, moving in a compressible medium (Fig. 9), at a fixed point $M(x, y, z)$. Since the motion of the body occurs along the axis, the change in the position of the center of the moving train relative to point M is taken into account by the introduction of a moving coordinate system $z = z_1 - v_0 t$, as shown in Fig. 3 [27].

The components of the velocities of the particles of the medium along the coordinate axes $0z$ and $0r$ are determined in terms of the velocity potential by the formulas $\varphi_1(r, z_1, t)$

$$v_z = \frac{\partial \varphi_1}{\partial z}, \quad v_r = \frac{\partial \varphi_1}{\partial r} \tag{1}$$

The equation for the propagation of an acoustic wave in the air can be written in the form

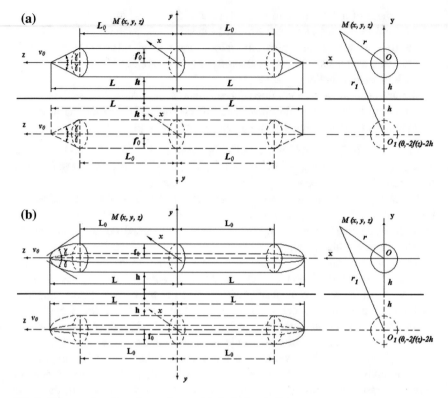

Fig. 9 Scheme of motion of an axisymmetric body in a half-space

$$\frac{\partial^2 \varphi_1}{\partial t^2} = a^2 \left(\frac{\partial^2 \varphi_1}{\partial r^2} + \frac{1}{r} \frac{\partial \varphi_1}{\partial r} + \frac{\partial^2 \varphi_1}{\partial z_1^2} \right). \tag{2}$$

Equation (2) can be reduced to the following form

$$\alpha^2 \frac{\partial^2 \varphi_1}{\partial z^2} - \frac{\partial^2 \varphi_1}{\partial r^2} - \frac{1}{r} \frac{\partial \varphi_1}{\partial r} = 0, \tag{3}$$

where, $\alpha - \alpha = \sqrt{1 - M^2}$, $M = v_0/a$—Mach number.

Equation (3) is integrated under the following boundary conditions:

1. The condition of the non-slipping of air particles on the surface of the body:

$$v_r = \frac{\partial \varphi_1}{\partial r} = v_0 \, tg \, \gamma \quad \text{at } r = f_1(z), \tag{4}$$

where $tg \, \gamma$ the equation of the surface of the body; the slope of the tangent to surfaces of a moving body.

The equation of the surface of a body can be determined by the formulas $tg\,\gamma = f_1'(z)$, $f_1(z) = f_1(-z)$. The value of the angle of inclination can vary within $0 \leq \gamma \leq 90°$.

2. The condition that the component along the axis $0y$ The velocity of the particles of the medium at the boundary of the half-space is determined from (5)

$$\frac{\partial \varphi_1}{\partial y} = 0 \quad \text{at } y = -h - f_1(z) \tag{5}$$

3. The symmetry condition with respect to the axis $0z$

$$\varphi_1 = 0 \quad \text{at } z = 0 \tag{6}$$

To find the solution of the equation, the method of sources [28]. Considering the function $\varphi(r, z)$, satisfying the Eq. (2) and the boundary condition (4), the solution can be represented in the form (7)

$$\varphi = -\frac{1}{4\pi} \int_{-L}^{L} \frac{q(\xi)d\xi}{\sqrt{(\xi - z)^2 + \alpha^2 r^2}} \tag{7}$$

where $q(z)$—the power of a source distributed over the surface of a moving body within $0 < r < f_1(z)$, $-L < z < L$.

For a thin axisymmetric body from formula (7), proceeding from [28], it can be asserted that:

at $r \to 0 \quad \frac{\partial \varphi}{\partial r} \to \frac{q(z)}{2\pi r}$;

at $0 < z < L \quad q(z) = 2\pi v_0 f_1'(z) f_1(z)$;

at $-L < z < 0 \quad q(z) = -2\pi v_0 f_1'(-z) f_1(-z)$.

Then the function $\varphi(r, z)$ becomes (8)

$$\varphi = -\frac{v_0}{2} \left(\int_{-L}^{0} \frac{f_1'(-\xi) f_1(-\xi)\, d\xi}{\sqrt{(\xi - z)^2 + \alpha^2 r^2}} + \int_{0}^{L} \frac{f_1'(\xi) f_1(\xi)\, d\xi}{\sqrt{(\xi - z)^2 + \alpha^2 r^2}} \right) \tag{8}$$

Given that $f_1(-\xi) = f_1(\xi)$, $f_1'(-\xi) = -f_1'(\xi)$ the Eq. (8) can be represented in the following form

$$\varphi = -\frac{v_0}{2} \left(\int_{0}^{L} \frac{f_1'(\xi) f_1(\xi)\, d\xi}{\sqrt{(\xi + z)^2 + \alpha^2 r^2}} - \int_{0}^{L} \frac{f_1'(\xi) f_1(\xi)\, d\xi}{\sqrt{(\xi - z)^2 + \alpha^2 r^2}} \right) \tag{8'}$$

In this case, the function $\varphi_1(r, z)$ can be represented in the following form (9)

$$\varphi_1 = \varphi(z, r) + \varphi(z, r_1) \tag{9}$$

where $r_1 = \sqrt{x^2 + [2f_1(z) + 2h + y]^2}$.

It can be shown that the function represented by (9) $\varphi_1(r, z)$ satisfies all the conditions of the problem under consideration. The components of the velocity vector of air particles in a Cartesian coordinate system at an arbitrary point in space are expressed by the following formulas (10)–(12)

$$v_x = \frac{\partial \varphi_1}{\partial x} = \left(\frac{\partial \varphi(z, r)}{\partial r} \frac{\partial r}{\partial x} + \frac{\partial \varphi(z, r_1)}{\partial r_1} \frac{\partial r_1}{\partial x} \right), \tag{10}$$

$$v_y = \frac{\partial \varphi_1}{\partial y} = \left(\frac{\partial \varphi(z, r)}{\partial r} \frac{\partial r}{\partial y} + \frac{\partial \varphi(z, r_1)}{\partial r_1} \frac{\partial r_1}{\partial y} \right), \tag{11}$$

$$v_z = \frac{\partial \varphi_1}{\partial z} = \frac{\partial \varphi(z, r)}{\partial z} + \frac{\partial \varphi(z, r_1)}{\partial z}. \tag{12}$$

As an example, we consider a body of revolution, the initial and final parts of which are cones and come to life. The shape of a moving body in the meridian plane is shown in Fig. 10. In this case, the accepted body dimensions are equal to the following values:

- distance L from the middle of the moving body Oz;
- distance from the middle of the moving body to the initial point of shape change (outlines) of the head parts of the moving body L_0;
- cross-section of an axisymmetric body R;

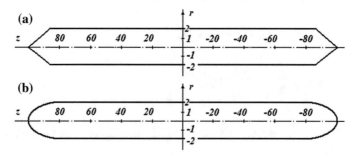

Fig. 10 Arrangement in the meridian plane of an axisymmetric body with the shape of the head and tail parts in the form: **a** cone; **b** came to life

- the cross section of the moving body is considered as a circle with a radius R of 2 m, the area of which is equal to the cross-sectional area of the train with a width of 3.0 m and a height of 4.0 m is 12 m^2;
- the angle of the cone of the head and tail is equal to $\gamma = 0.1538$ radius.

Functions $f_1(z)$ and $f_1'(z)$ for the conical shape of the head and tail parts of the body, when $\gamma_0 \approx tg\,\gamma_0 = R/(L - L_0)$ are represented as follows:

at $-L \le z \le -L_0$, $f_1 = \gamma_0(L + z)$, $f_1' = \gamma_0$;
at $-L_0 \le z \le L_0$, $f_1 = R$, $f_1' = 0$;
at $L_0 \le z \le L$, $f_1 = \gamma_0(L - z)$, $f_1' = -\gamma_0$.

For an axisymmetric body with a shape of the head and tail parts in the form, the functions $f_1(z)$ and $f_1'(z)$ have the following form:

at $-L \le z \le -L_0$, $f_1 = R[1 - (z + L_0)^2/(L - L_0)^2]$,
$\qquad\qquad\qquad\qquad f_1' = -2R(z + L_0)/(L - L_0)^2$;
at $-L_0 \le z \le L_0$, $f_1 = R$,
$\qquad\qquad\qquad\qquad f_1' = 0$;
at $L_0 \le z \le L$, $\quad f_1 = R[1 - (z - L_0)^2/(L - L_0)^2]$,
$\qquad\qquad\qquad\qquad f_1' = -2R(z - L_0)/(L - L_0)^2$.

In the meridian plane, the motion of the body is considered for the following instants of time:

1. t_1, when the beginning of the moving body (i.e. the head part of the train) is at a distance of 75.0 m to the origin of the coordinate axis (Fig. 11a);
2. t_2, when the body overcomes 75.0 m and its origin coincides with the axis of coordinates (Fig. 11b);
3. t_3, when the body crosses 125.0 m, i.e. ¼ part of its length and origin coincides with ¼ part of the body (Fig. 11c);
4. t_4, when the body crosses 175.0 m, i.e. ¼ part of its length and origin coincides with the center of the body (Fig. 11d);
5. t_5, when the body overcomes 225.0 m, i.e. ¾ part of its length and the origin of coordinates coincides with ¾ part of the body (Fig. 11e);
6. t_6, when the body overcomes 275.0 m, i.e. its entire length and origin coincide with the end (the tail part of the train) of the body (Fig. 11f);
7. t_7, when the body crosses 75.0 m and the origin is at a distance of 75.0 m behind the end (the tail part of the train) of the body (Fig. 11g).

Calculations of the velocity vector module $V = \sqrt{v_x^2 + v_y^2 + v_z^2}$ (theoretical) air flow caused by the movement of the body, for points located at a distance of 3.55, 6.00, 8.0, 10 m at $y = 0$ from the axis of the moving body (high-speed train) were carried out for the case. When the body moves at a constant speed (160, 200, 250, 350, 400 km/h) at altitude $h = 2$ m from the surface of the earth.

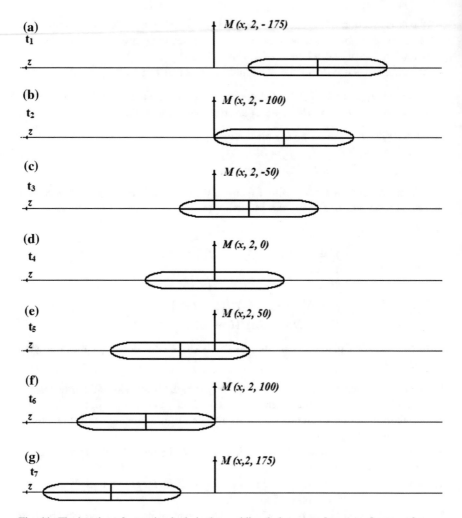

Fig. 11 The location of a moving body in the meridional plane: **a** t_1; **b** t_2; **c** t_3; **d** t_4; **e** t_5; **f** t_6; **g** t_7

The speed of the air flow formed by the movement of the body, for points located at a distance of 3.55, 6.00, 8.0 and 10 m from the axis of the moving body (high-speed train), is determined by formula (9). Based on the results of calculations using information technologies, graphs are constructed of the change in the velocity of the air flow along a moving body at different distances from it (Figs. 12 and 13).

As an example, Fig. 14 shows the graphs for the speed of 200 km/h.

Analysis of the airflow velocity along the moving body at a speed of 200 km/h (Fig. 14) shows that the general nature of the graphs for velocities of 160, 200, 250, 350, 400 km/h is identical. As the moving body approaches, there is a slight perturbation of the air environment (zone I, Fig. 14).

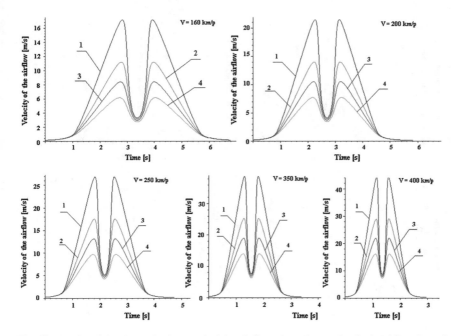

Fig. 12 Graphs of the change in the speed of the air flow along the moving body with a view of the head and tail parts in the form of a cone at a distance: 1–3.55 m; 2–6.0 m; 3–8.0 m; 4–10 m

The local maximum airflow velocity Vmax 1, that is, the maximum airflow velocity at the observation point M, located at a different distance from the moving body, reaches when the beginning of the moving body is located along the line connecting it with the observation point M (zone II, Fig. 14).

In the middle of the train, an extreme minimum air speed is required (zone III, Fig. 14).

The impulsive growth of velocity with an extremum of maximum Vmax 2 is observed when the end of the moving body is opposite point M (zone IV, Fig. 14).

After the passage of the point M, a gradual attenuation of the airflow velocity occurs (zone V, Fig. 14).

It should be noted that starting from the point A corresponding to the section AA in Fig. 14, while preserving its absolute value, the velocity vector changes in the opposite direction.

For example, at the point located at a distance of 3.55, 6, 8 and 10 m from the axis of the moving body (high-speed train) at a speed of 350 km/h, the maximum air speed is 41.5, 32.0, 26.2, 21.3 m/s (Figs. 11 and 12). At the points located at a distance of 3.55, 6, 8 and 10 m from the axis of the body moving at a speed of 200 km/h, the maximum speed of the air flow is respectively 21.4, 15.0, 10.5, 7.7 m/s.

Thus, the examination of a train as an axisymmetric body made it possible to obtain information revealing the features of the distribution of the air flow along

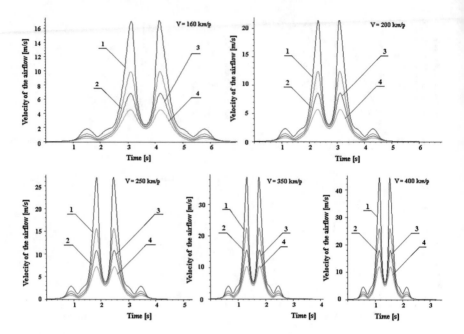

Fig. 13 The graphs of the change in the speed of the air flow along the moving body with the appearance of the head and tail parts in the form came to life at a distance: 1–3.55 m; 2–6.0 m; 3–8.0 m; 4–10 m

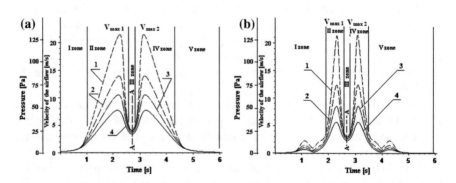

Fig. 14 Modification of the velocity vector of the air flow V(m/s) as a function of the position of the moving body at the point "M" at various distances from it 1–3.55 m; 2–6.0 m; 3–8.0 m; 4–10.0 m

and in the direction perpendicular from the axis of the moving train, and also the aerodynamic impact on the railway infrastructure objects and people [27, 29–31].

The resulting air flows have a negative impact on the environment, worsening the safe functioning of the system "high-speed railway—the environment (or surrounding objects)".

Providing traffic safety for trains and passengers; uninterrupted operation of the entire infrastructure of a high-speed railway is the main condition for the organization of high-speed and high-speed passenger train traffic. A high degree of security is usually provided at all stages of creating a high-speed highway, i.e. is laid during the design, is provided during construction and is realized during the operation of the infrastructure of high-speed railways. This task is relevant in the design of high-speed train traffic on existing railways, which were designed for the maximum speed of passenger trains of 120–160 km/h.

Since, on existing lines, high-speed traffic is possible after a large-scale reconstruction and modernization of permanent facilities and structures of the existing infrastructure, in designing the organization of high-speed traffic using the existing railway infrastructure, the maximum permissible speeds for a high-speed train for each facility should be set separately their technical condition.

To ensure the safe operation of a high-speed railway, it is necessary to consider the aerodynamic impact on people and railway infrastructure objects as one of the main safety criteria for high-speed passenger train traffic, since a train moving at high speed has an aerodynamic effect on each object of magnitude. The technical condition of the object allows it to perceive the impact with the maximum permissible value without reducing the safety level of high-speed trains.

In connection with the above, in order to ensure the safe operation of a high-speed railway at all facilities or structural elements of the existing path infrastructure must be condition.

$$P_{\max i} \leq P_{permi}, \tag{13}$$

where $P_{\max i}$ maximum pressure on the object; P_{permi} the maximum permissible value of an pressure that a given object can perceive.

Since, according to Bernoulli's law, the aerodynamic pressure varies directly in proportion to the speed of the air flow V_f created by the movement of the high-speed train, let us consider the theoretical aspects of establishing the value of the maximum train speed ensuring the fulfillment of condition (13).

The aerodynamic impact, strength and directivity of pressure on an object depend on the maximum speed and duration of the air flow, the spatial location, availability and proximity of the railway infrastructure objects relative to a moving high-speed train.

For each object i (or its constructive element) of the infrastructure of the existing railway, it is possible to draw up a design scheme of the effect of aerodynamic pressure on it (Fig. 15).

In all calculation cases, B for a known distance from a moving high-speed train to an object and the maximum permissible value of an impact that a given object can perceive, it is necessary to determine the speed of a high-speed train $V_{\max i}$ and hence the airflow V_f velocity that satisfies condition (13).

Thus, the problem of determining the aerodynamic effect on an i object is reduced to determining the velocity of the secondary airflow directly at the ith object when the high-speed train moves at a speed $V_{\max i}$.

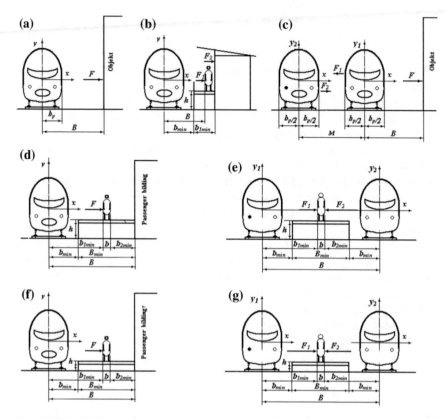

Fig. 15 Calculation schemes for the location of a high-speed train and objects: **a** train-object; **b** train-man-object; **c** train–worker; **d** train-passenger on a high platform-object; **d** train-passenger on a high platform-train; **e** train-passenger on a low platform-object; **g** train-passenger on a low platform-train

As the development of previously completed studies for a single solid body, the distribution of the air flow and the determination of its velocity along a moving high-speed train, was investigated on a model of a train consisting of a locomotive and 2n wagons [29].

To simplify the calculations, it is assumed that the locomotive and all cars in the cross section have the same shape, i.e. consist of a circular cylinder with identical head and tail shapes in the form of cones (Fig. 16).

The axisymmetric wave equation of the aerodynamic field near a high-speed train consisting of a locomotive and $2n$ wagons, can be decided as for a train consisting of one single wagon [30, 31]. In this case, the function $f(z)$ for each car can be represented in the form $f_{ij}(z)$, where the index i indicates the serial number of the wagon from the center of the train ($i = 0$ corresponds to the number of the middle car), the index j on the geometric shape of a part of the cars. If we assume

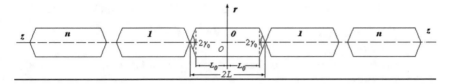

Fig. 16 Scheme of high-speed train with locomotive and wagons 2n

that the car consists of three geometric shapes, then $j = 1, 2, 3$ ($j = 1$ corresponds to the cylindrical part, $j = 2$ tail section, $j = 3$ the head).

The propagation of an acoustic wave in the air can be represented by Eq. (7) with the boundary conditions (8) and (10) [24, 27]. The condition that the component along the axis $0y$ the velocity of the particles of the medium at the boundary of the half-space, in contrast to Eq. (9) [27, 29], takes the following form

$$\frac{\partial \varphi_1}{\partial y} = 0 \quad \text{at } y = -h - R - f(z). \tag{14}$$

To find the solution of the equation, the method of sources was used [28]. Considering the function $\varphi(r, z)$ satisfying Eq. (2), the boundary conditions (8) and (10) [29], the solution of Eq. (14), can be represented in the form

$$\varphi_1 = -\frac{1}{4\pi} \int_{-(2n+1)L}^{(2n+1)L} \frac{q(\xi)\, d\xi}{\sqrt{(\xi - z)^2 + \alpha^2 r^2}}, \tag{15}$$

where $q(z)$—the power of a source distributed over the surface of a moving body within $0 < r < f_{ij}(z)$, $-(2n+1)L < z < (2n+1)L$.

For an axisymmetric body from formula (15) to [27], we can state that

$$\frac{\partial \varphi_1}{\partial r} \to \frac{q(z)}{2\pi r} \quad \text{at } r \to 0. \tag{16}$$

Since, the problem is symmetric about the axis $0z$, high-speed train consists of a locomotive and $2n$ the equation of the surface of the body $r = f_{ij}(z)$, as well as the power of the source $q(z)$ from each car and its components can be recorded separately.

For the cylindrical part of the car: at $2nL - L_0 < z < 2nL + L_0$ and $(2nL + L_0) < z < -(2nL - L_0)$ $f_{n1} = R$, $q = 0$.

For the head and tail parts of the wagon located up to the middle of the train:

at
$$2nL + L_0 < z < (2n+1)L \quad f_{n2} = \gamma_0[(2n+1)L - z],$$
$$q = -2\pi v_0 \gamma_0^2 f_{n,2};$$

and
$$-(2n+1)L < z < -(2nL + L_0) \quad f_{n2} = \gamma_0[(2n+1)L + z],$$
$$q = 2\pi v_0 \gamma_0^2 f_{n,2}.$$

For the head and tail parts of the wagon located behind the middle of the train:

at
$$(2n+1)L < z < (2n+1)L + L_0 \quad f_{n3} = \gamma_0[z - (2n+1)L]$$
$$q = -2\pi v_0 \gamma_0^2 f_{n3};$$

and
$$-[(2n+1)L + L_0[< z < -(2n+1)L \quad f_{n3} = -\gamma_0[(2n+1)L + z]$$
$$q = -2\pi v_0 \gamma_0^2 f_{n,3}.$$

For example, for a car $i = 1$ and its parts $j = 2$ as:

$$2L + L_0 < z < 3L; \ f_{12} = \gamma_0(3L - z); \ q = -2\pi v_0 \gamma_0^2 f_{12}$$

Taking into account the symmetry of the problem with respect to the variable z, power source from each car and its components $q(z)$, Eq. (15) can be represented in the following form

$$
\varphi_1 = -\frac{v_0 \gamma_0^2}{2} \left\{ \sum_{i=0}^{n} \left[\int_{2iL + L_0}^{(2i+1)L} \frac{(2i+1)L - \xi}{\sqrt{(z+\xi)^2 + \alpha^2 r^2}} d\xi - \int_{2iL + L_0}^{(2i+1)L} \frac{(2i+1)L - \xi}{\sqrt{(z-\xi) + \alpha^2 r^2}} d\xi \right] \right\}
$$
$$
+ \frac{v_0 \gamma_0^2}{2} \left\{ \sum_{i=1}^{n} \left[\int_{(2i-1)L}^{2iL - L_0} \frac{\xi - (2i-1)L}{\sqrt{(z+\xi)^2 + \alpha^2 r^2}} d\xi - \int_{(2i-1)L}^{2iL - L_0} \frac{\xi - (2i-1)L}{\sqrt{(z-\xi) + \alpha^2 r^2}} d\xi \right] \right\}.
$$
(17)

Introducing the variable $r_{ij} = r_{ij}(x, y, z)$ expressed by the formula

$$r_{ij} = \sqrt{x^2 + [2f_{ij}(z) + 2h + 2R + y]^2}.$$
(18)

The total potential presented in the form

$$\varphi_n = \varphi_1(r, z) + \varphi_1[r_{ij}(x, y, z), z].$$
(19)

Function $\varphi_n(x, y, z)$ satisfies the boundary condition (14), the function $\varphi_1 = \varphi_1[r_{ij}(x, y, z), z]$ satisfies Eq. (16) only when $\gamma_0 = 0$. Assuming γ_0 a small parameter and setting $f_{ij} = \gamma_0 f_{0ij}$ functions $1/r_{ij} = 1/\sqrt{x^2 + [2\gamma_0 f_{0ij}(z) + 2h + 2R + y]^2}$ can be expanded in powers of this parameter as

$$\frac{1}{r_{ij}} = \frac{1}{\sqrt{x^2 + [2\gamma_0 f_{0ij}(z) + 2h + y]^2}} = \frac{1}{r_1} + \gamma_0 \frac{2(2h+y)f_{0ij}(z)}{r_1^3} + \cdots \qquad (20)$$

where $r_1 = \sqrt{x^2 + (2h + 2R + y)^2}$.

If we substitute expression (20) into (17), then formula (19) takes the form

$$\varphi_n = \gamma_0^2 [\varphi_{01}(r, z) + \varphi_{01}(r_1, z) + \gamma_0 \varphi_{02} + \cdots], \qquad (21)$$

where

$$
\varphi_{01} = -\frac{v_0}{2} \left\{ \sum_{i=0}^{n} \left[\int_{2iL+L_0}^{(2i+1)L} \frac{(2i+1)L - \xi}{\sqrt{(z+\xi)^2 + \alpha^2 r^2}} d\xi - \int_{2iL+L_0}^{(2i+1)L} \frac{(2i+1)L - \xi}{\sqrt{(z-\xi) + \alpha^2 r^2}} d\xi \right] \right\}
$$
$$
+ \frac{v_0}{2} \left\{ \sum_{i=1}^{n} \left[\int_{(2i-1)L}^{2iL-L_0} \frac{\xi - (2i-1)L}{\sqrt{(z+\xi)^2 + \alpha^2 r^2}} d\xi - \int_{(2i-1)L}^{2iL-L_0} \frac{\xi - (2i-1)L}{\sqrt{(z-\xi) + \alpha^2 r^2}} d\xi \right] \right\}
$$
$$(22)$$

$$
\phi_{02} = v_0 \left\{ \sum_{i=0}^{n} \left[\int_{2iL+L_0}^{(2i+1)L} \frac{[(2i+1)L - \xi](2h+y)f_{0ij}(z)}{\sqrt{[\{(z+\xi)^2 + \alpha^2 r_1^2]^3}} d\xi \right. \right.
$$
$$
\left. - \int_{2iL+L_0}^{(2i+1)L} \frac{[(2i+1)L - \xi][2h+y]f_{0ij}(z)}{\sqrt{][z - \xi) + \alpha^2 r_1^2]^3}} d\xi \right]
$$
$$
+ v_0 \left\{ \sum_{i=1}^{n} \left[\int_{(2i-1)L}^{2iL-L_0} \frac{[\xi - (2i-1)L][2h+y]f_{0ij}(z)}{\sqrt{[(z+\xi)^2 + \alpha^2 r_1^2]^3}} d\xi \right. \right.
$$
$$
\left. - \int_{(2i-1)L}^{2iL-L_0} \frac{[\xi - (2i-1)L][2h+y]f_{0ij}(z)}{\sqrt{[(z-\xi) + \alpha^2 r_1^2]^3}} d\xi \right] \right\}
$$
$$(23)$$

In the sum of the potentials (21), the first approximation is the function φ_{01}, which satisfies Eq. (7) [27, 29] and the boundary condition (14).

The components of the velocity vector of air particles can be determined by the following formulas at $\gamma_0^3 \approx 0$:

$$\frac{\partial \varphi_n}{\partial x} = \gamma_0^2 \left(\frac{\partial \varphi_{01}(r,z)}{\partial r} \frac{\partial r}{\partial x} + \frac{\partial \varphi_{01}(r_1,z)}{\partial r_1} \frac{\partial r_1}{\partial x} \right)$$

$$= \gamma_0^2 \frac{v_0 x}{2r} \left\{ \sum_{i=0}^{n} \left[\int_{2iL+L_0}^{(2i+1)L} \frac{(2i+1)L - \xi}{(z+\xi)^2 + \alpha^2 r^2} d\xi - \int_{2iL+L_0}^{(2i+1)L} \frac{(2i+1)L - \xi}{(z-\xi)^2 + \alpha^2 r^2} d\xi \right] \right\}$$

$$- \gamma_0^2 \frac{v_0 x}{2r} \left\{ \sum_{i=1}^{n} \left[\int_{(2i-1)L}^{2iL-L_0} \frac{\xi - (2i+1)L}{(z+\xi)^2 + \alpha^2 r^2} d\xi - \int_{(2i-1)L}^{2iL-L_0} \frac{\xi - (2i-1)L}{(z-\xi)^2 + \alpha^2 r^2} d\xi \right] \right\}$$

$$+ \gamma_0^2 \frac{v_0 x}{2r_1} \left\{ \sum_{i=0}^{n} \left[\int_{2iL+L_0}^{(2i+1)L} \frac{(2i+1)L - \xi}{(z+\xi)^2 + \alpha^2 r_1^2} d\xi - \int_{2iL+L_0}^{(2i+1)L} \frac{(2i+1)L - \xi}{(z-\xi)^2 + \alpha^2 r_1^2} d\xi \right] \right\}$$

$$- \gamma_0^2 \frac{v_0 x}{2r_1} \left\{ \sum_{i=1}^{n} \left[\int_{(2i-1)L}^{2iL-L_0} \frac{\xi - (2i-1)L}{(z+\xi)^2 + \alpha^2 r_1^2} d\xi - \int_{(2i-1)L}^{2iL-L_0} \frac{\xi - (2i+1)L}{(z-\xi)^2 + \alpha^2 r_1^2} d\xi \right] \right\}$$

$$(24)$$

$$\frac{\partial \varphi_n}{\partial y} = \gamma_0^2 \left(\frac{\partial \varphi_{01}(r,z)}{\partial r} \frac{\partial r}{\partial y} + \frac{\partial \varphi_{01}(r_1,z)}{\partial r_1} \frac{\partial r_1}{\partial y} \right)$$

$$= \gamma_0^2 \frac{v_0 y}{2r} \left\{ \sum_{i=0}^{n} \left[\int_{2iL+L_0}^{(2i+1)L} \frac{(2i+1)L - \xi}{(z+\xi)^2 + \alpha^2 r^2} d\xi - \int_{2iL+L_0}^{(2i+1)L} \frac{(2i+1)L - \xi}{(z-\xi)^2 + \alpha^2 r^2} d\xi \right] \right\}$$

$$- \gamma_0^2 \frac{v_0 y}{2r} \left\{ \sum_{i=1}^{n} \left[\int_{(2i-1)L}^{2iL-L_0} \frac{\xi - (2i+1)L}{(z+\xi)^2 + \alpha^2 r^2} d\xi - \int_{(2i-1)L}^{2iL-L_0} \frac{\xi - (2i-1)L}{(z-\xi)^2 + \alpha^2 r^2} d\xi \right] \right\}$$

$$+ \gamma_0^2 \frac{v_0(y+2h+2R)}{2r_1} \left\{ \sum_{i=0}^{n} \left[\int_{2iL+L_0}^{(2i+1)L} \frac{(2i+1)L - \xi}{(z+\xi)^2 + \alpha^2 r_1^2} d\xi - \int_{2iL+L_0}^{(2i+1)L} \frac{(2i+1)L - \xi}{(z-\xi)^2 + \alpha^2 r_1^2} d\xi \right] \right\}$$

$$- \gamma_0^2 \frac{v_0(y+2h+2R)}{2r_1} \left\{ \sum_{i=1}^{n} \left[\int_{(2i-1)L}^{2iL-L_0} \frac{\xi - (2i-1)L}{(z+\xi)^2 + \alpha^2 r_1^2} d\xi - \int_{(2i-1)L}^{2iL-L_0} \frac{\xi - (2i-1)L}{(z-\xi)^2 + \alpha^2 r_1^2} d\xi \right] \right\}$$

$$(25)$$

$$\frac{\partial \varphi_n}{\partial z} = \gamma_0^2 \Big(\frac{\partial \varphi_{01}(r,z)}{\partial z} + \frac{\partial \varphi_{01}(r_1,z)}{\partial z} \Big)$$

$$= \gamma_9^2 \frac{v_0}{2} \left\{ \sum_{i=0}^{n} \left[\int_{2iL+L_0}^{(2i+1)L} \frac{[(2i+1)L - \xi](z+\xi)}{(z+\xi)^2 + \alpha^2 r^2} d\xi - \int_{2iL+L_0}^{(2i+1)L} \frac{[(2i+1)L - \xi](z-\xi)}{(z-\xi)^2 + \alpha^2 r^2} d\xi \right] \right\}$$

$$- \gamma_0^2 \frac{v_0}{2} \left\{ \sum_{i=1}^{n} \left[\int_{(2i-1)L}^{2iL-L_0} \frac{[\xi - (2i-1)L](z+\xi)}{(z+\xi)^2 + \alpha^2 r^2} d\xi - \int_{(2i-1)L}^{2iL-L_0} \frac{[\xi - (2i-1)L](z-\xi)}{(z-\xi)^2 + \alpha^2 r^2} d\xi \right] \right\}$$

$$- \gamma_0^2 \frac{v_0}{2} \left\{ \sum_{i=0}^{n} \left[\int_{2iL+L_0}^{(2i+1)L} \frac{[(2i+1)L - \xi](z+\xi)}{(z+\xi)^2 + \alpha^2 r_1^2} d\xi - \int_{2iL+L_0}^{(2i+1)L} \frac{[(2i+1)L - \xi](z-\xi)}{(z-\xi)^2 + \alpha^2 r_1^2} d\xi \right] \right\}$$

$$- \gamma_0^2 \frac{v_0}{2} \left\{ \sum_{i=1}^{n} \left[\int_{(2i-1)L}^{2iL-L_0} \frac{[\xi - (2i-1)L](z+\xi)}{(z+\xi)^2 + \alpha^2 r_1^2} d\xi - \int_{(2i-1)L}^{2iL-L_0} \frac{[\xi - (2i-1)L](z-\xi)}{(z-\xi)^2 + \alpha^2 r_1^2} d\xi \right] \right\}$$

$$(26)$$

Absolute speed of air flow generated by the system of high-speed train cars when it moves with a steady speed at an arbitrary point M(x, y, z) can be defined as

$$v = \sqrt{v_x^2 + v_y^2 + v_z^2}. \tag{27}$$

The air flow pressure can be determined from the formula

$$\Delta p = -\rho_0 \frac{\partial \varphi_n}{\partial t} = \rho_0 v_0 \gamma_0^2 \left[\frac{\partial \varphi_{01}(r,z)}{\partial z} + \frac{\partial \varphi_{01}(r_1,z)}{\partial z} \right]. \tag{28}$$

In Fig. 17 airs of change in airflow. In calculations it is accepted $L = 25$ m, $L_0 = 20$ m, $R = 2$ m, $v_0 = 200$ km/h, $\rho_0 = 1.2$ kg/m^3, $\gamma_0 = 0.4$.

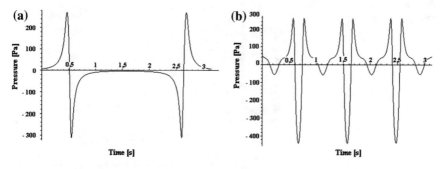

Fig. 17 Graphs of the change in aerodynamic airflow pressure versus time: **a** with closed and **b** open between car spaces

A comparison of the calculation results with the experimental data obtained in [16, 32] shows their qualitative agreement, the quantitative difference is 1.5–2.5 times. This difference can be explained by the fact that in the accepted calculation model the train is represented in the form of a long whole axisymmetric body whose initial and terminal sections have a kind of revived. The air is received by a compressible ideal gas, which is an essential approximation of the circuit to the real situation. In reality, the speed of the air flow is largely influenced by the shape of the moving train, the additional air flow arising in the space between the cars, and also the phenomenon of turbulence in the flow near a moving train. To take into account these factors, the correction factors can be introduced into the calculation formulas by the value of the airflow velocity.

Analysis of the graphs shows that in both cases the negative pressure is greater than the positive one. The reliability of these calculations is confirmed by the results of earlier experiments in the USA, Russia, Sweden [32, 33]. The effect between the car space on the magnitude of the negative (suction) aerodynamic pressure is clearly visible in the graph shown in Fig. 16b. On the railways of individual states, the movement of dual high-speed trains is practiced. With sufficient streamlining of the head and tail wagons, in places the pairing of trains produces a negative aerodynamic pressure, the value of which considerably exceeds the value of the excess pressure. Similar graphs can be constructed for other velocities and distances.

Thus, it can be argued that in order to ensure the safe operation of the railway infrastructure, it is necessary to take into account aerodynamic flows and pressures, regardless of their orientation. Using the results of calculations, it is also possible to construct a curve for the dependence of the magnitude of the aerodynamic pressure on the speed of trains and the distance to the point under consideration $P_{max} = f(V_{max}, B)$ (Fig. 18). With the help of these dependencies, it is possible to set the maximum permissible speed of passage of a high-speed train along the object.

As an example, solutions of the problems most often encountered in the practice of designing the high-speed movement of passenger trains, whose calculated schemes are shown in Fig. 15, are considered.

Fig. 18 Graphs of change in aerodynamic pressure

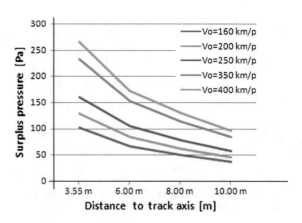

Ensuring the safety of passengers on the platform and workers of the railway in the immediate vicinity of the passing high-speed train is an actual task in connection with the current change in the aerodynamic field, due to the involvement of the air mass in motion, the speed of which and the pressure created depend on the speed and geometry of the high-speed train, the presence of the surrounding railway infrastructure facilities.

The strength of the impact on people and infrastructure depends not only on the maximum speed of the air flow, but also on its duration, location relative to the moving train, which should be considered as one of the main safety criteria. So far, the studies have been empirical. In particular, experiments conducted in Japan, France, Germany, the United States, Russia and other countries have experimentally established airspeed rates, the value of aerodynamic pressure around a high-speed train; its impact on people on the passenger platform and the construction of railway infrastructure facilities. Theoretically, the issue has not been studied enough.

For this purpose, an attempt was made to develop a design scheme with the following boundary conditions: a passenger platform of the "coastal" type adjoins, the passenger building; the edge of the platform is at a distance b_{min} from the axis of the path; The height of the platform is h (Fig. 15c); the passenger is on a platform width B_{min} on distance b_{1min} from the axis of the fence and distance b_{2min} from the passenger building; distance from the track axis to the edge of the platform b_{min} and platform height h are dimensioned "C_{250}" [8, 9]; a person is affected by force F, axial x and perpendicular to the vertical axis of the person. We can assume, since the surface (area) of the person on the platform is constant, the value of the force F, apparently varies directly in proportion to the value of the aerodynamic pressure P, formed during the passage of a high-speed train.

It is also clear that to ensure the safety of the person on the platform, the actual pressure on it should not exceed the regulatory threshold

$$P_f \leq P_n \qquad (29)$$

where

P_f the actual (excess) pressure (air flow) per person, Pa;
P_n the value of permissible pressure, regulated by the sanitary norms of the country, Pa.

For practical purposes, it is important to determine the point "M" with the coordinates M (x, y, z) where condition (29).

If a person is on a platform in a stationary state, at some point "M" his coordinates can be represented in the form $M(b_{1min}, y, z)$. In this case $x = b_{1min}$, $y = const$, $z = const$ (Fig. 19).

According to the Bernoulli law, the aerodynamic pressure is derived from the aerodynamic flow

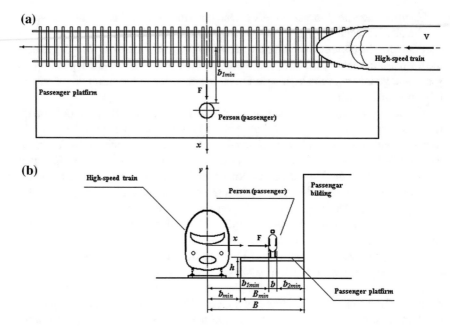

Fig. 19 The layout of the person (passenger) on the high coastal platform: **a** in plan, **b** in cross section

$$\Delta P = \frac{v^2 \rho}{2}. \tag{30}$$

From where you can determine the maximum value of the air speed

$$v_{\text{max.additional}} = \sqrt{\frac{2\Delta P_n}{\rho}}. \tag{31}$$

Thus, the solution of the problem reduces to determining the minimum distance $b_{1\min}$ or points $M(b_{1\min}, y, z)$, where the maximum value of air speed reaches $v_{\text{max.additional}}$ and condition (29).

Based on the results of previous calculations, graphs of the dependence of the maximum airflow velocity on the remoteness of the considered observation point can also be constructed (Fig. 20).

The resulting graphs allow us to establish:

- the minimum distance $b_{1\min}$, where the maximum values of the airflow velocity are equal $v_{\text{max.additional}}$;
- minimum width of the passenger platform $B_{\min}$ on sections of high-speed trains;
- the maximum speed of passage of high-speed passenger trains at the station (a detour) along a high passenger platform with a known width.

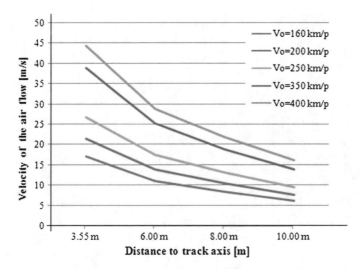

Fig. 20 The graph of the maximum airflow velocity versus the distance of the observed point of observation

In all cases, the condition (29) is satisfied.

Let us assume that the maximum permissible pressure value regulated by sanitary norms is equal to $P_n = 100$ Pa. According to the formula (31), the values of the maximum airflow velocity $v_{max.additional}$, which is 12.9 m/s, which corresponds to a distance of 4.44 m (Fig. 20). The maximum pressure on a person in a stationary state (standing still) at a distance $b_{1min} \geq 4.44$ m from the axis of the moving body with a speed of 160 km/h to make no more than 100 Pa, which corresponds to the sanitary standards.

The above methodology also allows one to investigate the effects of secondary air streams, formed by high-speed passenger trains, on the surrounding environment along the high-speed railway.

In the republics of Central Asia and Kazakhstan, about 3000 km of railways built in the sand desert zone are constantly exposed to sand drifts and are operated in extremely difficult conditions.

Currently, the Bukhara-Miskin line is being built, designed in the future to organize high-speed traffic through Urgench to Khiva station and separately from Miskin to Nukus station. The railway from Bukhara to Miskin crosses the massifs of the sands of southern Kyzylkum (Fig. 21).

These sands with a wind speed on the surface of the sand more critical (>4.1 m/s) come into motion and with a further increase in speed are transferred and create mobile forms of relief. This creates problems both during the construction period and during the operation of the railway. In particular, the roadbed is blown out, the drainage structures and the top structure of the road are filled up.

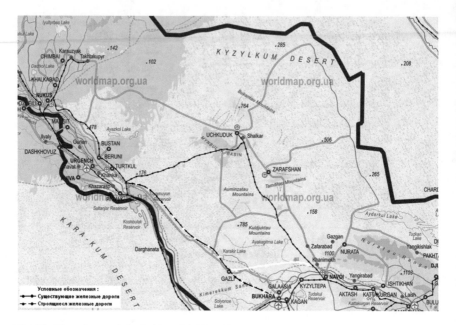

Fig. 21 The scheme of the Bukhara-Miskin railway

In the course of construction technogenic sands are formed, losing their natural composition and structure. As a result, technogenic sands grow after 6–8 years [34, 35].

When introducing high-speed traffic of passenger trains on this section, unlike other routes, additional questions arise that require their scientific and technical solutions. These include ensuring the stable operation of the roadbed, securing the slopes of the roadbed erected from sand-dune and the roadside strip.

The technology of construction and strengthening of the roadbed erected from sand-dune, fixing its slopes and adjacent to the railroad tracks are devoted to the work of A.I. Adylkhodzhaev, R.S Zakirov, M.M. Mirakhmedov, T.I. Fazylov and other scientists [34–38]. They also proposed various ways to protect the way from sand drifts and methods of fixing sand-dune carried by the wind.

In this study, the possibility of sand deflation of the adjacent strip, embankment slopes and excavation by a secondary airflow—a stream formed from the movement of a high-speed train—is considered.

The theoretical premise can be the transition of a grain of sand from a state of rest to a motion which, in the prevailing case, occurs at air flow velocities above 4.1 m/s, called critical. Above this threshold, the sand grains come into motion, at the beginning, by sliding, rolling, then spasmodically and in a suspended state.

The main form of motion is a jump, the parameters of which depend both on the size and shape of the sand, and the structure of the wind-sand flow, formed by the wind speed, in this case on the speed of the air flow created by the moving high-speed train. the negative impact of which on the state of the environment in the sand desert area should be included in the high-speed railway project. As measures, the following organizational and technical measures may be envisaged:

(1) fixation of susceptible sands:
(2) in the body of the roadbed;
(3) on the slopes of the roadbed;
(4) on the adjacent strip of width B (m);
(5) limiting the speed of the movement of high-speed trains to a level where the speed of the air flow created by them will be equal to or less than the critical one, i.e.

$$v_{af} \leq v_{cr} \tag{32}$$

The technology of erecting a roadbed from sand-dune, reinforcing the roadbed, fixing its slopes and adjacent to the railroad tracks are not the subject of this research.

Thus, the issue of selection and justification of measures preventing the negative impact of the movement of a high-speed train on the state of the environment in the area of sandy deserts is reduced to establishing:

1. Point locations n_i on the surface of the earth with coordinates (x_i, y_i, z_i), where condition (32), those at a known value of the speed of the distance train (x_i, y_i, z_i).
2. Train speeds v_{ni} at the point $n_i(x_i, y_i, z_i)$ on the surface of the earth, under which condition (32), those at a certain value of the physical and mechanical properties of the soil at the point n_i speeds v_{ni}.

In both cases, in order to solve the problem, it is necessary to establish the velocity of the secondary airflow v_{af}, which can be established by expression (27). The speed of a high-speed train can be established by traction calculations.

Using the graph of the maximum air speed as a function of the remoteness of the observed observation point (without taking into account the correction factor), which is shown in the figure with a thick line (Fig. 22), the minimum distance from the axis of the path on which the sand fixing works should be established.

These graphics allow:

– determine the nature of the distribution of air flow and determine its speed along a moving high-speed train;
– to build along the roadbed isolines with the same speed of air flow, formed by the movement of a high-speed train;
– establish the width of the strip of fastening work along the earthen canvas, sprinkled from sand-dune and exposed to wind-transported sands;
– optimize the amount of sand fixing work;
– choose the method (method) of fixing sand.

The results of the research can be used in the design of high-speed railways in the regions where loose sand is spread.

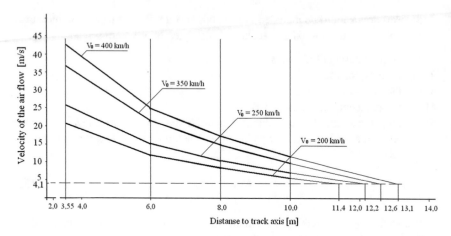

Fig. 22 The graph of the maximum airflow velocity versus the distance of the observed point of observation

As a result of a complex of theoretical studies of the aerodynamics of a high-speed train, as an axisymmetric body moving in a compressible (acoustic) medium,

1. The nature of the distribution of the air flow and the speed along the moving high-speed train, as well as the velocity distribution along the roadbed (isolines with the same speed) of the air flow formed during the movement of the high-speed train.
2. The maximum permissible speed of a high-speed train, taking into account the technical condition of permanent devices and structures of the existing railway infrastructure.
3. Technical parameters of individual objects and structural elements of the infrastructure of high-speed railways subject to the effect of aerodynamic pressure for a given maximum speed of high-speed trains.
4. The width of the strip of the roadbed (including slopes, cuvettes) spilled from sand-dune and to be fixed with appropriate justification (covering with heavy soil, impregnation of astringent, shelter from geotextile in combination with hydroseeding) provided that the velocity of the secondary wind generated from movement high-speed train exceeds the critical, as well as the minimum required sand-consolidation work.
5. The places of installation of fences on the runways, preventing unauthorized access to people and animals, the minimum safe distance of people in the vicinity of the passing train in accordance with the requirements of the national standard.
6. Rationally use the land fund allocated for the construction of high-speed railways.

The proposed methodological approach can be used in the practice of organizing high-speed and high-speed train traffic, in particular, when designing it on existing and newly constructed railways.

4 The Combined Movement of Passenger and Freight Trains to High-Speed Sections of Existing Railways

As an alternative solution to high-speed traffic, consider the organization of high-speed passenger traffic on existing lines with the combined movement of freight and passenger trains.

The presence of combined cargo and passenger traffic, the tendency to a known increase in freight traffic, in the event of an actual depletion of capacity reserves, create serious additional difficulties when introducing high-speed train traffic on single-track lines.

The introduction into circulation of several high-speed trains can lead to premature exhaustion of the carrying capacity of single-track rail and road and the emergence of a "capacity deficit". Therefore, the task of increasing the speed of passenger trains must be considered together with the task of eliminating the shortage of throughput and carrying capacity resulting from the accelerated removal of freight trains by accelerated passenger trains.

Let us assume that the required carrying capacity of a single-track section varies linearly, those

$$G_r = G_p + a \cdot t \tag{33}$$

where,

G_p carrying capacity for the initial year of operation, mln.t.;
a the rate of annual growth in the required capacity, mln.t/year;
t billing year

Required carrying capacity $G_r = f(t)$ changing in time, as a rule, increases. Possible carrying capacity of the site up to $G_p^t = f(t)$ and after $G_p^a = f(t)$ the introduction of high-speed train traffic also varies with time. Since in most cases the dimensions of passenger traffic are increasing, the possible carrying capacity of the section decreases with time, i.e. curves $G_p = f(t)$ tend to decrease (Fig. 23).

The graphical representation of the change in the required and possible carrying capacity (Fig. 23) shows that an increase in speed after modernization leads to a premature exhaustion of capacity for a period Δt, those, there is a shortage of carrying capacity ΔG.

If the annual growth rate of the required capacity is known, then it is possible to establish a time limit Δt for the premature exhaustion of the possible carrying capacity.

Fig. 23 Graphs of possible
and required carrying capacity

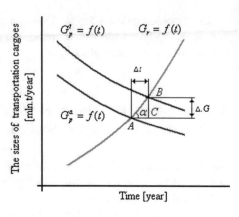

From the triangle ABC

$$tg\,\alpha = \Delta G/\Delta t, \tag{34}$$

as

$$tg\,\alpha = a, \tag{35}$$

then,

$$\Delta t = a\Delta G. \tag{36}$$

The magnitude of the deficit of carrying capacity ΔG there is a difference in the possible carrying capacity of the site up to $G_p^t = f(t)$ and after $G_p^a = f(t)$ introduction of high-speed train traffic, those

$$\Delta G = G_p^t - G_p^a \tag{37}$$

In turn, the possible carrying capacity of the site before $G_p^t = f(t)$ and after $G_p^a = f(t)$ putting into motion N_{h-s}^a high-speed trains, respectively, can be defined as

$$G_p^t = \frac{365 \cdot Q_c 10^{-6}}{\gamma} N_c^t == \frac{365 Q_c 10^{-6}}{\gamma}$$
$$\cdot \left\{ \frac{N_p}{1+\beta} - N_{pas}^t \varepsilon_{pas} - N_{as}^t (\varepsilon_{as} - 1) - N_e^t \varepsilon_e - N_{emp}^t \right\} \tag{38}$$

$$G_p^a = \frac{365 \cdot Q_c 10^{-6}}{\gamma} N_c^a == \frac{365 Q_c 10^{-6}}{\gamma}$$
$$\cdot \left\{ \frac{N_p}{1+\beta} - N_{h-s}^a \varepsilon_{h-s} - N_{pas}^a \varepsilon_{pas} - N_{as}^a (\varepsilon_{as} - 1) - N_e^a \varepsilon_e - N_{emp}^a \right\} \tag{38'}$$

The carrying capacity of the haulage in cargo traffic can be determined by the following formula [39]

$$N_c = \left\{ \frac{N_p}{1+\beta} - N_{pas} \varepsilon_{pas} - N_{as} (\varepsilon_{as} - 1) - N_e \varepsilon_e - N_{emp} \right\} \tag{39}$$

Introduction of high-speed (or high-speed) traffic of passenger trains N_{h-s}^a on existing single-track lines can be carried out:

1. Due to newly introduced high-speed trains;
2. By transferring a number of existing trains into the category of high-speed trains.

In the second variant, the number of passenger trains after the transfer to the category of high-speed N_{h-s}^a trains can be defined as,

$$N_{pas}^a = N_{pas}^t - N_{h-s}^a \tag{40}$$

We assume that the number of prefabricated, economic, empty, accelerated freight trains before and after the introduction of high-speed traffic of passenger trains does not change. Substituting the expressions (38), (38') and (40) into the formula (37) after some transformations, it is not difficult to establish the magnitude of the deficit of the carrying capacity ΔG after the introduction of high-speed traffic of passenger trains:

- with the introduction of newly high-speed trains

$$\Delta G = \frac{365 \cdot Q_c 10^{-6}}{\gamma} N_{h-s}^a \varepsilon_{h-s} \tag{41}$$

- when moving N_{h-s}^a the number of existing trains in the category of high-speed trains

$$\Delta G = \frac{365 \cdot Q_c 10^{-6}}{\gamma} N_{h-s}^a (\varepsilon_{h-s} - \varepsilon_{pas}) \tag{42}$$

Expression analysis (37), (41), (42) shows that the term of premature exhaustion of the possible carrying capacity Δt directly depends on the number of high-speed trains N_{h-s}^a, ratio of speeds of high-speed and passenger trains $\varepsilon_{pas}, \varepsilon_{h-s}$, as well as the rate of annual growth in the required carrying capacity—a.

Compensation of the deficit of carrying capacity resulting from the introduction of high-speed passenger train traffic on single-track sections is possible due to:

1. Increase in the norms of the mass of the train while maintaining the possible throughput N_{h-s}^a, N_p;
2. Increase in possible bandwidth N_p due to the transition to a smaller inter-train interval I on a single-track line.

In the first case, in order to preserve the possible carrying capacity before and after the introduction of the high-speed traffic of passenger trains, the condition

$$G_p^t = G_p^a. \tag{43}$$

Thus equality (38) and (38′) can be written down in a following kind

$$G_p^t = \frac{365 \cdot Q_c^t 10^{-6}}{\gamma} N_c^t, \tag{44}$$

$$G_p^a = \frac{365 \cdot Q_c^a 10^{-6}}{\gamma} N_c^a. \tag{45}$$

Substituting (44), (45) into (43) and solving the resulting equation for Q_c^a

$$Q_c^a = Q_c^t \frac{N_c^t}{N_c^a}. \tag{46}$$

The ratio of the number of freight trains before and after the introduction of high-speed trains will be denoted by

$$k_{inc} = \frac{N_c^t}{N_c^a}. \tag{46′}$$

Then, Eq. (46) takes the following

$$Q_c^a = Q_c^t k_{inc}, \tag{47}$$

where, k_{inc}—coefficient of increase in the mass of the freight train after the introduction of high-speed traffic of passenger trains.

Thus, to preserve the possible carrying capacity after the introduction of high-speed traffic N_{h-s}^a passenger trains it is necessary to increase the mass of each freight train in accordance with (47) by the amount k_{inc}. The value of the weight increase coefficient of the composition depends on the quantity, the maximum speed of movement of high-speed trains in circulation, and also on the speed of freight trains. The graphical representation of the change in the values of the mass increase coefficient of the composition is shown in Fig. 24.

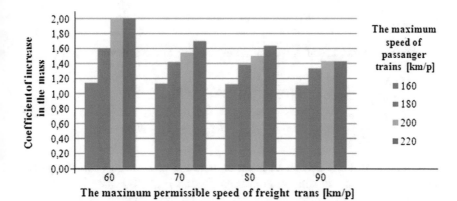

Fig. 24 Change in the value of the mass composition increase coefficient

Analysis of the results of the research shows that if in the section of a single-track railway in question passenger trains with speeds of up to 120, 140, 160 km/h were in circulation, then a sharp increase in the value of the conditional mass increase of the freight composition occurs when moving to higher speed levels. At speeds above 180 km/h, the value of the conditional coefficient of increase in the mass of the composition does not actually change, i.e. of the speed does not depend. To a large extent, it depends on the number of high-speed trains being introduced (or accelerated) and the speed of freight trains. With the reduction in the difference in the ratio of the speeds of freight and freight trains, the value of the coefficient decreases. The increase in the maximum permissible speed of freight trains from 60 km/h to 70, 80, 90 km/h, depending on the maximum speed (180, 200, 220 km/h) will reduce the value of this coefficient respectively by 10–11%, 14% and 20–27%. The maximum value of the coefficient of increase in the mass of the composition for the case under consideration is 1.43.

Assuming that before the introduction of high-speed passenger trains, the weight of a freight train at a maximum speed of 60 km/h was 3800 ton, in order to maintain the existing capacity after the introduction of one high-speed passenger train, it is necessary to increase the weight of the freight train to 5400 ton. The train must first be checked for traction constraint and the useful length of the receiving and sending tracks.

With an increase in the speed of freight trains from 60 to 70, 80, 90 km/h, the coefficient value k_{inc} respectively, to 1.30 and 1.25 and 1.13. The weight of the freight train, which can ensure the preservation of the existing capacity, respectively, to make 4950, 4750, 4300 ton.

Consider the second case of maintaining or increasing the possible band width N_p due to the transition to a smaller inter-train interval I on the single-path line

$$N_c^t = N_c^a \tag{48}$$

Taking (39) into account, we can represent (48) in the following form

$$\left\{ \frac{N_p}{1+\beta} - N_{pas}^t \varepsilon_{pas} - N_{as}^t (\varepsilon_{as} - 1) - N_e^t \varepsilon_e - N_{emp}^t \right\}$$
$$= \left\{ \frac{N_p}{1+\beta} - N_{h-s}^a \varepsilon_{h-s} - N_{pas}^a \varepsilon_{pas} - N_{as}^a (\varepsilon_{as} - 1) - N_e^a \varepsilon_e - N_{emp}^a \right\} \tag{49}$$

Assuming that the calculated intervals between passing trains in the packet, respectively in the odd and even direction I_{cal}' and I_{cal}'' before and after the introduction of high-speed traffic of passenger trains are equal, those $I_{cal}' = I_{cal}'' = I_{cal}^t$, as well as the period of the unpaired train schedule does not change, then Eq. (49) can be written in the following form

$$\frac{2(1440 - t_{tech})\alpha_r}{(2 - \alpha_{pac})T_{r.d.} + 2I_{cal}^t \alpha_{pac}}$$
$$- \left\{ \frac{N_p}{1+\beta} - N_{pas}^t \varepsilon_{pas} - N_{as}^t (\varepsilon_{as} - 1) - N_e^t \varepsilon_e - N_{emp}^t \right\}$$
$$= \frac{2(1440 - t_{tech})\alpha_r}{(2 - \alpha_{pac})T_{r.d.} + 2I_{cal}^a \alpha_{pac}}$$
$$- \left\{ \frac{N_p}{1+\beta} - N_{h-s}^a \varepsilon_{h-s} - N_{pas}^a \varepsilon_{pas} - N_{as}^a (\varepsilon_{as} - 1) - N_e^a \varepsilon_e - N_{emp}^a \right\} \tag{49'}$$

Proceeding from (49'), we can obtain an equality at which compensation of the deficit of the carrying capacity

$$I_{cal}^a = \frac{1}{2\alpha_{pac}}$$
$$\cdot \left\{ \frac{1}{\frac{1}{(2-\alpha_{pac})T_{r.d.} + 2I_{cal}^t \alpha_{pac}} - \frac{(N_{pas}^t - N_{pas}^a)\varepsilon_{pas} - N_{h-s}\varepsilon_{h-s}}{2(1440 - t_{tech})\alpha_r}} - (2 - \alpha_{pac})T_{r.d.} \right\} \tag{50}$$

To study the possible throughput in the freight traffic of a single-track section before and after the introduction of high-speed train traffic, calculations were made for the initial data given in Table 4 and reflecting the conditions for the movement of trains.

The change in the calculated interval between passing trains for the case when one or two high-speed passenger trains is put into motion on a single-track section (Table 4) under boundary conditions that the calculated interval between passing trains before the introduction of high-speed traffic is 15 min. Then, in order to maintain the existing carrying capacity of a single-track section with the

Table 4 Initial data for determining the throughput

No.	Index	Unit of measurement	Amount	
			Before	After
1	Number of trains	pp/day		
	– Passenger		2	2
	– National teams		1	1
	– Economic		1	1
	– Empty		–	–
	– Accelerated		1	1
	– Accelerated passenger		–	$1(2)^a$
	– High-speed		–	$1 (2, 4)^a$
2	Speed of trains	km/h	60	60
	– Passenger		120	120
	– Assembled		60	60
	– Economic		60	60
	– Empty		60	60
	– Accelerated		90	90
	– Accelerated passenger		–	160
	– High-speed		–	$180^a (200^a, 220^a)$
3	Estimated interval between trains in the package	min	20	18(16, 14, 12, 10, 8)

[a]In perspective

introduction of high-speed train traffic, it is necessary to reduce the calculated interval between trains to 11 and 7 min respectively.

For the main single-track lines of Uzbekistan, Tashkent-Pape-Andijan, Samarkand-Karshi-Termez, Samarkand-Navoi-Bukhara, Bukhara-Urgench, it is characteristic that the freight traffic prevails in them, the number of regular passenger trains does not exceed 4–6 pairs of trains per day (or 6–8 pairs of trains in the summer train schedule). In the short term, these lines can introduce 2–4 pairs of high-speed trains per day, while maintaining the existing dimensions in passenger transportation.

The study of the change in the carrying capacity after the introduction of high-speed train traffic was carried out for the electrified section of the single-track railroad Samarkand-Karshi, a length of 156 km. The section of the railway consists of 10 stretches, the length of which is within 7.6–22 km. In this section, freight trains are driven by electric locomotives UZel yuk, passenger UZel yo'l. The maximum speed of freight trains is limited to a speed of 60 km/h, a passenger speed of 160 km/h. To establish the limiting distillation, traction calculations were carried out, the results of which are summarized in Table 5. Analysis of the calculations showed that the distances No. 7, 8, 9 are limiting.

Table 5 Travel time of freight and passenger trains by limiting the stretch in section C-K

Category of trains	Speed trains (km/h)	Number of distillation									
		1	2	3	4	5	6	7	8	9	10
Cargo	60	20.82	24.14	40.72	34.59	35.54	34.02	46.62	46.55	45.13	29.86
	70	19.26	22.92	36.16	31.15	33.72	33.75	40.82	41.67	40.79	27.39
	80	18.44	22.22	32.75	29.87	32.52	28.97	36.8	38.6	38.68	25.74
	90	17.98	21.95	30.73	29.53	31.93	27.93	33.19	35.6	35.38	23.88
Passenger's	120	9.25	10.33	19.35	10.15	15.70	15.54	21.59	21.33	22.38	13.52
	160	7.63	9.30	15.58	9.01	14.41	13.69	16.64	16.64	18.88	11.96
	180	7.25	9.19	14.65	8.85	13.34	13.58	15.07	14.85	17.76	11.67
	200	7.1	9.19	14.06	8.82	14.34	13.58	13.97	13.8	17.25	12
	220	7.08	9.19	13.75	8.82	14.34	13.58	13.35	13.74	17.07	12.12

In the study of changes in the carrying capacity, it was assumed that the maximum permissible speed of freight trains was limited to 70, 80, 90 km/h; passenger trains up to 180, 200, 220 km/h. The results of calculations for determining the travel time of freight, passenger and high-speed trains are summarized in Table 5.

Further studies were carried out for the haul-outs No. 9 and for haul-off No. 3, at which the time of the freight train's journey was 56 min according to the order.

Intensive growth, when the speed of a passenger train increases from 120 to 180 km/h (Fig. 25). Further growth slows down, but at speeds above 200 km/h does not change. This is explained by the fact that the length of the distances can't be accelerated by high-speed passenger trains up to the maximum speed, since there are speed limits for switch points (120, 140 km/h) on separate points. Withdrawal of speed limits on the line 3 of this section of the railway, the admission of load-lifting trains from 2.1–2.6 to 1.6–1.8.

The possible throughput capacity of a single-track section of S-K equipped with auto-blocking, with a partial-package schedule for the hauls No. 9 and No. 3, was investigated for the maximum speed of freight trains 60, 70, 80, 90 km/h; values of the calculated intervals between trains 8, 10, 12, 14, 16, 18, 20 min. In this case, the station interval is taken equal to 3 min.

Analysis of calculation results showed that reducing the estimated interval between trains in the package from 20 to 8–10 min allows to increase the possible throughput capacity of the S-K railway section to 20% (Fig. 26). The existing temporary speed limit on the No. 3 line reduces the possible throughput capacity of the S-K railway section to 20–25%.

To study the possible throughput and carrying capacity in the freight traffic of the section of the CK before and after the introduction of high-speed passenger trains, calculations were made using the initial data given in Table 4.1. At the same

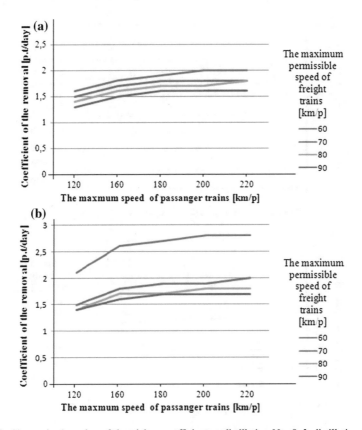

Fig. 25 Change in the value of the pickup coefficient: **a** distillation No. 9, **b** distillation No. 3

time, the maximum permissible speed of freight trains varied between 60–90 km/h in increments of 10 km/h, the estimated interval between trains in the package was 8, 10, 12, 14, 16, 18, 20 min, the duration of the technical operations was adopted 120 min calculations.

Based on the results of calculations of the possible throughput capacity of the S-K railway section in cargo traffic, with a calculated interval between trains in a package equal to 20 min and formula (46′) the values of the conditional coefficient of increase in the mass of the freight composition are determined, the graphical representation of which is shown in Fig. 27.

For a comparative assessment of the change in the possible carrying capacity of the section of the railway S-K, depending on the level of the maximum permissible speeds of freight and passenger trains, the estimated interval between trains in the package, the values of the carrying capacity are determined.

Taking into account the initial technical condition of the railway section, graphic images of the change in the possible carrying capacity for different levels of the maximum permissible speeds of freight and passenger trains (Fig. 28) are

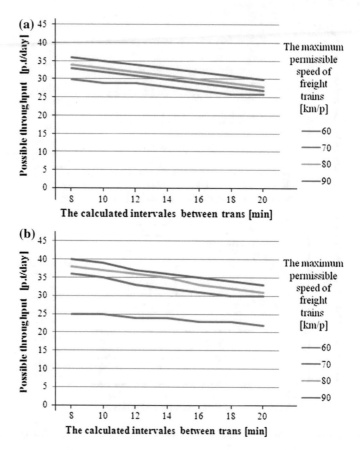

Fig. 26 Change of possible carrying capacity (at t_{tech}=120 min): **a** haulage No. 9, **b** haul No. 3

constructed. Similar graphic images can be construct for different calculation intervals between trains in a package, the duration of technological "windows", the mass of the freight train.

If the histogram is to plot a horizontal line of the required carrying capacity, then it is possible to establish the magnitude of the deficit of the carrying capacity and the measures for its compensation (Fig. 29).

Carrying out a horizontal line of the required carrying capacity on a histogram reflecting the possible carrying capacity of this section, at a level of 18 million ton per year, it is not difficult to establish, with the existing technical condition (the maximum permissible speed of freight trains is 60 km/h), there is some reserve of

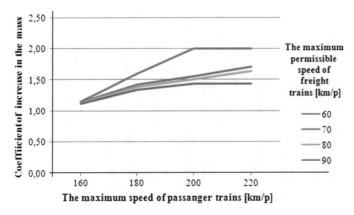

Fig. 27 The values of the conditional coefficient of increase in the mass of the freight composition in the section of C-K

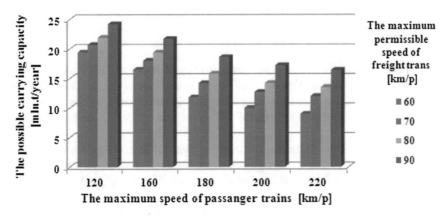

Fig. 28 The possible carrying capacity of the section of C-K

carrying capacity. A reserve of capacity is also available when organizing high-speed train traffic at speeds of 160 km/h. At the same time, the maximum permissible speed of freight trains should be more than 70 km/h. Further increase in the speed of passenger trains will lead to a shortage of carrying capacity, there is a need to develop measures to increase the capacity of the railway section in question.

Summarizing the results obtained, we can draw the following conclusion.

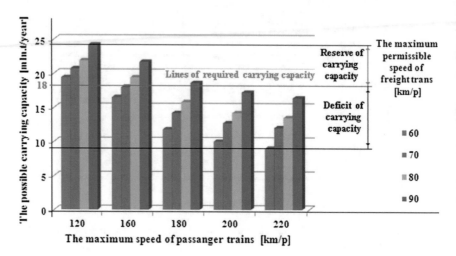

Fig. 29 Possible and required carrying capacity

5 Conclusion

1. An increase in the speed of passenger trains from 120 to 160, 180, 200, 220 km/h at different speeds of freight trains leads to an increase in the coefficient of removal from 1.3 to 2.6. On average, the pick-up ratio increases from 22 to 62%. At the same time, the maximum value of the takeoff coefficient can be more than 3.4.

2. The duration of the technological window does not have a significant effect on the throughput. Increase in the duration of technological windows from 75 up to 150 min or from 90 to 180 min reduces the possible throughput by an average of 5%.

3. When every high-speed train is put into motion, a sudden decrease in the possible carrying capacity takes place, which depends on the number of high-speed trains and their maximum speed. An increase in the speed of high-speed trains exceeding 180 km/h does not lead to a significant reduction in the carrying capacity.

4. The introduction into circulation of several high-speed trains can lead to a premature exhaustion of the carrying capacity of a single-track railway and the appearance of a "capacity shortage". The term of premature exhaustion of the possible carrying Δt capacity directly depends on the number of high-speed trains N_{h-s}^a, the ratio of the speeds of high-speed and passenger trains ε_{h-s}, ε_{pas} and the rate of annual growth in the required carrying capacity—a.

5. Compensation of the deficit of carrying capacity resulting from the introduction of high-speed passenger train traffic on single-track sections is possible due to:

   - increase in the mass of the train while maintaining the available capacity N_p;

- increasing the possible through put by N_p switching to a smaller inter-train interval I on a single-track line.

6. To maintain the possible carrying capacity after the introduction of high-speed N_{h-s}^a passenger trains, it is necessary to increase the mass of each freight train in accordance with (47) by an amount k_{inc}. The value of the weight increase coefficient of the composition depends on the quantity, the maximum speed of movement of high-speed trains in circulation, and on the speed of freight trains. An increase in the speed of freight trains from 60 to 70, 80, 90 km/h will significantly reduce the value of the weight gain coefficient by k_{inc} 10, 14, 27%, respectively.

7. Reducing the estimated interval between trains in the package from 20 to 8–10 min allows increasing the possible capacity of the section of the railway to 20%.

Bibliography

1. Commission Recommendation of 21 March 2001 on the basic parameters of the trans-European high-speed rail system referred to in Article5(3)(b) of Directive 96/48/EC. http://eur-lex.europa.eu/legal-content/EN/TXT/PDF/?uri=CELEX:32001H0290&from=EN

2. Сотников ЕА (1996) Мировой опыт создания ВСМ. Инженер путей сообщения. ВСМ: Спец. выпуск. Санкт-Петербург, pp 20–23 [In Russian: Sotnikov EA (1996) World experience of creating high-speed highways. Railway Engineer]

3. Киселев ИП (2005) Краткий обзор истории европейских высокоскоростных поездов. Часть 1. Железные дороги мира. No. 12, pp 20–36 [In Russian: Kiselev IP (2005) Brief review of the history of European high-speed trains. Part 1. Railways of the world]

4. Перспективы высокоскоростных пассажирских перевозок (2015) Железные дороги мира. No. 12, pp 9–20 [In Russian: Prospects for high-speed passenger transportation (2015) Railways of the world]

5. Сушков ЮС (2013) Проблемы и закономерности развития скоростных железных дорог в мире. Градостроительство. No. 1, pp 75–81 [In Russian: Sushkov YuS (2013) Problems and patterns of development of high-speed railways in the world. Urban Development]

6. World Speed Survey (2013) China sprints out in front. Railway Gazette International. http://www.railwaygazette.com/news/high-speed/single-view/view/world-speed-survey-2013-china-sprints-out-in-front.html

7. Тархов СА (2010) Расширение сети железных дорог Китая в 2000-е годы. Региональные исследования. No. 1, pp 89–102 [In Russian: Tarkhov SA (2010) Expansion of the network of railways of China in the 2000s. Regional studies]

8. ВСН 333-Н (2015) Инфраструктура скоростной железнодорожной линии Самарканд-Карши. Общие технические требования. Ташкент: АО«УТЙ». [In Russian: Departmental Building Norms 333-N (2015) Infrastructure of the high-speed railway line Samarkand-Karshi. General technical requirements. Tashkent]

9. ВСН 448-Н (2010) Инфраструктура скоростной железнодорожной линии Ташкент - Самарканд. Общие технические требования. Ташкент: ГАЖК«УТЙ». [In Russian: Departmental Building Norms 448-N (2010) The infrastructure of the high-speed railway line Tashkent–Samarkand. General technical requirements. Tashkent]

10. Винокуров БЮ (2009) Международные коридоры ЕврАз ЭС: быстрее, дешевле, больше: Отраслевой обзор. Алма-Аты. http://transtec.transtec-neva.ru/files/File/eurozec.

pdf [In Russian: Винокуров БЮ (2009) International corridors of EurAs ES: faster, cheaper, more: Industry review. Alma-Ata]

11. Волчок ЮГ (2011) Формирование единого транспортного пространства Евразийского экономического сообщества. Транспорт Российской Федерации. No. 1(32), pp 4–7 [In Russian: Volchok YuG (2011) Formation of the common transport space of the Eurasian Economic Community. Transport of the Russian Federation]

12. Djabbarov S (2016) Prospects for raising passenger train speed on the reconstructed section of the Uzbekistan railways. Transp Probl 11(4):103–110

13. Djabbarov S, Mirakhmedov M (2015) Features of the organization movements of high-speed passenger train on Tashkent–Andijan line (of the Uzbekistan railway). In: VII Int. Sci. Conf. "Transport Problems". Silesian University of Technology, Katowice, pp 355–360

14. Karimova F, Djabbarov S, Mirakhmedov M (2015) The organisation of high-speed movement of passenger trains on the international transport corridors of the Central Asia. In: 8th International Symposium for Transportation Universities in Europe and Asia, Nanjing

15. Правила тяговых расчетов для поездной работы (1985) Москва: Транспорт, 287 p [In Russian: Rules of traction calculations for train operation (1985) Transport, Moscow]

16. Raghunathan RS, Kim H-D, Setoguchi T (2002) Aerodynamics of high-speed railway train. Prog Aerosp Sci 38(6–7):469–514

17. Diedrichs B (2006) Studies of two aerodynamic effects on high-speed trains: crosswind stability and discomforting car body vibrations inside tunnels. Doctoral thesis. KTH, Stockholm, 74 p. http://kth.diva-portal.org/smash/get/diva2:11067/FULLTEXT01.pdf

18. Holmes S, Schroeder M, Toma E (2000) High-speed passenger and intercity train aerodynamic computer modeling. In: International Mechanical Engineering Congress & Exposition. November 5–10, 2000, Orlando, Florida

19. Khayrullina A, Blocken B, Janssen W, Straathof J (2015) CFD simulation of train aerodynamics: train-induced wind conditions at an underground railroad passenger platform. J Wind Eng Ind Aerodyn 139:100–110

20. Quinn AD, Hayward M, Baker CJ, Schmid F, Priest J, Powrie W (2009) A full-scale experimental and modelling study of ballast flight under high speed trains. J Rail Rapid Transit 224(2):61–74

21. Jing GQ, Zhou YD, Lin J, Zhang J (2012) Ballast flying mechanism and sensitivity factors analysis. Int J Smart Sens Intell Syst 5(4):928–939

22. Baker CJ, Dalley SJ, Johnson T, Quinn A, Wright NG (2001) The slipstream and wake of a high speed train. In: Proceedings of the IMechE, Part F: Journal of Rail and Rapid Transit, vol 215(2), pp 83–99

23. Sterling M, Baker CJ, Jordon SC, Johnson T (2008) A study of the slipstreams of high-speed passenger trains and freight trains. In: Proceedings of the IMechE, Part F: Journal of Rail and Rapid Transit, vol 222, pp 177–193

24. Baker C (2010) The flow around high speed trains. J Wind Eng Ind Aerodyn 98:277–299

25. Қурилиш меъёрлари ва қоидалари. ҚМҚ 2.01.05-96 (1998) Тошкент, Ўзбекистон республикаси давлат архитектура ва қурилиш қўмитаси. [In Uzbek: Building codes and regulations 2.01.05-96 (1998) State Committee of the Republic of Uzbekistan on Architecture and Construction. Tashkent]

26. ВСН-450 (2010) Ведомственные технические указания по проектированию и строительству. Железные дороги колеи 1520 мм. Ташкент: ГАЖК. 48 p. [In Russian: Departmental Building Norms 450 (2010) Departmental technical instructions for design and construction. Railways of gauge 1520 mm. Tashkent]

27. Djabbarov S, Mirakhmedov M, Mardonov B (2016) Aerodynamic field model of high-speed train. In: VIII Conference International "Transport Problems". Katowice: Silesian University of Technology, pp 107–115

28. Ламб Г (1947) Гидродинамика. Москва: ОГИЗ, 929 p [In Russian: Lamb G (1947) Hydrodynamics. OGIZ, Moscow]

29. Джаббаров СТ, Мирахмедов М, Мардонов БМ (2016) К вопросу безопасности пассажира на высокой железнодорожной платформе при прохождении высокоскоростного поезда. Инновационный транспорт. No. 3(21), pp 39–44 [In Russian: Jabbarov ST, Mirakhmedov M, Mardonov BM (2016) On the question of the safety of the passenger on a high railway platform when high-speed train is passing. "Innotrans" Journal]

30. Джаббаров СТ (2015) Исследование поля скоростей частиц воздуха вблизи вагонов при движении скоростного поезда в равнинной местности. Проблемы механики. Ташкент. No. 2. Р. 80–84 [In Russian: Djabbarov ST (2015) Investigation of the velocity field of air particles near wagons during the movement of a high-speed train in a flat terrain. Problems of mechanics. Tashkent]

31. Джаббаров С, Мирахмедов М, Марданов Б (2017) Определение границы зоны безопасности высокоскоростного поезда. Вестник ТашГТУ. Ташкент. No. 1. Р. 68–74. [In Russian: Jabbarov S, Mirakhmedov M, Mardanov B (2017) Determination of the boundary of the safety zone of a high-speed train. Bulletin of the Tashkent State Technical University. Tashkent]

32. Лазаренко ЮМ, Капускин АН (2012) Аэродинамическое воздействие высокоскоростного электропоезда "Сапсан" на пассажиров на платформах и на встречные поезда при скрещении. Вестник Научно-исследовательского института железнодорожного транспорта. No. 4, pp 11–14 [In Russian: Lazarenko YuM, Kapuskin AN (2012) Aerodynamic impact of high-speed electric train "Sapsan" on passengers on platforms and on oncoming trains at crossing. Bulletin of the Scientific Research Institute of Railway Transport]

33. Orellano A (2012) Aerodynamics of High Speed Trains. Vehicle Aerodynamics Lecture, 79 p. https://www.mech.kth.se/courses/5C1211/Orellano_2012.pdf

34. Закиров РС (1986) Железные дороги в песчаных пустынях: Проектирование, сооружение земляного полотна и эксплуатация пути. Москва: Транспорт. 221 p. [In Russian: Zakirov RS (1986) Railways in sandy deserts: Design, construction of the road bed and operation of the track. Transport, Moscow]

35. Закиров РС (1983) Предупреждение песчаных заносов железных, автомобильных дорог и ирригационных сооружений. Москва: Медицина, 166 p [In Russian: Zakirov RS (1983) Prevention of sandy drifts of iron, roads and irrigation facilities. Medicine, Moscow]

36. Фазылов ТИ, Мирахмедов М, Адылходжаев АИ и др (1984) Рекомендации по закреплению подвижных песков вяжущими веществами. Москва: Транспорт, 32 p [In Russian: Fazylov TI, Mirakhmedov MM, Adilkhodzhaev AI et al (1984) Recommendations for fixing mobile sands with astringents. Transport, Moscow]

37. Мирахмедов ММ (1991) Технология и организация работ по закреплению подвижных песков. Ташкент: Фан, 143 p [In Russian: Mirakhmedov MM (1991) Technology and organization of works on fixing mobile sands. Fan, Tashkent]

38. Фазылов ТИ (1987) Закрепление подвижных песков вяжущими веществами. Ташкент: Фан, 104 p [In Russian: Fazylov TI (1987) Fixing mobile sands with astringents. Fan, Tashkent]

39. Инструкция по расчету наличной пропускной способности железных дорог (2011) Москва: ОАО«РЖД», 305 p [In Russian: Instruction on the calculation of the available railway capacity (2011) JSC Russian Railways, Moscow]

40. Сагомонян АЯ (1974) Проникание. Москва: МГУ, 299 p [In Russian: Sagomonyan AY (1974) Penetration. Moscow State University, Moscow]

Printed in the United States
By Bookmasters